Aftermath

JAPANESE SOCIETY SERIES
General Editor: Yoshio Sugimoto

Lives of Young Koreans in Japan
Yasunori Fukuoka

Globalization and Social Change in Contemporary Japan
J.S. Eades, Tom Gill and Harumi Befu

Coming Out in Japan: The Story of Satoru and Ryuta
Satoru Ito and Ryuta Yanase

Japan and Its Others:
Globalization, Difference and the Critique of Modernity
John Clammer

Hegemony of Homogeneity: An Anthropological Analysis of Nihonjinron
Harumi Befu

Foreign Migrants in Contemporary Japan
Hiroshi Komai

A Social History of Science and Technology in Contemporary Japan, Volume 1
Shigeru Nakayama

Farewell to Nippon: Japanese Lifestyle Migrants in Australia
Machiko Sato

The Peripheral Centre:
Essays on Japanese History and Civilization
Johann P. Arnason

A Genealogy of 'Japanese' Self-images
Eiji Oguma

Class Structure in Contemporary Japan
Kenji Hashimoto

An Ecological View of History
Tadao Umesao

Nationalism and Gender
Chizuko Ueno

Native Anthropology: The Japanese Challenge to Western Academic Hegemony
Takami Kuwayama

Youth Deviance in Japan: Class Reproduction of Non-Conformity
Robert Stuart Yoder

Japanese Companies: Theories and Realities
Masami Nomura and Yoshihiko Kamii

From Salvation to Spirituality: Popular Religious Movements in Modern Japan
Susumu Shimazono

The 'Big Bang' in Japanese Higher Education:
The 2004 Reforms and the Dynamics of Change
J.S. Eades, Roger Goodman and Yumiko Hada

Japanese Politics: An Introduction
Takashi Inoguchi

A Social History of Science and Technology in Contemporary Japan, Volume 2
Shigeru Nakayama

Gender and Japanese Management
Kimiko Kimoto

Philosophy of Agricultural Science: A Japanese Perspective
Osamu Soda

A Social History of Science and Technology in Contemporary Japan, Volume 3
Shigeru Nakayama and Kunio Goto

Japan's Underclass: Day Laborers and the Homeless
Hideo Aoki

A Social History of Science and Technology in Contemporary Japan, Volume 4
Shigeru Nakayama and Hitoshi Yoshioka

Scams and Sweeteners: A Sociology of Fraud
Masahiro Ogino

Toyota's Assembly Line: A View from the Factory Floor
Ryoji Ihara

Village Life in Modern Japan: An Environmental Perspective
Akira Furukawa

Social Welfare in Japan: Principles and Applications
Kojun Furukawa

Escape from Work: Freelancing Youth and the Challenge to Corporate Japan
Reiko Kosugi

Japan's Whaling: The Politics of Culture in Historical Perspective
Hiroyuki Watanabe

Gender Gymnastics: Performing and Consuming Japan's Takarazuka Revue
Leonie R. Stickland

Poverty and Social Welfare in Japan
Masami Iwata and Akihiko Nishizawa

The Modern Japanese Family: Its Rise and Fall
Chizuko Ueno

Widows of Japan: An Anthropological Perspective
Deborah McDowell Aoki

In Pursuit of the Seikatsusha:
A Genealogy of the Autonomous Citizen in Japan
Masako Amano

Demographic Change and Inequality in Japan
Sawako Shirahase

The Origins of Japanese Credentialism
Ikuo Amano

Pop Culture and the Everyday in Japan: Sociological Perspectives
Katsuya Minamida and Izumi Tsuji

Japanese Perceptions of Foreigners
Shunsuke Tanabe

Migrant Workers in Contemporary Japan:
An Institutional Perspective on Transnational Employment
Kiyoto Tanno

The Boundaries of 'the Japanese', Volume 1:
Okinawa 1868–1972 – Inclusion and Exclusion
Eiji Oguma

International Migrants in Japan: Contributions in an Era of Population Decline
Yoshitaka Ishikawa

Globalizing Japan: Striving to Engage the World
Ross Mouer

Beyond Fukushima: Toward a Post-Nuclear Society
Koichi Hasegawa

Japan's Ultra-Right
Naoto Higuchi

The Boundaries of 'the Japanese', Volume 2:
Korea, Taiwan and the Ainu 1868–1945
Eiji Oguma

Creating Subaltern Counterpublics: Korean Women
in Japan and Their Struggle for Night School
Akwi Seo

Social Stratification and Inequality Series

Inequality amid Affluence: Social Stratification in Japan
Junsuke Hara and Kazuo Seiyama

Intentional Social Change: A Rational Choice Theory
Yoshimichi Sato

Constructing Civil Society in Japan: Voices of Environmental Movements
Koichi Hasegawa

Deciphering Stratification and Inequality: Japan and Beyond
Yoshimichi Sato

Social Justice in Japan: Concepts, Theories and Paradigms
Ken-ichi Ohbuchi

Gender and Career in Japan
Atsuko Suzuki

Status and Stratification: Cultural Forms in East and Southeast Asia
Mutsuhiko Shima

Globalization, Minorities and Civil Society:
Perspectives from Asian and Western Cities
Koichi Hasegawa and Naoki Yoshihara

Fluidity of Place: Globalization and the Transformation of Urban Space
Naoki Yoshihara

Japan's New Inequality:
Intersection of Employment Reforms and Welfare Arrangements
Yoshimichi Sato and Jun Imai

Minorities and Diversity
Kunihiro Kimura

Inequality, Discrimination and Conflict in Japan:
Ways to Social Justice and Cooperation
Ken-ichi Ohbuchi and Junko Asai

Social Exclusion: Perspectives from France and Japan
Marc Humbert and Yoshimichi Sato

Global Migration and Ethnic Communities:
Studies of Asia and South America
Naoki Yoshihara

Stratification in Cultural Contexts: Cases from East and Southeast Asia
Toshiaki Kimura

Advanced Social Research Series

A Sociology of Happiness
Kenji Kosaka

Frontiers of Social Research: Japan and Beyond
Akira Furukawa

A Quest for Alternative Sociology
Kenji Kosaka and Masahiro Ogino

Modernity and Identity in Asia Series

Globalization, Culture and Inequality in Asia
Timothy S. Scrase, Todd Miles, Joseph Holden and Scott Baum

Looking for Money:
Capitalism and Modernity in an Orang Asli Village
Alberto Gomes

Governance and Democracy in Asia
Takashi Inoguchi and Matthew Carlson

Liberalism: Its Achievements and Failures
Kazuo Seiyama

Health Inequalities in Japan: An Empirical Study of Older People
Katsunori Kondo

Aftermath

Fukushima and the 3.11 Earthquake

Edited by

TSUJINAKA Yutaka and INATSUGU Hiroaki

Kyoto University Press
Kyoto

Trans Pacific Press
Melbourne

This English edition published in 2018 jointly by:

Kyoto University Press
69 Yoshida Konoe-cho
Sakyo-ku, Kyoto 606-8315, Japan
Telephone: +81-75-761-6182
Fax: +81-75-761-6190
Email: sales@kyoto-up.or.jp
Web: http://www.kyoto-up.or.jp

Trans Pacific Press
PO Box 164, Balwyn North
Victoria 3104, Australia
Telephone: +61-(0)3-9859-1112
Fax: +61-(0)3-8611-7989
Email: tpp.mail@gmail.com
Web: http://www.transpacificpress.com

Designed and set by Sarah Tuke, Melbourne, Australia.

Printed by Focus Print Group, Melbourne, Victoria, Australia.

Distributors

Australia and New Zealand
James Bennett Pty Ltd
Locked Bag 537
Frenchs Forest NSW 2086
Australia
Telephone: +61-(0)2-8988-5000
Fax: +61-(0)2-8988-5031
Email: info@bennett.com.au
Web: www.bennett.com.au

USA and Canada
International Specialized Book
Services (ISBS)
920 NE 58th Avenue, Suite 300
Portland, Oregon 97213-3786
USA
Telephone: 1-800-944-6190
Fax: 1-503-280-8832
Email: orders@isbs.com
Web: http://www.isbs.com

Asia and the Pacific (except Japan)
Kinokuniya Company Ltd.
Head office:
3-7-10 Shimomeguro
Meguro-ku
Tokyo 153-8504
Japan
Telephone: +81-(0)3-6910-0531
Fax: +81-(0)3-6420-1362
Email: bkimp@kinokuniya.co.jp
Web: www.kinokuniya.co.jp
Asia-Pacific office:
Kinokuniya Book Stores of Singapore Pte., Ltd.
391B Orchard Road #13-06/07/08
Ngee Ann City Tower B
Singapore 238874
Telephone: +65-6276-5558
Fax: +65-6276-5570
Email: SSO@kinokuniya.co.jp

ISSN 1443–9670 (Japanese Society Series)

ISBN 978–1–925608–96–0 (paperback)

National Library of Australia Cataloguing-in-Publication entry

Title: Aftermath : Fukushima and the 3.11 earthquake / Tsujinaka Yutaka, Inatsugu Hiroyuki editors.
Edition: First edition.
ISBN: 9781925608960 (paperback)
Series: Japanese Society series.
Notes: Includes bibliographical references and index.
Subjects:
Fukushima Nuclear Disaster, Japan, 2011 Nuclear accidents--Japan--Fukushima-ken.
Nuclear power plants--Accidents--Investigation--Japan--Fukushima-ken.
Emergency management--Japan.
Disaster relief--Japan--Fukushima-ken.
Radioactive pollution--Health aspects--Japan.
Other Creators/Contributors: Tsujinaka, Yutaka, editor. Inatsugu, Hiroyuki, editor.

Front cover photo: A tsunami hits Miyako, Iwate Prefecture. Photo taken from the city hall. Courtesy of Miyako City/Shutterstock/Aflo.
Spine photo: A pine tree standing firm at the coast of Rikuzentakata, Iwate Prefecture, after the tsunami. Courtesy of Aflo.

Contents

Figures

Tables

Contributors

Hamamoto Shinsuke
Associate Professor, Graduate School of Law and Politics, Osaka University

Inatsugu Hiroaki
Professor, School of Political Science and Economics, Waseda University

Itō Masatsugu
Professor, Graduate School of Social Sciences, Tokyo Metropolitan University

Itō Yasushi
Professor, Faculty of Humanities and Social Sciences, Chiba University of Commerce

Kamikawa Ryūnoshin
Associate Professor, Graduate School of Law and Politics, Osaka University

Kubo Yoshiaki
Associate Professor, Department of Political Science and International Relations, The University of the Ryukyus

Matsui Nozomi
Professor, Faculty of Urban Liberal Arts, Tokyo Metropolitan University

Nishide Junro
Professor, Faculty of Policy Studies, Iwate Prefectural University

Ōkura Sae
Assistant Professor, Faculty of Humanities, Law and Economics, Mie University

Tsujinaka Yutaka
Professor, Faculty of Humanities and Social Sciences, The University of Tsukuba

Tsunekawa Keiichi
Senior Professor, National Graduate Institute for Policy Studies

Wada Akiko
Deputy Mayor, Sakata City, Yamagata Prefecture

Yamamoto Hidehiro
Associate Professor, Yamagata University

With respect to Japanese name order, this book generally follows the Japanese convention that surname precedes given name.

Acknowledgments

The Great East Japan Earthquake and the Fukushima Daiichi nuclear accident resulted in the deaths, mainly by tsunami, of more than 18,000 victims and the destruction of more than 400,000 buildings. The peak number of evacuees exceeded 400,000, and in 2017, even after six years, numbered more than 120,000 people. This is the biggest postwar crisis we have faced, and the devastating impact continues to this day. Behind these figures are countless tragedies, sacrifice and pain, and still many people suffer from the traumatic experience. The editors cannot but wish for the victims' swift return to peaceful daily life.

When the powerful magnitude 9.0 earthquake hit the Tōhoku area on March 11, 2011, one of the editors was in Shinjuku, Tokyo, and the other was in the city of Tsukuba, Ibaraki Prefecture. The maximum seismic intensity was 'lower 5' (meaning moderately strong on a scale used in Japan) in Shinjuku, and 'lower 6' (moderately severe) in Tsukuba. Aftershocks of magnitude 5.0 or higher have occurred 850 times in East Japan since the day of the powerful earthquake. The distance from Fukushima Daiichi Nuclear Power Station to Shinjuku and Tsukuba is approximately 220 kilometers and 170 kilometers respectively. To recall how the crisis evolved back then, one of the editors studied the old newspapers he kept at hand.

The factors that led to such a complex disaster, the pattern of the damage it caused and the process of recovery all have their origins in the social world in its broadest sense. The causes of the disaster and the initial response by national and local governments, and the provision of disaster relief by both domestic and international actors, as well as the forms of risk communication that governments used with the populace, involve themes of broad concern to social sciences such as institutions, policy, conflict resolution and leadership. In the long-term, disaster recovery will require not only technical and engineering solutions, but also the forging of a consensus about how to design and implement a reconstruction plan and a mechanism for sharing the work and the responsibility for implementing it. We realized that political, policy, economic and social analysis of these processes was necessary in order to contribute to recovery and reconstruction. Through comprehensive analysis of the crisis and the recovery, social scientists are called on to contribute to the development of a social system that can minimize the impact of this kind of disaster.

In 2011 the Japan Society for the Promotion of Science initiated the Social Scientific Study of the Great East Japan Earthquake. This project brought together 90 social scientists devoted to understanding the disaster. An eight-volume series, *Daishinsai ni Manabu Shakaikagaku* (Lessons for the Social Sciences from the 2011 Great East Japan Earthquake), published by Tōyō Keizai Shinpōsha, is one of the fruits of this joint effort. This present collection contains a number of revised and translated essays from the original Japanese series.

In publishing this book, we again received full support from the Japan Society for the Promotion of Science. Professors Muramatsu Michio and Tsunekawa Keiichi, who organized this project, gave us valuable advice on the contents of this book and on signing a publishing contract. Professor Tsunekawa also contributed one chapter to this book.

Mr. Suzuki Tetsuya of Kyoto University Press and Professor Yoshio Sugimoto of Trans Pacific Press helped us publish this book in English. We are grateful for their efficient editorial support. Without their guidance, this book would not have been completed.

Finally, we are especially grateful to the residents, local government officials and many others in the stricken area who cooperated with us in this research on earthquake disaster.

Tsujinaka Yutaka
Inatsugu Hiroaki
Spring 2017

Introduction: 3.11 and Fukushima

Tsujinaka Yutaka and Inatsugu Hiroaki

On March 11, 2011, a magnitude 9.0 earthquake struck off the Pacific coast of north-east Japan. The earthquake's fracture energy was equivalent to dozens of times that of the Great Kanto Earthquake in 1923 and more than 1,000 times the Great Hanshin–Awaji Earthquake in 1995. The earthquake triggered a massive tsunami, which swept across north-east Japan's Pacific coast. The number of dead and missing reached some 20,000, mostly caused by the tsunami. The disaster also caused enormous damage to property. At the Fukushima Daiichi Nuclear Power Station, the earthquake and tsunami led to a total loss of external power, leading to a severe nuclear accident.

The factors that led to such a complex disaster, the pattern of the damage it caused and the process of recovery all have their origins in the social world in its broadest sense. The causes of the disaster and the initial response by national and local governments and the provision of disaster relief by domestic and international agents, as well as the forms of risk communication that governments used with the populace, involve themes of broad concern to social sciences, such as institutions, policy, conflict resolution and leadership. In the long term, disaster recovery will require not only technical and engineering solutions but also the forging of a consensus about how to design and implement a reconstruction plan and a mechanism for sharing the work and the responsibility for implementing it. We realized that political, policy, economic and social analysis of these processes was necessary in order to contribute to recovery and reconstruction.

Social scientists are called on to contribute to the development of social mechanisms and structures that can minimize the impact of this kind of disaster by conducting a comprehensive analysis of the crisis and the recovery. The Japan Society for the Promotion of Science initiated the Social Scientific Survey of the Great East Japan Earthquake in 2011. The project brought together 90 social scientists, who threw themselves into the task of trying to understand the disaster. The publication of an eight-volume series, *Daishinsai ni Manabu Shakaikagaku* (Lessons for the Social Sciences from the 2011 Great East Japan Earthquake), by Tōyō Keizai Shinpōsha is one of the fruits of this joint effort. This collection contains a number of essays from the series that have been revised and translated into English. Tsunekawa Keiichi's 2016 edited volume *Five Years After: Reassessing Japan's Responses to the Earthquake, Tsunami, and the Nuclear Disaster* was conceived on a similar basis and can be considered as the companion to this collection.

Many books that analyze the Great East Japan Earthquake from a social science perspective have already appeared in English. As Tsunekawa (2016a: 1) points out, however, while civil society and local government responses to the disaster have been lauded in the existing literature in English, there is a tendency to emphasize the ineptitude of the national government's response to the disaster (in terms of both politicians and the bureaucracy)[1] or to take the skeptical view that even in the face of such a major disaster, Japan is incapable of implementing genuine reform (Samuels 2013). These are important observations and are considered in this volume. However, an analysis of the existing literature in English leads us to conclude that the government's proactive role in responding to the disaster has not yet been sufficiently examined. A considered examination of the disaster would reveal that the interventions by government were both important and effective. What are the reasons for this perception gap? The chapters chosen for inclusion in this volume go some way towards answering this question.

Nevertheless, we have no intention of defending the actions of government during the disaster. Rather, we argue that when looked at from the point of view of the research fields in which the two editors have worked for some time, we can see a different side to the government's disaster response. Tsujinaka Yutaka's research has focused on the way government is entangled in social networks and on the issue of governance. Inatsugu Hiroaki's research concerns Japan's integrationist system of local government, which resembles continental European systems, and the institutional and non-institutional horizontal and vertical relationships that bind central and local governments. Our research themes clarify why we have chosen the term 'network governance' for the title of Part II of this volume.[2] Many people both inside and outside the Diet were undoubtedly angered by the appearance of politicians from both the government and opposition parties constantly bickering with one another in the midst of a national emergency. However, if indeed there was more to the government's response than this, how can we close the gap between the general public perception and what was really going on? What are the findings of the authors of the chapters in this volume? In the next section, we provide an outline of the chapters in this book, emphasizing those points of greatest interest to us as editors.

Structure and chapter outline

We have divided the 14 chapters in this volume into three parts:

- Part I: The Failure of Political Actors
- Part II: The Triumph of Network Governance
- Part III: Picturing Fukushima from the Data.

As mentioned above, the chapters in this collection have been selected from the eight volume *Daishinsai ni Manabu Shakaikagaku* and then revised and translated into English. In making the selection for this volume, we chose chapters that discuss the specific actions undertaken by the political actors (such as the Diet and the political parties), as well as those that analyzed the relationship between TEPCO and the other actors involved in the nuclear accident and the issue of nuclear power more generally from a political science perspective. We also chose chapters that discuss cooperation between central and local governments in local administration and touch on the legal procedures for compensating victims of the nuclear accident, a topic that has not received adequate attention in the existing literature. Finally, we chose chapters that examine the data on the extent of nuclear damage and radiation exposure, public opinion and social movements, and the impact of the disaster on Japan's foreign relations. Rather than attempting a comprehensive argument about political governance and the disaster, we chose chapters with a narrower focus that highlight a number of key points.

In the first chapter of Part I, Hamamoto discusses the response in the National Diet to the disaster. The relationship between political parties both before and after the disaster was shaped by the realities of a divided Diet. The Democratic Party of Japan (DPJ) government had a majority in the lower house but not in the upper house. This made the government's task of managing Diet business more complicated. What influence did the crisis and the divided Diet have on coalition formation and the legislative process? Hamamoto proposes a number of hypotheses drawn from coalition theory. Having done so, he analyzes the process of coalition formation both before and after the disaster and the way each party engaged with the legislative process during the DPJ's term in office. He discovers that while the crisis created by the disaster suggested the need for parties to form the grand coalition, the government's poor showing in the polls and the lack of institutional incentives for coalition formation in the National Diet obstructed coalition formation. The opposition refused to join a coalition with the DPJ. Instead, while Kan was leader, it actively tried to force him to resign and call a general election. The DPJ therefore had no choice but to try and govern with a minority in the House of Councillors.

Hamamoto's analysis of the passage of legislation and budgetary bills through the Diet shows that, even without a stable majority, most political parties cooperated with the government to pass disaster-related legislation. In other policy areas, the legislative process was characterized by a three-party consensus. However, the Kan government had to endure a protracted standoff over its second supplementary budget with the opposition, which wanted Kan to resign and to thereby force a dissolution of the lower house. Overall, by moving forward with legislative amendments proposed by the opposition Liberal Democratic Party (LDP) and Komeito, passing the consumption tax hike and formulating nuclear energy policy, the DPJ ultimately cooperated with the opposition to a large degree.

Of course, it was the nuclear accident at Fukushima that attracted the greatest attention after the 3.11 disaster. In Chapters 2 and 3, Kamikawa analyzes TEPCO's political and economic power. In Chapter 2 he analyzes TEPCO's relationship with a range of other political actors before the March 2011 disaster. Kamikawa shows how TEPCO created the myth of nuclear safety and a power structure that enabled the company to wield enormous influence over energy policy. TEPCO was able to create the myth of nuclear safety by using its political resources to successfully contain opposition to nuclear power. He argues that the source of TEPCO's power was its financial clout, itself a product of the comprehensive cost-accounting method enshrined in the *Electricity Business Act*. As a major procurer of goods and services, TEPCO had enormous influence in the business world. The company created a cooperative relationship with the Ministry of Economy, Trade and Industry (METI), the ministry responsible for supervising its activities, by finding positions for former METI officials both inside TEPCO and in related companies in the practice known as *amakudari* (descent from heaven, or parachuting into a related organization). This is the origin of the so-called iron triangle discussed in theories of regulatory capture.

On a number of occasions since the 1990s, there have been moves within METI to try to liberalize the electricity sector. TEPCO was successful in opposing these plans thanks to the assistance of the LDP, to whom it provided both votes and political donations. The company also had influence within the DPJ, then in opposition, which was also obtained by the same method. TEPCO effectively gutted the nuclear regulatory regime by establishing collusive relationships with staff at the regulator and with the academics and nuclear energy researchers who sat on its committees. The company dispensed research funding liberally to researchers and entertained officials involved in the regulatory process.

Local governments in nuclear host communities received subsidies from the national government. The utilities supplemented these payments through property taxes, nuclear fuel taxes and anonymous donations. This helped them to gain the consent of local communities. In order to shift public opinion in favor of nuclear energy, TEPCO won over the mass media by purchasing huge amounts of advertising. While a few media outlets criticized the nuclear industry when troubles occurred, very few subjected it to any serious scrutiny. Numerous documentary sources also tell of how TEPCO intervened in education by asking for elementary and junior high school students to be taught about the safety and usefulness of nuclear energy. The judiciary, too, accepted the myth of nuclear safety and refused to suspend operations at nuclear power plants when local residents brought legal actions against plant operators.

In Chapter 3 Kamikawa develops his argument by discussing the fate of nuclear energy policy after the Fukushima accident. While it made some effort to reduce Japan's reliance on nuclear power, the DPJ government ultimately failed to effect significant change. It lacked the power to pacify the powerful and diverse network of interests that profited from the nuclear energy policy regime constructed under the previous LDP administration. While the DPJ made much of its plan to phase out nuclear power by 2040, it failed to implement any concrete policy changes that would enable it to realize that goal. The LDP–Komeito coalition under Prime Minister Abe Shinzō, which returned to power at the end of 2012, tried to restart Japan's nuclear reactor fleet. However, the restart has not proceeded as the government had hoped. A now highly independent regulator, the Nuclear Regulation Authority (NRA), has been conducting close inspections of existing power stations based on a much stricter set of safety standards developed after the Fukushima accident. Two district courts have also made orders that suspended operations at a nuclear power station. Some prefectural governors and municipal mayors have also been more cautious in their approach to reactor restarts. These factors all constitute obstacles to restarting the reactor fleet in the future. Furthermore, the Abe government, wary of a popular backlash, has avoided dealing with the issue of new reactor construction or the revision of the nuclear fuel cycle project. While the DPJ and the LDP–Komeito governments had different stances on nuclear energy policy on the outside, both have been mindful of the negative attitude the public has towards nuclear power. They have tended to avoid making an issue out of nuclear power and have deferred making important policy decisions necessary to the long-term future of the industry.

In Chapter 4 Kubo investigates the competition between political parties that emerged after 3.11 in order to understand the mechanisms behind this stalemate on nuclear power. As an example, he looks at the process of building opinion within parties (intraparty), within political blocs (intrabloc) and between political blocs (interbloc) to set the agenda in both party leadership elections and the 2012 general election. Kubo's investigation reveals a number of points. First, because a consensus on disaster reconstruction developed through the cooperation of the two major political blocs, there was little apparent difference between the position the government and opposition parties took to the 2012 lower house elections. While there are strong differences between the DPJ and the LDP on a number of fronts, on disaster reconstruction they tended to proceed based on a basic consensus. Second, while three smaller parties tried to turn nuclear and energy policy into an election issue, the lack of consensus within the other three political blocs meant that the issue did not become a major one in the election as a whole. However, an analysis of voting patterns in the 2012 lower house election shows that neither of the major blocs achieved a particularly favorable evaluation from voters. The Japan Restoration Party, whose entry into the political arena was a challenge to the two major parties, received a similar degree of support to the DPJ. The Tomorrow Party of Japan, though it only secured a small number of seats, also received some support. The 3.11 disaster, along with other factors, provided the stimulus for a new type of politics. The agenda that formed after 3.11 contributed to the emergence of a third force that challenged the two major political blocs. However, the eruption of so many different issues onto the political stage effectively prevented the battle over 3.11 from becoming a national issue.

How did ordinary citizens react to the actions of the political parties and the politicians? How did they respond as political actors themselves? In Chapter 5 Yamamoto discusses the development of the issue of nuclear power from the perspective of issue-attention cycle theory. The Fukushima accident triggered the rapid development of a new style of antinuclear movement. Yamamoto points out that in the background to the emergence of these new movements was a growing popular opposition to nuclear energy among the population at large. Politicians who actively opposed nuclear power within institutional arenas also served as movement allies. In other words, the political opportunity structure and the discursive opportunity structure were both open to the emergence of an antinuclear movement.

However, while popular opinion was firmly opposed to nuclear power, reaching a peak in the summer of 2012, it did not have a substantive influence on policymakers. After the summer of 2012, divisions within the DPJ over a proposed consumption tax hike deepened and many antinuclear politicians who had acted as movement allies split from the party and no longer had a role in government. For the Japanese people as a whole, the consumption tax issue was of greater concern than nuclear energy and this led to a dampening of the antinuclear mood. The political opportunity structure and the discursive opportunity structure for the antinuclear movement began to close.

Nevertheless, support for phasing out nuclear power is still firmly entrenched in popular opinion. Opposition to the technology has become pervasive throughout society since the Fukushima accident.[3] Furthermore, popular opposition to nuclear energy is not merely the result of the dominance of sectional interests but is accompanied by widespread demands for a more democratic and impartial process. Yamamoto shows that the antinuclear movement has helped to create a new set of values that is becoming more deeply ingrained in society at large.

The political implications of the legal system had tended to be overlooked, but since 3.11 litigants opposing the restart of nuclear reactors have been successful in Fukui Prefecture, where they obtained a provisional disposition stopping the restart of the Ōi Nuclear Power Station (May 21, 2014). With legal action taking place all over the country aimed at stopping the operation of nuclear power stations, there is a growing awareness of the importance of legal decisions.[4] Chapter 6, the final chapter in Part I, looks at the legal process in its broadest sense as a final recourse for dispute resolution. In this chapter, Ōkura and Kubo discuss the compensation issues that arose after Fukushima. They examine the processes that were created for reconciling the conflicting interests of victims and perpetrators (TEPCO and sometimes the government) and the impact that the process can have on the relief that victims receive. They follow three specific procedures by which victims can receive compensation based on documentary evidence and the result of a survey of legal teams involved in representing disaster victims. These include a direct process managed by TEPCO, a mediation scheme managed by the Nuclear Damage Compensation Dispute Resolution Center and the final recourse of legal action in the courts. The authors find that, in general, the compensation framework was constructed on an ad hoc basis after the disaster and reacted to the claims for compensation received from victims. This led to delays in the commencement of legal proceedings in the courts and to considerable delays in some victims receiving compensation.

One factor that contributed to the delay was the Justice System Reform. The Justice System Reform was supposed to increase the efficiency of the court system by increasing the legal workforce. However, only a few additional judges were recruited and public sector cuts undertaken by the Koizumi government meant that the number of administrative staff working in the court system remained static. However, there was widespread expectation that the establishment of an alternative dispute resolution (ADR) process might be one way to address the problem of a sluggish court system. However, the ADR system was placed under the Ministry of Education, Culture, Sports, Science and Technology. As a cog in a larger bureaucratic machine, it lacked the budget and the human resources to meet the heavy demand for settlement mediation after 3.11. This also contributed to long delays in victims receiving nuclear damage compensation. Ōkura and Kubo conclude that both the government and civil society need to anticipate the kinds of disputes that might arise after a disaster. Regardless of whether they are to be resolved via legislation, the courts or an administrative agency, such systems need to be able to mobilize sufficient resources in a flexible manner.

Part II concerns the relationship between local governments and between local and national governments. Relationships between central and local governments and the work of local governments determined the way reconstruction progressed after the disaster. Unlike the separationist model that is characteristic of local governments in Britain and the United States, local administration in Japan is based on an integrationist model. Central, prefectural and municipal governments exchange information and collaborate on a daily basis. This kind of network comes into its own in a disaster. Administrative networks encompass all of Japan's social systems. In this volume, it is our assessment that this kind of network governance (or administrative network governance) performed well in response to the disaster. When compared with other countries, the characteristic feature of central–local government relations in Japan is interdependent vertical and horizontal networks formed between central and local governments (prefectures and municipalities). The chapters in this section show that these features enhanced the performance of network governance in the post-disaster context. This is true both of networks between governments (central–local and local–local) and those networks that link government agencies with the public via the quasi-private sector, such as Social Welfare Councils. Volunteer centers established by prefectural governments, as well as prefectural social welfare councils, were able to match volunteers and NPOs with the needs of disaster-affected municipalities.

In Chapter 7 Itō Masatsugu explains the way the Reconstruction Agency was set up from the perspective of rational choice theory. By looking at how the Reconstruction Agency carried out its function as a 'control tower' and provided a 'one-stop shop', he analyzes the process by which institutional mechanisms for reconstruction were developed. The Reconstruction Agency functioned as a control tower for reconstruction policy but was also meant to serve as a centralized one-stop shop where local governments in the disaster zone could go to obtain the assistance they needed. The Reconstruction Agency was established by Cabinet in February 2012 as an interim organization that would operate for ten years. It was headed by the Prime Minister and the Minister for Reconstruction. In addition to its role in formulating policy and planning for reconstruction, the agency was also intended as a centralized body for implementing policy. The agency was established during the divided Diet that existed after the 2010 upper house elections. As such, the ruling DPJ had to incorporate input from the opposition parties in designing the new institution.

The Reconstruction Agency began operating as an independent administrative agency ten months before the 2012 elections, when the LDP–Komeito coalition returned to power. With the election of a new government, many expected that the agency's functions as a control tower and one-stop shop would be strengthened in order to accelerate the pace of reconstruction. However, while the Reconstruction Agency and the regional Reconstruction Bureaus were meant to direct reconstruction efforts, they mainly served a coordinating role. The organization's ability to effectively coordinate the reconstruction effort depended on the interaction between agency staff and those from other government agencies. It also provided some support to local governments. The government originally intended that the Reconstruction Agency would carry out some reconstruction-related projects by itself where possible. However, its major task was to encourage cooperation between the relevant agencies. The work that was carried out by the Reconstruction Agency was successful in that it encouraged reconstruction while respecting local government autonomy in the affected areas.

In Chapter 8 Inatsugu analyzes the dispatch of local government personnel to the disaster zone. Immediately after the disaster, a large number of firefighters and police from all over Japan were mobilized to carry out search and rescue and other operations. The dispatch of these emergency services personnel was based on existing legislation. However, there was no existing legislation covering the dispatch of other government workers to disaster zones. Local government employees have a legal responsibility to work exclusively for the welfare of residents in the local government area in which they are employed. Local government employees are not expected to go off to work for distant administrative districts and there was no legal basis for such dispatches. Nevertheless, local governments throughout Japan sent large numbers of personnel to assist in the disaster zone. Why? Inatsugu finds that, in addition to the networks that already existed between some local governments, the disaster provoked a kind of competition between local governments over how many staff they would send to help in the disaster zone. A fully functional form of administrative network governance after 3.11 seems to have come out of this process.

In Japan local governments provide services to residents based on the profile of the residential community developed in the family registration and resident registration systems. After 3.11 some local governments began providing resident services to evacuees who were not registered in the region to which they had fled. In Chapter 9 Wada examines the circumstances that led to this departure from usual government practice and tries to discover the reason behind it. In Japan municipalities provide a range of administrative services to residents based on data from the resident registration system. However, after 3.11 some municipalities began providing these services to evacuees who were not registered with them. Furthermore, whereas in the past local governments have provided temporary housing to disaster victims who had a certificate from the local government jurisdiction in which they were registered, after 3.11 some prefectures provided housing even without such certificates. Wada investigates the reasons behind this expanded definition of who was eligible for resident services by looking at the example of Yamagata Prefecture, which bordered the disaster zone, and at the cities of Yamagata and Yonezawa located in that prefecture. She speculates that prefectures and municipalities expanded their usual criteria for service provision because they were conscious that the central government wanted them to do so. While the fact that financial incentives stimulated local governments to provide government services to unregistered evacuees seems natural, the example is important in that it highlights the nature of the relationship between central and local governments in Japan. Wada argues that while local governments ordinarily refuse to deviate from existing legislation and regulation, in a time of disaster they took into account the attitudes of numerous stakeholders, including evacuees, residents and municipal councils, in extending these services to non-residents.

In Chapter 10 Matsui analyzes how local government employees who were involved in reconstruction on the ground perceived the reconstruction process. This chapter is based on the results of a mail survey of 1,352 section heads in the three disaster-affected prefectures (Iwate, Miyagi and Fukushima) and 37 coastal municipalities within these prefectures that suffered heavy damage from the tsunami and nuclear accident. From the results of this survey, Matsui clarifies a number of issues. First, local government employees respond to the daily ups and downs of reconstruction work based on the organizational and administrative systems that were in use before the disaster. Second, with regard to relations between government agencies, many commentators on the 3.11 disaster have concluded that while the central government played a major role, the municipalities' relationships with prefectural governments have become increasingly diluted. However, by looking at the frequency of interactions reported by the section heads who responded to this survey, Matsui concludes that far from becoming weaker, there were actually more points of contact between cities and prefectures than with the central government, casting doubt on the idea that the relationship between these two tiers has been diluted. The survey paints a picture of networks that are built between governments on a daily basis being activated for the purpose of reconstruction. Third, the average local government employee working on reconstruction projects in the disaster-affected areas needs a strong source of psychological support to continue his or her work, as well as a degree of specialist knowledge.

In Chapter 11 Nishide examines the situation of volunteering in the disaster zone. Immediately after the 3.11 disaster, numerous volunteer centers were established all over the disaster zone. The total number of volunteers exceeded 1.35 million. The volunteers and the volunteer centers were supported by a number of actors, including NGOs, NPOs and for-profit companies, and these actors worked hard to ensure that volunteers would be available in the disaster zone in the long term. After analyzing the NGOs, NPOs and professional organizations that worked independently in the disaster zone, Nishide investigates some of the issues raised by volunteering after the 3.11 disaster and some of the proposed solutions. This chapter points out that social networks played an important role in volunteers' relationships with local governments. Prefectures established volunteer centers that were at the center of these networks but vertical and horizontal networks linking government officials with private citizens were also critical. Tōno, a municipality in Iwate Prefecture, had established a logistical support center before the disaster. Iwate Prefecture established its own logistical support center immediately after 3.11. These centers were located at some distance from Iwate's major cities in the coastal areas that were devastated by the 3.11 disaster. This enabled them to serve as relay points for getting disaster relief to where it was needed.

Part III comprises three chapters that paint a picture of the Fukushima nuclear accident. The arguments in this section are based primarily on the data. Although many questions remain about the Fukushima nuclear accident, the reality of what has happened is steadily becoming clearer. These chapters aim to present this data clearly.

In Chapter 12 Itō Yasushi collates the important data on the health impacts of the nuclear disaster. This information is essential if we are to understand the extent of the damage caused by the Fukushima nuclear accident. Itō begins with an overview of the results of the Fukushima Health Management Survey and the radiation doses that workers at the Fukushima Daiichi Nuclear Power Station engaged in the clean-up have been exposed to. Following the nuclear accident, national and local governments ordered people living in the region of the power station to evacuate in an attempt to prevent their exposure to radiation and the consequent damage to their health. Itō presents the data on population movements following these directives. He also provides information on the scale of the compensation payments TEPCO has had to make to cover the healthcare costs and preventative health measures for people who were forced to evacuate and those in neighboring areas that are now suffering from depopulation. He also looks at the figures for public spending on disaster relief, mediation and recovery, and discusses what the government considers as disaster-related damage and how it has proceeded with reconstruction. Itō demonstrates that the extent of the damage is far greater than has been communicated in the mainstream media. Furthermore, TEPCO is liable for an enormous compensation bill that it will be unable to pay without assistance. The government is helping the company to meet its compensation obligations but the mechanism by which it is providing this assistance is complex and difficult to understand. Itō explains the ways in which the government subsidizes TEPCO's compensation scheme. This chapter raises a number of questions that will have to be examined in the future.

In Chapter 13 Itō Yasushi turns his attention to the cost effectiveness of decontamination and tries to highlight some of its problems. The Fukushima nuclear accident triggered the release of a large quantity of radioactive particles, forcing many people living near the power station to evacuate. The government's basic policy is that evacuees ought to be able to return to their former homes. However, in some cases even decontamination work does not lower radiation levels sufficiently. The cost of these procedures is enormous and there has been growing criticism of the decontamination policy. At the root of these issues is the government's adherence to a policy of 'decontamination followed by return'. This chapter therefore discusses the process of how the government adopted such a policy. Having explained the background to decontamination, Itō conducts a cost-benefit analysis to determine whether this is the best method for helping evacuees to regain their former standard of living. He analyzes the implementation of the decontamination policy thus far and compares it with an alternative option, focusing in particular on decontamination in the evacuation instruction zone. As a result, Itō shows that if the government was to allow residents who would like to relocate to an area where radiation levels are lower to do so, this would not only give the evacuees more choice but would also be a far more cost-effective means of restoring quality of life.

In the final chapter in this volume, Tsunekawa analyzes the impact of the radioactive fallout from Fukushima on Japan's international relations. An effective response to the radioactive fallout from the Fukushima Daiichi Nuclear Power Station is important not only to ensure the lives and the health of people in Japan but for Japan to regain the trust of the international community. Tsunekawa examines this issue by focusing on three examples: overseas visitor numbers to Japan, Japanese exports and the discharge of radioactive effluent into the Pacific Ocean.

There was a dramatic decline in the number of overseas visitors to Japan in the immediate aftermath of the nuclear crisis. While this did have an impact on Japan's tourism industry, the fact that only a limited area was affected meant that by 2013 international visitor numbers had returned to pre-disaster levels. Thereafter, they began to climb steadily. The disaster impacted Japanese exports as a number of countries imposed important bans and restrictions on Japanese products. However, like international visitor numbers, exports had recovered to pre-disaster levels by 2013. Nevertheless, a number of countries still have import restrictions in place affecting Japanese products. One of the reasons for this is the continuing discharge of radioactive effluent from the Fukushima Daiichi site into the Pacific Ocean. The machinery that has been designed to treat radioactive effluent keeps breaking down and there have been delays in setting up systems to capture and treat the groundwater that is flowing into the reactor housing. These problems have forced TEPCO to construct an enormous number of storage tanks. Like the need to decommission the damaged reactor, the effluent problem is one of the major lingering effects of the Fukushima crisis. These issues continue to exert a harmful influence on Japan's standing in the international community.

Questions raised in this collection

The chapters in this collection address the question of why, while the relative success of network governance in brokering a government response to both the disaster response and to reconstruction (Part II), politics failed either to prepare for the disaster or to respond adequately to it (Part I).

Tsunekawa (2016a: 27–8) concludes:

> Looking back at Japan's responses to the earthquake/mega-tsunami/ nuclear accident from five years after the disaster makes it clear that, despite its notorious vertically divided structure, [the] Japanese government, in fact, responded quite well, especially in areas where it had been well prepared, such as emergency rescue and relief, infrastructure restoration, reopening of schools and medical services, and preparation of residential spaces for disaster victims. Coordination of activities across government bureaus was also smoothly carried out with the help of staff members who were seconded to the emergency disaster management headquarters as well as the Reconstruction Agency on the basis of the pre-March 11 experiences nurtured through mutual personnel exchanges and loan of personnel to the Cabinet Office. In contrast, not only TEPCO but also the national government was powerless against the nuclear disaster, for which prior preparedness had been overly inadequate.

This observation raises another important question. Why were some components of the same Japanese political and social system well prepared while others (when it came to the nuclear issue) were very poorly prepared? Tsunekawa (2016a: 28) argues that 'the very act of placing high hopes on the Great East Japan Earthquake to function as a critical juncture [in terms of provoking reform] is misguided'. As Samuels (2013: 200) points out, 'the 3.11 catastrophe was not the "game changer" many policy entrepreneurs desired'. Nevertheless, it did clearly contrast the positive and negative aspects of the Japanese political and social system and lead to the formation of new political movements.

There is a need to explain these two clearly contrasting components in terms of the logic of a single system (Tsunekawa 2016a: 13).[5] Our basic hypothesis is as follows.

There are two sides or faces of Japan's postwar political system. The first side comprises interdependent vertical and horizontal networks formed between central and local governments (prefectures and municipalities); quasi-governmental organizations such as Social Welfare Councils, district welfare officers and probation officers; quasi-private organizations such as chambers of commerce and industry and crime prevention associations; social organizations such as industry groups, labor unions and farmers cooperatives; professional organizations such as medical associations and civil society organizations (including the more than 50,000 NPOs/NGOs currently operating in Japan); innumerable regional residents groups that organize around the 300,000 neighborhood associations (*Jichikai* or *Chōnaikai*); and the more than two million for-profit companies from large corporations to small businesses (Tsujinaka et al. 2009: 24–5, Figure 1.2). It is these organizations that enable what we call 'network governance'.

For example, local municipalities are connected with one another not only through formal organizations such as the six peak federations of local officials and bodies like National Governors Association, but also through the thousands of administrative associations involving municipalities that are responsible for more than 10,000 different operational issues (Inatsugu 2013: 205–9).

The diverse organizations, associations, unions etcetera mentioned above, whether they are set up as juridical persons or not, operate primarily at the municipal level. They provide public services and municipal activities and engage in advocacy through their local governments.

This aspect of Japan's political system facilitated public, collective and individual efforts to assist disaster victims in a time of crisis. Because administrative bodies, including central government agencies and local governments, often act like hubs, they are referred to as administrative networks. In these networks there is a degree of hierarchy in the vertical relationship between central government agencies and the municipalities. However, after a series of decentralization reforms that took place up until the early 2000s, the relationship between the central and local governments became much more equal. Overall, the majority of these networks are horizontal and open. They are characterized by flexibility and mutual respect. We might call this the pluralistic face of the Japanese system.

On the other hand, as can be seen in Kamikawa's two chapters in particular, since 1955 a structure of political and economic power that unites politicians and the bureaucracy with the major corporations has been a core part of the Japanese political system. A system for promoting nuclear power, with TEPCO at the center, stood at the heart of this system. This network incorporates not only the LDP and the business world but even extends to the opposition and some labor unions. It had its political-economic basis in the legal framework designed to promote nuclear power: the comprehensive cost-accounting method that was specified in the *Electricity Business Act*. It took the form of the iron triangle, which, despite a degree of internal conflict, linked TEPCO (business), the LDP and bureaucratic structures such as METI (formerly MITI, the Ministry of International Trade and Industry), and the Science and Technology Agency. TEPCO was the financial power that gave the network the resources it needed to exercise its influence. While they may have been less prominent, other writers have already shown how in other industries, too, similar political influence networks existed in related policy areas (Ōtake 1996; Tsujinaka (ed.) 2016).

In the electric power industry, an opaque and closed world was created that was entirely encircled by a hierarchical network and economic gain. Its prototype is the policy community known as the 'nuclear village'. The village is a closed circle built on power and profit that limited the creativity of multiple different actors and pushed inconvenient possibilities, which should have been considered, outside of consideration. This created a situation in which preparations for a nuclear accident were inadequate. For example, there was no comprehensive manual detailing how to respond in a severe emergency and no training for such an eventuality (Saitō 2015: 170–252). This is the negative face of the power elite. Their organizations are irresponsible in the sense that they failed to prepare adequately for a nuclear accident, depending on closed-mindedness and an addiction to profit.

These two faces are actually two sides of the same coin, a system of governance established by the LDP over a period of more than half a century. Its positive face is connected with the LDP's long reign and efficiency and creativity, as well as the welfare systems that earned the party the support of voters. Its negative face seems to have supported this long reign by providing money and human resources to help the LDP to keep winning elections. Furthermore, because the LDP remained in power for so long, the mechanisms by which it ruled were never exposed to public scrutiny and the party deferred difficult public policy decisions in order to remain in power.

It is important to note that the 3.11 disaster happened not under the LDP, which built these two networks over more than half a century, but under the DPJ coalition. The DPJ was formed in 1996. After contesting a number of elections, the victory it achieved in the autumn of 2009 was almost without precedent since 1955 and led to a change of government (that is, the party obtained the largest number of seats in the lower house). However, the party was still new to government and displayed little mastery of the art of administration. While the responsibility for what happened during the 3.11 disaster really lay with the power elite centered on the LDP, this fact was obscured because the disaster occurred under a different government. This, along with the DPJ's clumsy handling of the situation, led to widespread criticism of the new government. There was no opportunity to adequately express the LDP's real responsibility for the disaster.

In fact, when disaster struck, the formerly opposition DPJ leadership had been in power for a little over a year. The party had little understanding of the structure of political power in Japan and lacked the ability to overcome its own ignorance. The leaders of both major parties considered forming a grand coalition after the disaster, but neither could come up with a creative political solution to make it happen.

The issues raised by the existence of these two networks involve knowledge and ignorance, institutionalization and deinstitutionalization, and open and closed networks. The pluralistic system that was cultivated by the LDP over half a century in power had already become more horizontal in nature and gave the party the ability to listen to the grassroots in each social sector in all sorts of creative ways. This was the system that responded effectively to the disaster. It provided a solid political and social infrastructure to support reconstruction. On the other hand, the opaque and closed LDP power network, which avoided facing up to the consequences of some of its policies for half a century, was unprepared for the nuclear disaster and failed to pass on the information the new government needed to deal with the disaster.

The issues associated with reconstruction policy are directly connected with the problems arising from the first network, while the failure of the response to the nuclear crisis is representative of the latter. Saitō Makoto (2015: ii), the social scientist who has devoted the greatest attention to studying this particular disaster, has concluded that 'in terms of reconstruction policy, there is a degree of over-preparedness while the response to the nuclear crisis was the product of a decided lack of preparedness. Unfortunately, this has led to an excessive concentration on reconstruction policy while nuclear policy has been completely inadequate.' These differences between the two responses may be related to the degree of legitimacy possessed by each network. Both the government and the opposition can agree on a disaster reconstruction budget to invigorate pluralistic administrative networks that are open to civil society and are taking the lead in disaster reconstruction in the name of 'creative reconstruction'. On the other hand, when it comes to the closed power networks of the nuclear village, it might be necessary to make TEPCO, the company that caused the accident, bear much more of the responsibility.

A number of issues are indicative of the continuing problems posed by the politics of nuclear energy. At the end of November 2016, METI announced that 'the cost of decommissioning the Fukushima Daiichi Nuclear Power Station, compensating the victims, decontaminating the fallout zone and an interim storage facility for the waste will be double previous estimates' – in other words, 22 trillion yen opposed to the previous estimate of 11 trillion yen.[6] The ministry suggested that most of the balance, which exceeds TEPCO's ability to pay, should be covered by the government and by raising electricity prices. In 2016 the government decided to decommission the Monju fast breeder reactor. At the same time, it also decided to maintain the fast breeder development program.[7] As at the end of 2016, a state of nuclear emergency under the *Act on Special Measures Concerning Nuclear Emergency Preparedness* remains in place at the Fukushima Daiichi Nuclear Power Station.[8]

A public opinion survey on disaster prevention and energy was conducted five years after the disaster in December 2015. Ninety-two per cent of respondents to the survey said that the situation at the Fukushima Daiichi Nuclear Power Station was 'of concern' (45% said they were ill at ease and 47% said they were more uneasy than not). Another 85% said they were concerned about the possibility of future nuclear accidents. Forty-nine per cent said that Japan ought to reduce the number of nuclear reactors and 22% said that all nuclear reactors should be shut down. On restarting the reactor fleet, 40% were opposed and 42% answered 'can say neither way'. Only 17% were in favor of reactor restarts. Sixty-five per cent of respondents said that they doubted whether nuclear energy could ever be made safe (the question asked was, 'Can nuclear power be controlled?'). When compared with similar surveys conducted one and three years after the disaster, these results demonstrate that there has been very little change in popular attitudes towards nuclear energy (Kōno et al. 2016). However, surprisingly, in the four national elections that have been held since 2012, nuclear energy policy has not been a major election issue.

Keeping in mind the discussion of the previous research at the beginning of this introduction, we would like to close with the following points.

1. The success of civil society and local government responses to the 3.11 disaster was due to these actors cooperating with the central government via institutions and networks discussed above (integrationist relations between central and local governments).
2. Where institutionalized mechanisms or other preparations existed, government responses, whether by politicians or the bureaucracy, were generally appropriate. However, where the government had to respond to a new issue for which it was

not prepared, as was the case with the nuclear disaster, its ineptitude was marked.

Examples of appropriate responses include the government's conduct of diplomatic relations with the United States and other countries that offered disaster relief. The policing response to assist with search and rescue, recovering the corpses of the deceased, and the rapid mobilization of the fire brigades and 100,000 Self-Defense Forces were all positive. The government did successfully pass a number of supplementary budgets that enabled it to finance the disaster response. The government's secondary crisis management, such as in avoiding a financial crisis and providing the necessary finances to repair roads, ports and other infrastructure, as well as the rapid restoration of educational services for children, can all be considered to have been successful. Where central and local governments worked together successfully in response to the disaster, this was always the result of being prepared (in terms of institutions, networks and transitional activities). On the other hand, the myth of safety meant there was no preparation for the nuclear accident, not even a severe-incident manual or a crisis response drill.

3. In responding to the first major crisis, caused by the nuclear accident, since the end of the Second World War, politicians lacked creativity and were unprepared to come up with new solutions, such as forging a grand coalition and putting aside their differences in order to deal with the crisis.
4. In the political realm, few have sought to question why they were so unprepared. The greater part of the political responsibility for the lack of preparation in regards to both TEPCO's crisis management and the government's handling of it lies with the LDP, the party that has held onto power almost exclusively for more than half a century (and with the various LDP Cabinets during this period). However, not even the media have dared to pursue those responsible.
5. In dealing with complex issues such as nuclear power, where they anticipate a public backlash, the major political parties (DPJ, LDP, Komeito) have chosen avoidance and governments have made opaque decisions or postponed making tough decisions that supposedly address the issue. There is an ongoing tendency for politicians to avoid making difficult decisions. The Fukushima Daiichi Nuclear Power Station is

still legally considered to be in a state of emergency. Only a vague timeline exists for decommissioning the facility. There is little discussion among leading parties of the important issues such as who will pay the cost of decommissioning and decontamination or whether to restart the nuclear reactor fleet.

6. In many ways, it still remains to be seen whether the disaster will lead to long-term change. In terms of energy policy, important decisions – such as coming up with a detailed timeline for the alternative sources of energy that will be needed as the Fukushima Daiichi Nuclear Power Station is decommissioned and the overall number of reactors is reduced – have not been articulated.
7. The question of whether the disaster will lead to major reform is not limited to energy policy. It also depends upon how shifts in public opinion and civil society activism with regard to other contentious policy areas, such as national security and international relations, are articulated with the political parties.
8. Rather than engaging in a dialogue with the Japanese people about the important issues, the LDP-led government that returned to power after the disaster adopts an ambiguous stance on key issues while doing its best to project an image of governability. It seems to regard hanging onto power as its only purpose.

Part I
The Failure of Political Actors

1 Coalition Formation and the Legislative Process in a Divided Diet

Hamamoto Shinsuke

The scale of the Great East Japan Earthquake and the nuclear accident it caused at the Fukushima Daiichi Nuclear Power Station were unprecedented. The Japanese government responded with a number of legislative measures designed to promote recovery and reconstruction. By 2013 the Diet had enacted 41 government (Cabinet) bills.[1] Four supplementary budgets had also been passed by February 2012. A significant number of private members' bills were also enacted, including the *Basic Act on Reconstruction in Response to the Great East Japan Earthquake* (hereafter, the Basic Act on Reconstruction).

These laws were enacted in the context of a divided Diet. Nine months before the earthquake, the government had lost control of the upper house, creating a divided Diet. However, minority governments, in which the ruling party does not control the balance of power in the first house, exist in 30% of countries that have parliamentary governments. This means that the situation in Japan is far from exceptional.

In this chapter, I analyze coalition formation and the legislative process during the administration of the Democratic Party of Japan (DPJ) by asking two questions about the divided Diet. First, how did coalition negotiations between the government and the opposition parties proceed? Second, what were the features of the legislative process for disaster recovery and reconstruction? I focus on the period of the DPJ administration because of its importance for understanding nuclear energy policy after the earthquake. As I discuss below, numerous attempts were made to form a coalition during the divided Diet. The government had three possible policy options and four potential coalition partners. Depending on which coalition the DPJ formed, different policy options would have been available for addressing nuclear energy, restarting Japan's nuclear reactors and reforming the electric power industry. The period of the DPJ administration marked a turning point in nuclear energy policy in Japan. My study of this period therefore contributes to a broader examination of nuclear energy policy after Fukushima.

Analyzing the crisis may also prove to be a profitable means of evaluating the usefulness of existing coalition theory. Existing coalition theories leave a number of questions unanswered. In a significant number of cases of actual coalition formation, they fail to provide an adequate explanation. Martin and Stevenson (2001) developed 21 hypotheses about coalition formation drawing on the existing research in the field. They divided these hypotheses into three categories – number of seats, policy and institutions – and subjected them to statistical analysis. They found that even a model that takes all these factors into account can only accurately predict coalition formation about half the time. It is therefore difficult to accurately predict coalition formation using existing theories. Existing theories also tend to focus on coalitions that have already formed. Little attention has been given to the process of coalition building (Andeweg 2011). There is therefore a need to develop more effective models of coalition formation that take this process into account, as well as the various factors that can influence coalition formation.

In this chapter I focus on the effect of situational factors on coalition formation, specifically a situation of crisis. This is not a new idea and recently scholars have examined coalition formation in the context of economic crisis (Moury and De Giorgi 2015). Crisis forces opposition parties to make a choice about whether to cooperate with the government or to take advantage of the situation to seize power for themselves.

This study also contributes to the discussion of the explanatory power of institutions in coalition formation. The institutional approach is the central approach used to explain minority governments (Strom 1990). In the context of fierce electoral competition, where there are a large number of marginal seats, opposition parties tend not to participate in coalition governments. In institutional settings where opposition parties are able to influence the policy-making process, they are also less likely to join in a coalition government. Minority governments are formed when opposition parties make rational choices about the value of participating in government. However, the crisis brought about by the earthquake disaster created a situation that favored cooperation and coalition building between political parties rather than conflict. By examining coalition formation, we can discover whether institutional factors or situational factors have a greater influence.

This chapter is structured as follows. In the first part I explain the institutional settings of the Japanese National Diet. I then propose three hypotheses regarding coalition formation. In the second section I examine the DPJ's three attempts to form a coalition in the aftermath of the earthquake disaster. In the third section I discuss how each party responded to the bills introduced to the Diet by the DPJ government. Finally, I summarize the discussion and discuss its implications for nuclear policy and coalition theory.

Coalition theory and the divided Diet

The structure of the Japanese National Diet favors the development of legislation by majorities rather than enabling minorities to influence the legislative process. Döring (1995) studied how agenda-setting takes place in 18 European parliaments. He looked at seven stages in the legislative process in order to compare the different systems. Masuyama Mikitaka applied the same measurement to the Japanese National Diet. After comparing the Diet with parliaments around the world using multivariate analysis, he found that it favors majorities more than other systems. Unlike the consensus-oriented models typical of parliaments in northern Europe, the institutional mechanisms of the Japanese National Diet ensure a more confrontational model similar to that of the British parliament (Masuyama 2003).

However, the Japanese National Diet incorporates a bicameral system in which both houses have relatively equal powers. The House of Representatives does have two features that make it superior to the House of Councillors. The House of Representatives can pass legislation that has been rejected by the House of Councillors (Constitution of Japan, Article 59, Paragraph 2). If the decisions of the two houses differ on a given bill, it can still become law if it is passed a second time by a two-thirds majority of the House of Representatives. However, other than in the period 2005–09, the government has rarely been able to command a two-thirds majority in the House of Representatives, making this a particularly steep hurdle.

The superiority of the House of Representatives is also recognized in the Constitution. With regard to the budget (Article 60), the nomination of the prime minister (Article 67, paragraph 2) and the ratification of treaties (Article 61), the Constitution gives the House of Representatives the power to overrule a decision of the House of Councillors. However, the superiority of the House of Representatives is limited to these items. The House of Councillors has equal power to that of the House of Representatives when it comes to making laws. A particularly important feature is that while the House of Representatives may have greater powers when it comes to making budgets, this power does not extend to money bills more generally. This means that if the government is in a minority in the House of Councillors, while it has the ability to pass the budget, it cannot force through other related bills, such as those authorizing the issue of special deficit-financing bonds.

Therefore, while the House of Representatives is invested with greater powers than the House of Councillors, they are not particularly strong and in practice there is little difference between the powers of the two houses. Lijphart (2012) compared two aspects of bicameral systems: the powers granted to them under their respective constitutions and their electoral procedures. He gave Japan's House of Councillors three points on a four-point scale, placing it in a category of moderately powerful bicameral systems.

The House of Councillors also has significant autonomy in terms of its relationship with the prime minister and Cabinet. First, the prime minister does not have the power to dissolve the House of Councillors. Members of the House of Councillors are elected for a six-year term, with half of the seats declared open for election every three years. Therefore, the prime minister cannot use the threat of dissolution as a means of pressuring members of the House of Councillors. Opposition members in the House of Councillors also have the right to propose censure motions against Cabinet ministers. Censure motions have no basis in the Constitution or in legislation and are therefore not legally binding. Nevertheless, they are politically effective (Kawato 2015). The passage of a censure motion against a Cabinet minister will typically move the prime minister to force the minister in question to resign. Otherwise, the prime minister risks Diet business grinding to a halt in the committee process. This first occurred in the divided Diet of 1998, when the Liberal Democratic Party's (LDP) Nukaga Fukushirō, director-general of the Defense

Agency, resigned following a censure motion against him in the House of Councillors. Since then, while censure motions against the prime minister are usually ignored, prime ministers have not reappointed Cabinet ministers who have been censured. In effect, this means that while it may lead to a breakdown in trust between the councillors and the prime minister, the upper house does have the power to stop bills in committee, vote them down in a general sitting and, via censure motions, to force a mass resignation or dissolution of the Cabinet.

Third, while the Constitution provides for the holding of a conference committee made up of members from both houses, this does not occur in practice. According to the Constitution, a conference committee can be triggered when the two houses vote differently on a bill. The Constitution specifies that a conference committee is required in cases of disagreements over constitutional matters, the nomination of the prime minister and the ratification of international treaties. The conference committee is made up of ten members of each house. In order to pass a bill, the joint committee needs the agreement of a two-thirds majority. However, no conference committee has ever successfully passed a budget or nominated a prime minister. There are only a very few examples of legislation ever having been passed by such a committee.[2]

Comparatively speaking, the way the Japanese National Diet favors majorities in each house, combined with the relative equality between the houses, incorporates both centralized and decentralized decision making. The mechanisms available to governments for resolving conflict between the two houses are (1) to form a coalition in order to secure a majority or (2) to depend upon the cooperation of the opposition when formulating policy.

Coalition theory: models and hypotheses

In a divided Diet, parties frequently attempt to form coalitions. How are we to understand the choices available to political parties when they seek to enter into a coalition? In this section, I examine the framework suggested by coalition theory.

Coalition theory proposes two basic models of coalition formation. The first model, the office-seeking model, suggests that parties seek to form coalitions in order to secure government posts. The second model, the policy-seeking model, suggests that parties form coalitions in order to influence policy. In the office-seeking model, the main benefit of gaining a majority is seen to be gaining positions such as Cabinet posts. In order to become a majority, potential coalition partners have to form a winning coalition that commands a majority of seats. Riker (1962) argues that parties seek to avoid forming oversized coalitions that contain more parties than is absolutely necessary. Instead, they try to form a minimum winning coalition, thereby reducing the number of people who have to share in the spoils of victory. Leiserson (1966) suggests parties seek to include the least number of parties possible in order to reduce the costs of negotiation between the partners in a minimum winning coalition.

The policy-seeking model suggests that parties also consider their potential coalition partners' policy preferences, as well as the number of seats they hold. According to this model, parties that share similar policy positions are able to form coalitions more readily. Axelrod's (1970) notion of a minimal connected winning coalition, in which the members of a winning coalition share similar policy preferences, follows on from this. De Swaan's (1973) notion of a closed minimal range coalition is another variant in which the potential choice of coalition partners is based on the distance between the partners in terms of policy. Policy considerations can sometimes lead to the formation of an oversized winning coalition, which includes more parties than is necessary to form a majority.

In addition to the number of seats and the policy positions held by potential coalition partners, some researchers emphasize the impact of institutional factors on coalition formation. For Strom, political parties are motivated by the pursuit of three objectives: office, policy and votes. In addition to gaining government posts, participation in an administration can enable parties to realize their policy objectives and to increase their vote in future elections. However, it is difficult for parties to realize all three of these objectives at the same time. Participating in a coalition might force a party to compromise on its policy agenda, or even to accept policies it opposes. They also run the risk of losing their existing support base.

By analyzing minority governments, researchers have identified two factors that govern the choices parties make. The first is the policy influence differential (Strom 1990: 42). Political parties cannot realize their goal of obtaining official posts unless they participate in an administration. However, their ability to influence policy also depends on institutional factors. By participating in an administration, parties do gain more influence over policy than they would if they remained in opposition. However, in parliamentary systems where committees have wide powers, such differences are reduced. In institutional settings where opposition parties can easily exert an influence on policy, they have little incentive to participate in the administration. This makes it more difficult to form a broad coalition government.

As discussed above, the two houses of the Japanese National Diet have relatively equal powers. As a result, in a divided Diet, if opposition parties are able to form a majority in one house, they can influence the legislative process on everything except budgetary and treaty bills. Based on these factors, I propose the following hypothesis:

> *The parliamentary system hypothesis* – in institutional settings where opposition parties are able to participate in the policy-making process, they will not seek to join a coalition government.

The second factor depends upon the competitiveness and decisiveness of the upcoming election (Strom 1990: 45–9). There are a number of electoral considerations that political parties take into account when deciding whether or not to participate in a coalition government. The first factor is the degree to which pre-electoral commitments have been made by the relevant parties. The second is electoral volatility. The third factor is the influence that their participation in the government is likely to have on their showing in the next election. Finally, there is the question of whether or not coalition realignment occurs outside the electoral cycle. Where the performance of a coalition government while in office has a significant influence on electoral results and there is a chance that the participating parties might lose seats, opposition parties have less incentive to participate in a coalition government.

The elections for both the House of Representatives and the House of Councillors are particularly volatile and are influenced by public opinion about the government's performance. In the House of Representatives, 300 of the total 480 members were elected from single-member constituencies. In the House of Councillors, 146 of the total 242 members are elected from prefectural constituencies, while 96 are elected via proportional representation from a single, nationwide constituency.[3] This system tends to produce a great degree of electoral volatility. Furthermore, elections to the House of Representatives are held about every three years on average. House of Councillors elections also take place every three years but with only half the seats up for election each time. This results in a short electoral cycle, meaning that parties tend to avoid participating in a coalition government unless they have a lot to gain.

> *The election hypothesis* – where electoral competition is strong and the government's actions tend to effect electoral results significantly, opposition parties prefer not to participate in coalition governments.

Other researchers have investigated the effect of situational factors such as economic crises on coalition formation. Economic crises present opposition parties with a dilemma. They have to decide whether to cooperate with the government to deal with the crisis or to take advantage of the situation to try and win government (Moury and De Giorgi 2015). However, in the case of a natural disaster, reconstruction and recovery becomes an important issue that all parties agree must be addressed. In an unprecedented crisis, public opinion demands that the government and the opposition cooperate in order to carry out restoration and recovery programs. This may produce a situation favorable to coalition formation.

> *The crisis hypothesis* – after a crisis, government and opposition parties need to cooperate and opposition parties will seek to participate in a coalition government.

These hypotheses drawn from coalition theory suggest a number of possible avenues for examining coalition formation: the number of seats, the policy positions held by political parties, and institutional and situational factors. In the next section I analyze three attempts to form a coalition that took place during the DPJ administration and examine each of these factors in turn.

Coalition formation before and after the disaster[4]

June 2010–March 2011

The LDP has governed Japan for much of the past 60 years (1955–93, 1994–09 and 2012–present). It was the largest party in the House of Representatives from 1955 until 2009, and in the House of Councillors from 1955 until 2007. However, since the 1990s it has often failed to win an outright majority and has entered into coalition governments. Since 1999, when the LDP has been in government it has been in a coalition with Komeito.

In 2007 the LDP and Komeito suffered a catastrophic defeat in the House of Councillors elections and lost control of the house. The opposition parties, centered on the DPJ, commanded a majority in the upper house, creating a divided Diet. In 2007 the DPJ entered into a coalition with the People's New Party (PNP). The party also had an existing electoral alliance with the Social Democratic Party (SDP). However, the government still controlled a two-thirds majority in the House of Representatives, enabling it to pass important bills.

The DPJ won a decisive victory in the 2009 general elections and established a governing coalition with the PNP and the SDP. The coalition was able to command a majority in both houses. However, an electoral funding scandal involving Prime Minister Hatoyama Yukio and controversy surrounding the relocation of Marine Corps Air Station Futenma in Okinawa undermined the coalition. The SDP left the coalition as a result of the DPJ's failure to resolve the Futenma issue and the government's approval ratings plummeted. In June 2010 Hatoyama was forced to resign.

In June, Kan Naoto stood for the leadership of the DPJ. Kan, who already had leadership experience, won a majority of votes from Diet members and was elected. Following Kan's assumption of the leadership, his personal approval rating and that of the government improved. However, suggestions by Kan that it might be necessary to raise the consumption tax cost the party support in the next House of Councillors elections, where it suffered a major defeat. The DPJ's defeat in the House of Councillors elections led once again to a divided Diet. While the DPJ remained the largest party in the upper house, the ruling coalition no longer had a majority. Without a two-thirds majority in the lower house, the government now had to rely on the support of opposition parties in order to pass anything other than supply bills.

The DPJ's potential coalition partners included the LDP and a number of smaller parties. Table 1.1 shows the number of seats held by each party from 2004 to 2012. In July 2010 the DPJ and the PNP needed another 12 seats to form a majority in the upper house. If they were to enter into a coalition with Komeito, they would gain the required seats. Alternatively, if the SDP was to rejoin the coalition, the DPJ would have a two-thirds majority in the lower house and be able to pass bills that were rejected by the councillors.[5]

Table 1.1: Number of seats by party in both houses (2004–12)

Diet session	Government	Prime Minister	Period			LDP	Komeito	DPJ	PNP	SDP	YP	JCP	Other	Vacant	Total seats
						House of Representatives									
160	LDP/Komeito	Koizumi	2004	7	30	**249**	**34**	178		6		9	4		480
163	LDP/Komeito	Koizumi	2005	9	21	**296**	**31**	114	6	7		9	17		480
165	LDP/Komeito	Abe	2006	9	26	**292**	**31**	113	6	7		9	20	2	480
167	LDP/Komeito	Abe	2007	8	7	**306**	**31**	113	6	7		9	8		480
168	LDP/Komeito	Fukuda	2007	9	10	**305**	**31**	113	6	7		9	9		480
170	LDP/Komeito	Asō	2008	9	24	**303**	**31**	114	7	7		9	8	1	480
172	DPJ/SDP/PNP	Hatoyama	2009	9	16	119	21	**312**	**3**	**7**	5	9	4		480
175	DPJ/PNP	Kan	2010	7	30	116	21	**307**	**4**	7	5	9	9	2	480
176	DPJ/PNP	Kan	2010	10	1	116	21	**307**	**4**	6	5	9	10	2	480
181	DPJ/PNP	Noda	2012	10	29	118	21	**248**		6	5	9	71	2	480
182	LDP/Komeito	Abe	2012	12	26	**294**	**31**	57		2	18	8	70		480
						House of Councillors									
160	LDP/Komeito	Koizumi	2004	7	30	**114**	**24**	83		5		9	7		242
163	LDP/Komeito	Koizumi	2005	9	21	**112**	**24**	82	3	6		9	5	1	242
165	LDP/Komeito	Abe	2006	9	26	**111**	**24**	83	5	6		9	4		242
167	LDP/Komeito	Abe	2007	8	7	**84**	**20**	112	4	5		7	10		242
168	LDP/Komeito	Fukuda	2007	9	10	**84**	**21**	115	4	5		7	6		242
170	LDP/Komeito	Asō	2008	9	24	**83**	**21**	118		5		7	8		242
172	DPJ/SDP/PNP	Hatoyama	2009	9	16	85	21	**118**		**5**		7	4	2	242
175	DPJ/PNP	Kan	2010	7	30	83	19	**106**	**3**	4	11	6	10		242

176	DPJ/PNP	Kan	2010	10	1	83	19	**106**	**3**	4	11	6	10		242
181	DPJ/PNP	Noda	2012	10	29	87	19	**90**		4	11	6	25		242
182	LDP/Komeito	Abe	2012	12	26	**83**	**19**	87	3	4	11	6	23	6	242

Note: The number of seats given is the number held by each party at the opening of the Diet session. The bold numbers indicate the seats held by the ruling coalition. Source: Compiled by the author.

Next let us examine the relationship between the different parties in terms of policy. Figure 1.1 represents the positions held by candidates and voters in each party in a number of policy areas. The data used were produced from a survey of in which candidates and voters were asked the same set of questions. After analyzing their responses using principal component analysis, they were aggregated along two axes.[6] The average principal component score for each party was calculated and displayed as a graph.

These two axes are based on policy areas that have been continuously observed in Japanese politics since the 1980s. The horizontal axis represents positions on the constitution and national security. A high score indicates a tendency to support constitutional revision and remilitarization. The vertical axis represents positions on economic policy and structural reform. A high score indicates a favorable attitude towards neoliberalism. The highlighted area indicates the position held in common by the DPJ and its coalition partners.

Based on this figure, we can see that after the upper house elections, the DPJ had four options. It could either resurrect its original coalition with the SDP, enter into a coalition with Komeito, enter into a coalition with the LDP or form a coalition with Your Party (YP). Depending on which coalition partner the DPJ was able to secure, its potential future policies on nuclear energy and on raising consumption tax would vary considerably. If it revived its old coalition with the SDP, which was campaigning to abolish nuclear power and opposed raising the consumption tax, then the DPJ would have difficulty in pursuing a pronuclear agenda or increasing the tax. If it entered into a coalition with either Komeito or the LDP, there would likely be little change from the previous government's policy settings. The LDP had also gone into the House of Councillors elections promising to increase consumption tax to 10% and it had not changed its position. On the other hand, if the DPJ cemented a coalition with YP, it would probably have to pursue regulatory reform of the electric power industry and oppose a consumption tax hike.

Looking at Figure 1.1, we can see that Komeito was closest to the DPJ in terms of policy. Komeito's position was closer to the DPJ's own position than either the PNP or the SDP and the position of each party's supporters was also closer than that of any other party. Clearly, Komeito would have been the DPJ's ideal coalition partner both in terms of the number of seats it held and its policy position.

Figure 1.1: Policy positions of political parties and their supporters (based on a 2009 survey of candidates and voters)

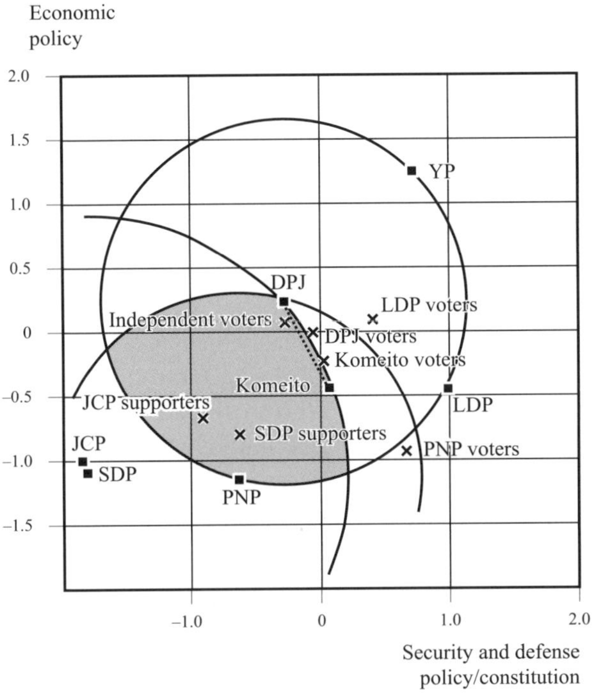

Source: Takenaka 2010.

After the House of Councillors election, Kan entered into negotiations with Komeito to form a coalition and end the divided Diet. Thanks to the cooperation of Komeito, 11 pieces of legislation were passed between September 2009 and December 2010 where the DPJ did not have the support of the LDP. Komeito did not always vote with the LDP.[7] The chair of Komeito's Diet Affairs Committee stated that 'our basic position is to fight from opposition. However, we will consider each piece of legislation on its merits…there is no need for us to always follow the LDP. We are free to cooperate with the government on drafting bills and proposing amendments' (*Asahi Shimbun*, July 21, 2010, morning edn).

However, the climate for negotiations began to take a turn for the worse. The DPJ government was facing constant criticism over a number of incidents. In September 2010 an incident occurred off the Senkaku Islands in which the captain of a Chinese fishing boat was arrested by Japan Coast Guard officials. However, the district prosecutor in Naha declined to prosecute the captain, announcing that he would be released without charge. Footage of a collision between the fishing boat and the Japan Coast Guard vessel was subsequently released over the internet, leading to criticism of the Kan government.

In October the Kan government suffered another blow when the mandatory indictment of senior DPJ lawmaker Ozawa Ichirō was made public. Then in November, Russian President Dmitry Medvedev visited the disputed Northern Territories, the first such visit by a Russian head of state. Justice Minister Yanagida Minoru was also forced to resign over a statement in which he seemingly made light of his duties in the Diet. The LDP also successfully censured Chief Cabinet Secretary Sengoku Yoshito and Land, Infrastructure and Transportation Minister Mabuchi Sumio in the House of Councillors.

These incidents caused the party's support in the polls to fall sharply for approximately two months. Figure 1.2 shows support for the Kan government, the DPJ and the LDP. While support for both Kan's government and the DPJ had increased after Kan's election as DPJ leader in June 2010, between September and November support for his government fell from 59% to 27%. These results were mirrored in other polls. In polling carried out by the *Mainichi Shimbun*, this was the most dramatic two-month fall in popularity ever recorded over such a period (Maeda 2015).

As support for the government and the DPJ fell, Komeito's attitude towards entering into a coalition hardened. The party did not regard the DPJ government as having much chance of survival. At the Komeito conference in October, party leader Yamaguchi Natsuo had stated that he would not seek to topple the current government. However, by November, amidst falling popular support for the government, the party adopted a more confrontational stance (*Mainichi Shimbun*, December 6, 2010, morning edn). It distanced itself from a DPJ whose popularity was declining due to dissatisfaction over the government's clumsy management of Diet business (Yomiuri Shimbun Seijibu 2011: 26). Komeito voted with the LDP to censure government ministers, further distancing itself from the Kan government. While there was support within Komeito for a coalition, including from party leader Yamaguchi, it never eventuated.

*Figure 1.2: Support for the government, the DPJ and the LDP (*Asahi Shimbun*)*

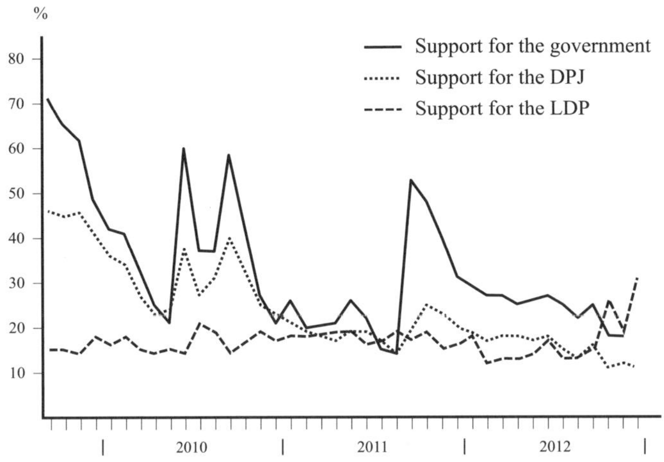

Source: Compiled by the author based on surveys conducted by the *Asahi Shimbun.*

As the prospect of a coalition with Komeito faded, the DPJ looked once more to its old coalition partner, the SDP. The SDP had indicated that it was willing to work with the DPJ, such as in an electoral alliance formed in September. Furthermore, it supported the government's supplementary budgets and opposed the LDP's censure motions. However, following the indictment of Ozawa Ichirō, internal conflict within the DPJ intensified. The party decided to suspend Ozawa's party membership until his case was concluded. Sixteen DPJ members, centered on Ozawa's faction, announced that they would leave the party's parliamentary group in protest at the decision. Open conflict broke out between them and the party leadership. The departure of the 16 members meant that the possibility of achieving a two-thirds majority in the lower house via a coalition with the SDP was now lost. The DPJ's internal conflict and its failure to respond to the SDP's policy proposals led the SDP to stop cooperating with the government on supply (*Asahi Shimbun*, February 18, 2011, morning edn). Then, on March 6, Foreign Minister Maehara Seiji resigned over an electoral funding scandal. The investigation even threatened to implicate the prime minister.

Kan's attempts to form a coalition, whether with Komeito or the SDP, or by bringing the smaller parties together, all failed.[8] Coalition theories that emphasize number of seats and policy similarity cannot explain this failure to form a coalition with parties that otherwise possessed optimum numbers of seats and policy compatibility. The election hypothesis provides a better explanation of the position of Komeito, with whom the DPJ could have formed a minimum winning coalition. Komeito decided not to participate in a coalition with the government due to the DPJ's declining popularity.

March–April 2011

In the midst of the DPJ's failed attempts to resolve the problem of the divided Diet, the Great East Japan Earthquake occurred. Before the disaster, the investigation into the abovementioned political donations scandal threatened to envelop Prime Minister Kan. After the disaster, the situation changed overnight and the political infighting halted. On the evening of March 11, the LDP even declared that 'if we were asked to form a coalition, then we could not refuse' (Tanigaki 2013: 86).

Widespread public support also existed for either party realignment or a grand coalition. There were numerous calls for the government and the opposition to come together to deal with the crisis. Public opinion polls (yes–no format) in the national newspapers concerning a grand coalition found 60% support for such a move and 30% opposed. Figure 1.3 shows the results of a public opinion poll on the composition of the government. Respondents were asked to choose their preferred form of government out of four possible choices. Support for a new type of administration, such as a grand coalition or party realignment, far exceeded support for a coalition centered on either the DPJ or the LDP.

However, at a leaders' summit held from March 11 to 13, Kan did not ask the opposition to join in a grand coalition nor to cooperate on budget-related bills. The DPJ and PNP leaderships proposed that Kan form a government of national unity and appoint members of the opposition to the Cabinet. Kan tried to organize a meeting with LDP president Tanigaki Sadakazu, but was unable to secure a one-on-one meeting (Kan 2012: 136–7). On March 19 Kan telephoned Tanigaki and invited him directly to join the DPJ in a grand coalition. He asked the LDP leader 'to share responsibility for dealing with this crisis affecting the nation', and offered to

Figure 1.3: Public opinion on the composition of government

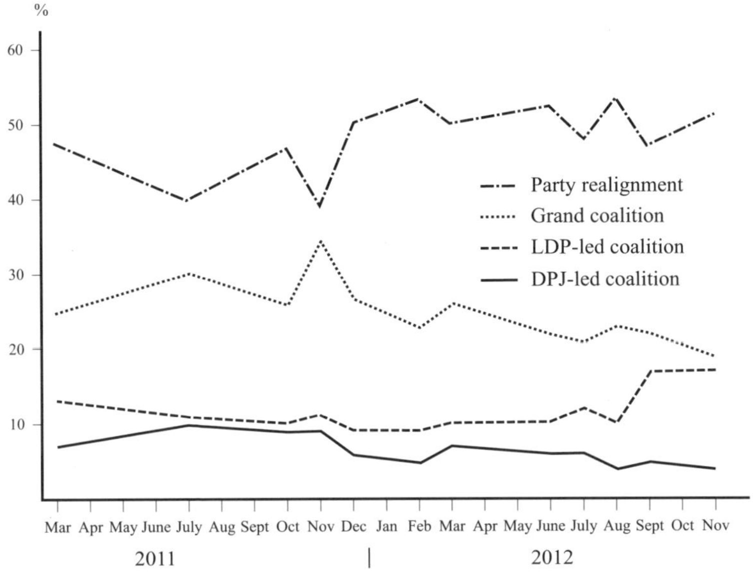

Source: Compiled by the author from opinion polling carried out by the *Yomiuri Shimbun*.

make Tanigaki deputy prime minister and Minister for Disaster Reconstruction as part of the grand coalition. However, by March 19 the favorable conditions for a grand coalition had dissipated. The LDP already had a sense that Kan wanted to make Tanigaki deputy prime minister and reconstruction minister. It seems the party was highly suspicious of any attempt to make it share responsibility for such a serious nuclear disaster (*Jiji Tsūshin*, March 20, 2011) and so decided not to participate in the Cabinet. Tanigaki formally refused Kan's invitation.

However, a grand coalition was still possible. Tanigaki indicated that his party was willing to work with the government to build trust and to compromise on policy. He told a press conference that the disaster had brought about 'a national crisis' and that his party had 'every intention of cooperating with the government where possible...I think we need to develop mutual trust by working together...The real problem is what kind of trust, what kind of mutual cooperation.'[9] He also stated that 'if there is a proposal for a new coalition, we need to have a frank discussion about policy'

(*Mainichi Shimbun*, March 20, 2011), signaling that the DPJ would have to compromise on policy if there was to be a grand coalition.

Later, Sengoku and then LDP vice-president Ōshima Tadamori held high-level talks, as did Okada Katsuya and LDP secretary-general Ishihara Nobuteru. In these talks, the LDP made it a condition of any potential grand coalition that an LDP prime minister be appointed temporarily, and called on the DPJ to modify its policies. It asked the DPJ to jettison some of its core policies, including the child allowance, the abolition of fees for using the national expressway network, social security for farming households and the abolition of fees for high school education. The funds saved by abandoning these policies could then be channeled into reconstruction. On April 1 Kan announced a partial freeze on expenditure from the 2011 budget, thereby indicating his willingness to compromise with the LDP. However, on April 5 Tanigaki announced that the LDP planned to 'fulfil [its] proper role as an opposition', extinguishing the possibility of a grand coalition.

We can see from this discussion that while negotiations about cooperation between the two parties and a possible grand coalition did take place, they never eventuated. This invalidates the crisis hypothesis. Amid increasing criticism of Kan's clumsy handling of the negotiations and the administration's handling of the crisis, the LDP decided not to join in a governing coalition. These facts can be best explained via the election hypothesis.

April–August 2011

While the LDP and Komeito refused to participate in a coalition government, they did become more involved in the formation of policy. On April 11 the DPJ presented a draft of its first supplementary budget for the year to the LDP, Komeito and the PNP. The budget included 4 trillion yen for earthquake reconstruction. Tanigaki indicated his support for the supplementary budget, saying that his party would 'cooperate fully wherever we can in responding to the disaster'. However, as the debate on how to fund recovery and reconstruction developed, the LDP immediately called on the DPJ to drop some of its major policies. The LDP proposed the following modifications: (1) a partial freeze on the reduction in fuel excise that had been brought in to deal with sudden price hikes, (2) a partial freeze on the trial abolition of fees for using the expressway network, (3) postponing the introduction of a new 2000 yen flat charge for using expressways on weekdays, (4) a freeze on the child allowance supplement (worth 200 million yen) and (5) a proposal to divert 2.5 trillion yen that had been put aside to maintain the government's future pension contributions for other purposes. On April 29 the DPJ, LDP and Komeito policy chiefs reached a compromise at a joint meeting and the supplementary budget was passed unanimously on May 2.

On May 13 the government introduced a bill for a Basic Act on Reconstruction to the Diet. The LDP and Komeito also presented their own versions of this bill. The main difference between the three bills was the kind of agency that would be created to supervise reconstruction and the powers that it would have. The government's bill limited the reconstruction authority to a planning and coordination role, while the LDP and Komeito bills gave it strong powers to implement policy. The LDP's bill also provided for the issuing of reconstruction loans to fund the plan, while Komeito's included a proposal for a system of designated disaster zones.

The Kan government presented an amended bill to the LDP and Komeito. The new bill would establish a Reconstruction Agency with the power to put forward plans and coordinate the disaster response. On April 29 the three parties agreed that the bill would enable the government to issue reconstruction bonds, create a method for managing the cost of reconstruction and establish a system of designated reconstruction zones. On June 1 the parties reached a final agreement. The government and the LDP withdrew their original bills and the amended bill was introduced to the Diet as a private member's bill. The Basic Act on Reconstruction was passed, though YP and the JCP voted against it.

During the political truce that followed the disaster, the opposition cooperated with the government on recovery and reconstruction. Nevertheless, it continued to criticize Prime Minister Kan's response to the disaster. With approximately 20 disaster response centers working to deal with the situation, there was confusion about the chain of command, leading to criticism over delays in the disaster response. Kan's ability to manage the crisis came under attack.

On May 27 Ozawa told an American newspaper, 'the sooner Kan is replaced as prime minister the better', signaling his intention to topple the prime minister. Former LDP Prime Minister Mori Yoshirō later revealed how he and Ozawa had plotted to pass a motion of no confidence in the Cabinet with the support of Ozawa's parliamentary group. They then intended to form a grand coalition with Tanigaki as prime minister.[10]

On June 1 the LDP, Komeito and the Sunrise Party of Japan put a motion of no confidence in the Cabinet to the Diet. They argued that 'Kan's refusal to lay out a clear plan of action is only adding to the confusion and hindering attempts at reconstruction'. Support for the motion, centered on Ozawa's group, continued to grow. Five senior vice-ministers and parliamentary secretaries affiliated with the group announced their resignations. The no-confidence motion looked as if it might pass. Seeking to avoid a split in the party, Kan met with Hatoyama and other party leaders. He promised to resign on condition that the Diet passed a number of bills, including budget measures. Following Kan's announcement, the no-confidence motion was defeated, with only two DPJ members voting in favor. Kan's conditions included the passage of the second supplementary budget for 2011, a bill authorizing the government to issue special deficit-financing bonds and another to establish a feed-in tariff to encourage renewable energy.

In the negotiations over Kan's conditions, the DPJ continued to backtrack on its other major policies. The party acceded to most of the LDP's demands and declared that it would alter some of its most important policies. The compromise reached included the introduction of an assets test for the child allowance and the abolition of the child allowance for 2011 (and restoration of the previous childcare allowance system). The parties also signed a memorandum of understanding to revisit the abolition of expressway fees and high school fees, and social security for farming households. Once the DPJ had accepted these changes, the opposition agreed to pass the bill authorizing the government to issue special deficit-financing bonds that had been at the center of the negotiations between the government and the opposition.

September 2011–December 2012

Following Kan's agreement to resign, the DPJ held internal party leadership elections. The five candidates who stood for the post differed over the kind of coalition they would seek to form; whether to revise the party's electoral manifesto to include a set of rules requiring the party to find a consensus with the DPJ, the LDP and

Komeito; if and when to raise consumption tax; and whether or not to expel Ozawa Ichirō from the party. Candidate Noda Yoshihiko indicated his desire for a consensus with the major opposition parties and the creation of 'a Cabinet of national salvation' based on a grand coalition. Noda was also emphatic on the need to raise consumption tax, which he described as essential (*Mainichi Shimbun*, August 18, 2011, evening edn). Noda was elected party leader and therefore became prime minister. The LDP and Komeito reacted positively to Noda's election. While Tanigaki was reserved about supporting the government on matters other than reconstruction, he praised Noda's policies and his political stance. Komeito leader Yamaguchi also discussed the need for the three parties to build a consensus and called for the maintenance of their special relationship.

The other opposition parties were critical of the way talk of a grand coalition dominated the DPJ leadership elections. YP leader Watanabe Yoshimi stated that 'in practice, there is already a grand coalition between the DPJ, the LDP and Komeito. Diet business is being conducted behind closed doors and the bill to bailout TEPCO is barely being debated in committee' (*Mainichi Shimbun*, August 17, 2011, morning edn).[11] JCP chairperson Shii Kazuo also distanced his party from the DPJ following Noda's election. He criticized Noda's position on raising consumption tax and forming a grand coalition, stating that 'the DPJ is essentially already in a grand coalition with the LDP having decided to substantially increase the consumption tax' (*Mainichi Shimbun*, August 30, 2011, morning edn).

Nevertheless, while Noda asked the LDP and Komeito to form a coalition, it never eventuated. While the DPJ leadership ballot was taking place, Tanigaki doused the idea of a coalition, saying that 'in almost all of the single-member constituencies we are in competition with the DPJ…in this context, entering into a grand coalition has to be thought of as the exception to the exception'.[12] On top of all this, the Noda government's popularity reached a three-month low, leaving the other parties with little enthusiasm for joining it in a coalition.

While the LDP and Komeito refused to participate in a coalition with the DPJ, they did become more involved in formulating policy. The increasing cooperation between the DPJ, the LDP and Komeito can also be observed by identifying patterns in the legislation that was passed. Table 1.2 shows how the parties cooperated on

Table 1.2: Patterns of cooperation on legislation (plenary session, first referred house)

Group	Hatoyama 173 Oct–Dec 2009	Hatoyama/Kan 174 Jan–Jun 2010	Kan 176 Oct–Dec 2010	Kan 177 Jan–Aug 2011	Noda 179 Oct–Dec 2011	Noda 180 Jan–Sep 2012	Noda 181 Oct–Dec 2012	Total
1. Unanimous	40.0	47.2	50.0	57.3	46.2	37.7	0.0	46.6
	4	17	7	47	6	23	0	104
2. Opposed by one party	40.0	22.2	7.1	26.8	15.4	16.4	0.0	21.1
	4	8	1	22	2	10	0	47
3. Cooperation with a third party	20.0	27.8	42.9	1.2	0.0	6.6	0.0	10.3
	2	10	6	1	0	4	0	23
4. Cooperation with the LDP	0.0	2.8	0.0	14.6	38.5	39.3	100.0	22.0
	0	1	0	12	5	24	7	49
No. of government-sponsored bills passed	10	36	14	82	13	61	7	223

Note: Diet sessions in which no bills were passed are omitted.
Source: Compiled by the author.

legislation in each Diet session. In this table, a vote by a party's parliamentary group for a piece of government-sponsored legislation is counted as an instance of cooperation. If we consider only seven parties – DPJ, PNP, LDP, Komeito, JCP, YP and SDP – there are 15 possible patterns whereby different combinations of opposition parties could cooperate to pass government-sponsored bills.

In Table 1.2 these patterns are classified into four groups. Group 1, 'Unanimous', refers to bills that passed with the support of all seven parties. Group 2, 'Opposed by one party', refers to bills that were opposed by only one of the seven parties.[13] Groups 1 and 2 include almost all the bills that passed with the consent of the government and the major opposition parties. Bills that do not fall under this category (e.g. those bills where two or more opposition parties voted against or abstained) are divided into a further two groups. These are Group 3, 'Cooperation with a third party', where the government obtained the support of a party other than the LDP, and Group 4, 'Cooperation with the LDP', which includes bills that had the support of the LDP.

By looking at Table 1.2, we can examine the changing patterns of cooperation on legislation during the DPJ administration. During the Hatoyama Cabinet, most bills were passed either unanimously or were opposed by a single party. In the 176th Diet session, when the Diet first became divided, the most common pattern was cooperation with a party other than the LDP, covering 42.9% of legislation passed. After the disaster, there was a greater degree of cooperation between the government and the opposition and up to 57.3% of bills were passed unanimously. Following Noda's election to the DPJ leadership, about 40% of bills were passed with the support of the LDP, indicating a significant departure from previous patterns. This changing pattern of legislative cooperation mirrors the trends discussed in the previous sections.

The Noda government's main source of cooperation in the Diet was the LDP and Komeito. His government worked to maintain this three-party consensus. Despite significant opposition from within his own party, Noda pushed the consumption tax bill through the House of Representatives with the support of the LDP and Komeito. The Ozawa group, which staunchly opposed the bill within the DPJ, split from the party over the issue.

Despite the government's attempts to secure a coalition with various parties in the context of a divided Diet, it was never able to secure a stable majority. This situation continued when the earthquake struck in 2011. Despite significant public pressure on the government and the opposition to cooperate with one another to deal with the crisis, a grand coalition never eventuated. Both Kan and Noda asked the LDP to participate in a governing coalition, but the LDP chose to remain in opposition. Nevertheless, the LDP and Komeito did influence the formulation of bills. Following Noda's takeover of the DPJ leadership, there was increased cooperation between the three parties.

In terms of coalition theory, neither number of seats nor policy position alone can account for these developments. The prime minister's personal qualities (such as his connections and negotiating skills) also cannot explain the DPJ's failure to form a coalition. Unlike Kan, Noda had a good working relationship with Tanigaki. However, though both leaders asked the LDP to form a grand coalition, neither was successful in doing so. Looking at the hypotheses I proposed earlier, the election hypothesis and the parliamentary system hypothesis appear to have greater explanatory power than the crisis hypothesis. The disaster did produce calls from the public for the government and the opposition to cooperate and there was progress in discussions about forming such a coalition. However, while the government's approval ratings fell, the LDP and Komeito chose not to join a coalition. Instead, they become more involved in making policy. As a result, the two parties were able to participate in drafting the Basic Act on Reconstruction and to force amendments to some of the DPJ's signature policies. They also obtained agreement on the need to increase consumption tax. Thinking ahead to the next election, the opposition parties focused on policy making in the Diet and decided to remain in opposition.

Disaster-related legislation and the legislative process

While the DPJ's failure to form a coalition meant that it was still in a minority in the upper house, its continued cooperation with the LDP and Komeito led to the development of a three-party consensus. By examining the legislative process, we can see how the three parties maintained this cooperative relationship. In this section I look at the government bills legislation during the DPJ administration from the perspective of legislative productivity and the Diet's committee process.

Legislative productivity

Both the government and individual members of both houses of the Diet have the right to introduce bills. However, a proponent of a private member's bill must obtain the support of 20 members if the bill is introduced to the House of Representatives (50 in the case of bills affecting the budget) and ten if it is first introduced in the House of Councillors (20 in the case of budget-related bills). Figure 1.4 shows the legislative productivity of the National Diet since 1956, including government bills and private members' bills. The shaded sections indicate years when the two houses were divided. This figure suggests two points. First, the DPJ's legislative productivity in terms of government bills was extremely low. The DPJ government introduced 295 bills to the Diet. Only 67.1% of these bills were enacted. The average legislative productivity from the end of the Second World War until August 2009 was 85.5%, meaning that the DPJ government's legislative productivity was 20% below average.

However, it is not true that legislative productivity is necessarily lessened during a divided Diet. Since 1955 there have been three periods with a divided Diet. Legislative productivity was 91.0% during the divided Diet of 1989–93, 87.9% during 2007–09 and 83.3% during 2007–09. The reasons behind the DPJ's lower legislative productivity lay in the party's failure to think broadly enough about how to cooperate with other parties and to refine its bills so they had a chance of passing. As a result, a large number of government bills failed to pass the Diet. One way of explaining this relative lack of productivity when compared with previous divided Diets is to see it as a result of stagnation caused by the majority doctrine.

By contrast, after the disaster the number of private members' bills introduced in the House of Representatives that were successfully enacted was the highest it has ever been. In 2011 legislative productivity for private members' bills in the lower house was 74.3%, about the same as for government bills. This reflects the degree of cooperation between the ruling and opposition parties. Legislative stagnation based on the majority doctrine seems therefore to reflect only one aspect of the legislative process. Compared with earlier

Figure 1.4: Legislative productivity for government and private members' bills

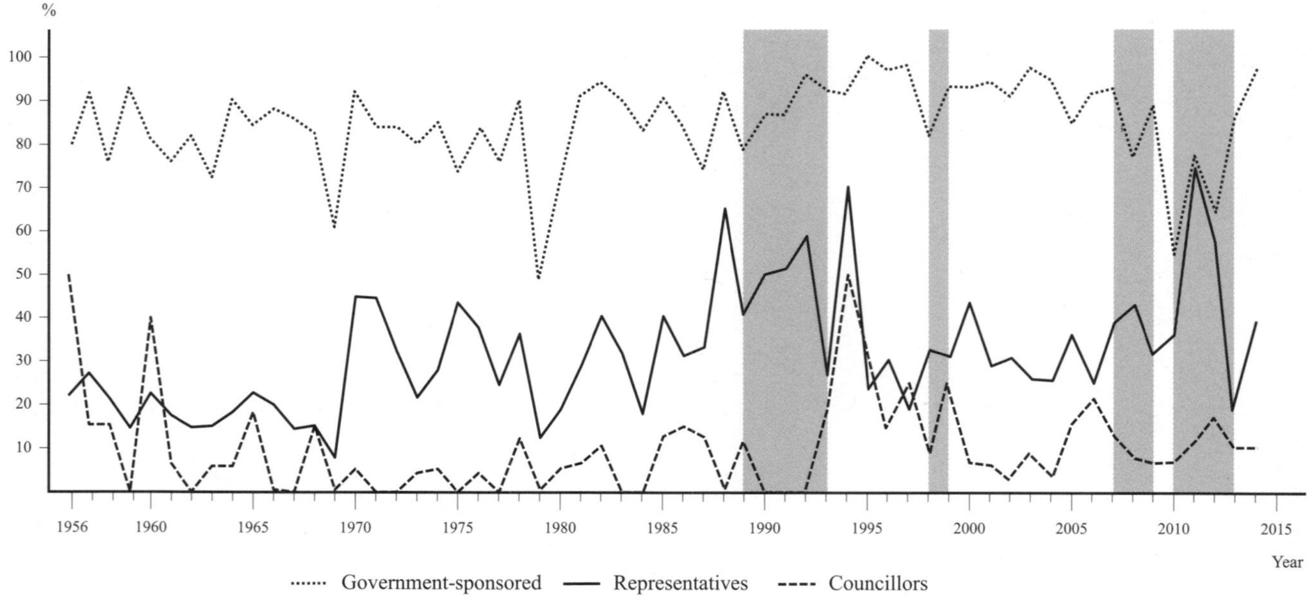

Note: The figure given for 2014 includes bills introduced up until the 186th Diet session. The shaded sections indicate years when the Diet was divided.
Source: Compiled by the author.

Table 1.3: Legislative productivity during the DPJ administration by type of bill

	Rejected	Enacted (1)	Enacted (2)
Disaster-related	19.1	80.9	87.2
	9	38	41
Other	35.5	64.5	73.4
	88	160	182

Note: The upper figure refers to the percentage, the lower to the number of bills. The Rejected and Enacted (1) categories include bills that were resolved during a single Diet session. The Enacted (2) category includes bills that were enacted after having been held over from a previous session.

divided Diets, legislative cooperation between the government and opposition parties proceeded by means of private members' bills.

How did the government and the opposition deal with disaster-related legislation? Table 1.3 shows legislative productivity for both disaster-related and other legislation.[14] It shows that 80.9% of disaster-related legislation was ultimately enacted. If we include bills that were enacted after a long period bogged down in committee, this figure rises to 87.2%. Legislative productivity for disaster-related bills was 15% higher than for other bills. The Diet also rejected nine disaster-related bills, including three related to nuclear energy.[15]

The committee process

How did disaster-related bills fare in committee? The Japanese National Diet is a committee-based legislature and all bills introduced to the parliament are referred to a committee for examination. Once they are introduced, it often takes some time for a bill to be referred to a committee. The length of the committee stage depends on the bill. Committees have wide powers to examine ministers and government representatives or to propose amendments. Once a bill has passed the committee stage, it is returned to the house, where members vote on it in a plenary session. Once a bill has gone through this same process in both houses, it finally becomes law.

Table 1.4 shows the average number of days each step in this process took by type of bill. By looking at how long each stage in the process took in each house, we can see how disaster-related

Table 1.4: Legislative process by type of bill (average)

	First house			Second house			Total		
	Referral	Committee	Plenary	Referral	Committee	Plenary	Referral	Committee	Enactment
Disaster-related	21.2	8.5	0.9	3.5	3.3	0.8	24.3	7.6	31.9
	44	38	38	38	38	38	38	38	38
Other	58.9	12.2	1.9	9.7	6.6	1.4	59.4	17.6	77.4
	207	166	166	163	160	160	166	160	160

Note: The top figure is the average number of days and the bottom figure is the number of bills.
Source: Compiled by the author.

Figure 1.5: The percentage of votes for government bills by party (plenary session, first referred house)

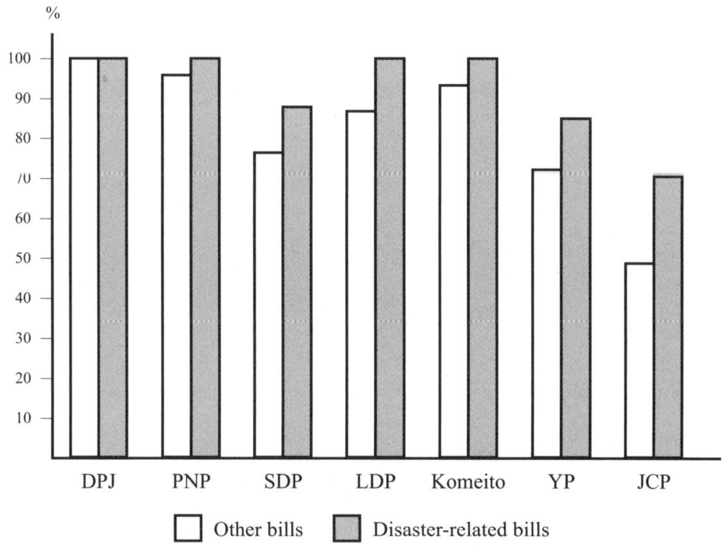

Source: Compiled by the author.

legislation progressed through the committee stage. On average it took 21.2 days for disaster-related bills that were introduced to one of the two houses to be referred to committee. This compares with 58.9 days for other bills. Disaster-related bills took much less time to go through committee and be voted on in a plenary session, making the entire process smoother in every respect.

Finally, let us examine how each party voted for each type of legislation. Figure 1.5 shows how each party voted in a plenary session of the first referred house where the bill was first introduced. These numbers are for the day the final vote took place. It should be noted that bills with little prospect of being enacted are unlikely to proceed to a full vote in a plenary session. Figure 1.5 shows that the PNP, the LDP and Komeito supported 100% of disaster-related bills. The SDP and YP each voted for the same bills about 85% of the time, while the JCP supported them just over 70% of the time. The fact that support for non-disaster-related bills was only about 5%–10% lower than for disaster-related bills for every party, with the exception of the JCP (which was 20%), reflects the degree of cooperation that existed in the Diet. These figures also demonstrate that it was the LDP and Komeito that were most active in their support for the government, rather than the SDP or YP.

Conclusion

In this chapter I have analyzed the following two points about the DPJ's attempts to form a coalition during the divided Diet. First, how did coalition negotiations between the government and the opposition parties proceed? Second, what were the features of the legislative process for disaster recovery and reconstruction?

During the divided Diet, the government tried to form a majority by forming a coalition with a third party. However, support for the government and for the DPJ as a party was falling and there were doubts about whether the DPJ could hang onto power. In this context, the cost of participating in a coalition with the DPJ increased and none of the opposition parties chose to participate. After the disaster, public opinion strongly favored cooperation between the government and the opposition. The LDP considered joining a coalition at one stage. However, it refused both Kan and Noda's overtures and a grand coalition never eventuated. The LDP chose to remain in opposition, while still offering support for disaster-related bills, and forced the DPJ to compromise on some of its signature policies.

Two legislative processes coexisted in the National Diet after the disaster. The first was a cooperative process centered on disaster-related bills. We can examine this process by looking at the political negotiations that took place and the way disaster-related bills passed through committees in the Diet. Of course, despite the general degree of cooperation that existed for disaster-related bills, not all were passed. Nor can the cooperation on disaster-related bills be entirely separated from what happened to other bills. While the two types of bills were treated differently by the opposition parties, the LDP traded its cooperation on disaster-related bills for amendments and even the wholesale abandonment of some of the DPJ's signature policies.

The second legislative process involved a three-party consensus. In the divided Diet, the DPJ administration had three possibilities for legislative cooperation. It could adopt a social-democratic position by aligning itself with its former coalition partner the SDP, continue the previous government's policies by aligning itself with Komeito and the LDP, or adopt a neoliberal position by aligning itself with YP. After the disaster the DPJ chose to accept the LDP and Komeito's amendments to its policies and to work with them to raise consumption tax and support nuclear energy. This is how the two processes – one based on cooperation between a large number of political parties to enact disaster-related bills and one based on a three-party consensus – coexisted.

This had two implications. First, the two major parties agreed on the need to raise consumption tax and restart the nation's nuclear reactors. However, by governing through a three-party consensus, Japan's two major parties narrowed the difference between themselves. Even though the two parties retain different bases of support in the electorate, the actual policy difference between them has shrunk. In practice, the LDP, the DPJ and Komeito all approve of restarting Japan's nuclear reactors, with only a few conditions. Furthermore, they all agree that nuclear energy should not be immediately abolished, but continued until at least 2040.

One way political parties can differentiate themselves from one another is by having a clearly different position on a particular issue. By coming to an agreement over a number of pressing questions, they lost one possible means of showing this to voters. As a result, consumption tax and nuclear energy became largely irrelevant as electoral issues. Had the DPJ cooperated instead with the SDP or YP, it is possible that consumption tax would have been kept at the existing level and a greater change might have come about in nuclear policy.

Second, looking at the three hypotheses I proposed in the introduction, the election and parliamentary system hypotheses seem to best capture the situation that developed in the divided Diet. The crisis hypothesis suggested that if public opinion clearly supported the formation of a broad coalition, then the political parties would seek to do so. However, when both Kan and Noda failed to form a grand coalition, this possibility disappeared. By contrast, both before and after the election, and even after changing its leader, the DPJ's popularity plummeted. This led the opposition parties to back away from any possible coalition. It therefore seems reasonable to conclude that the election hypothesis best explains what happened in the divided Diet. Furthermore, both after the disaster and the leadership change, the parliamentary system hypothesis also has a degree of explanatory power.

Acknowledgments

I would like to thank Katsumata Hiroto and Matsumoto Taku for assisting in preparing the data on the committee process. This research was supported by a Grant-in-Aid for Scientific Research (B) from the Japan Society for the Promotion of Science (Project Number 23330040).

2 TEPCO's Political and Economic Power before 3.11

KAMIKAWA Ryūnoshin

In this chapter I discuss the ways in which the Tokyo Electric Power Company (TEPCO) wielded political and economic power before the Great East Japan Earthquake. I focus on how the company used its authority to promote nuclear power. While TEPCO is the main subject of this chapter, in discussing the electric power industry's influence on nuclear policy I consider the activities of all ten electric utilities (Hokkaidō, Tōhoku, TEPCO, Hokuriku, Chūbu, Kansai, Chūgoku, Shikoku, Kyūshū and Okinawa) that make up the industry body known as the Federation of Electric Power Companies of Japan (FEPC). TEPCO is the most influential organization within the industry and the driving force behind the FEPC.[1] The Ministry of Economy, Trade and Industry (METI) – until 2000, the Ministry of International Trade and Industry (MITI) – also played a critical role, cooperating with the utilities to promote nuclear power. The roles of the government and TEPCO in the political struggle to pursue nuclear energy were intertwined in a network of overlapping interests that has been termed the 'nuclear village'.[2] In this chapter, I outline how the nuclear village operated in the spheres of the economy, government, the Liberal Democratic Party, the academy, the electric power industry unions, municipal and prefectural governments, the courts, the mass media and the education system.

Numerous authors have already pointed out that the accident at the Fukushima Daiichi Nuclear Power Station was not simply a natural disaster that 'exceeded expectations' but a manmade disaster that occurred because numerous warnings went unheeded. Soeda (2014) explains that after the 1993 Hokkaidō earthquake, seismologists re-evaluated existing predictions about tsunami risks at Japan's nuclear reactors. These investigations led scientists to point out the risk of a large tsunami, exceeding existing predictions, hitting the Fukushima Daiichi Nuclear Power Station (located in Ōkuma Town and Futaba Town, Futaba District, Fukushima Prefecture). Both the FEPC and the government's regulatory agencies were also aware of the risk that a large tsunami might damage reactor cores or cause a total loss of electricity to a nuclear power station. However, the FEPC did not modify its guidelines for tsunami preparedness to account for this new information. It neither checked the safety of the nuclear reactors (retrospective checking), nor made modifications to account for the risk of larger tsunamis (retrofitting). On the contrary, in a series of meetings held between 1999 and 2001, the Tsunami Evaluation Subcommittee of the Japan Society of Civil Engineers approved the underestimating of the tsunami level predictions. This subcommittee is fully funded by the utilities and more than half of its members work in the electric power industry. Despite repeated warnings from seismologists and regulators about the risk posed by tsunamis, TEPCO postponed the implementation of any countermeasures. Numerous people pointed out the possibility of a large-scale earthquake or tsunami that exceeded previous estimates. Others called on the utilities to plan for the possibility of a total loss of power to the station or a serious accident (Tokyo Shimbun Genpatsu Jiko Shuzaihan 2012: 167–219). The plaintiffs in a number of antinuclear lawsuits also made similar points (Kaido 2011: 58–79).

However, the utilities, the government and their academic advisers disregarded these warnings. Increasing safety measures would lead to increased costs. Nuclear advocates were also concerned that if they were to acknowledge the need for such safety measures, then they would likely face a backlash from residents of host communities concerned that such an accident might occur (Tokyo Shimbun Genpatsu Jiko Shuzaihan 2012: 191).[3]

TEPCO was able to ignore these warnings because it had the power to suppress opposition to nuclear power and to construct the 'myth of safety'. Since the disaster, journalists and insiders have begun to talk about aspects of TEPCO's power that were rarely spoken of in the past. The outline of TEPCO's power in relation to other political actors in this chapter is based on these reports.

The economy

TEPCO is one of Japan's largest and most important companies. In March 2010 (at the end of the Japanese financial year), TEPCO had 13.204 trillion yen worth of assets, making it the thirteenth largest

Figure 2.1: Map of Japan's nuclear power stations (May 2015)

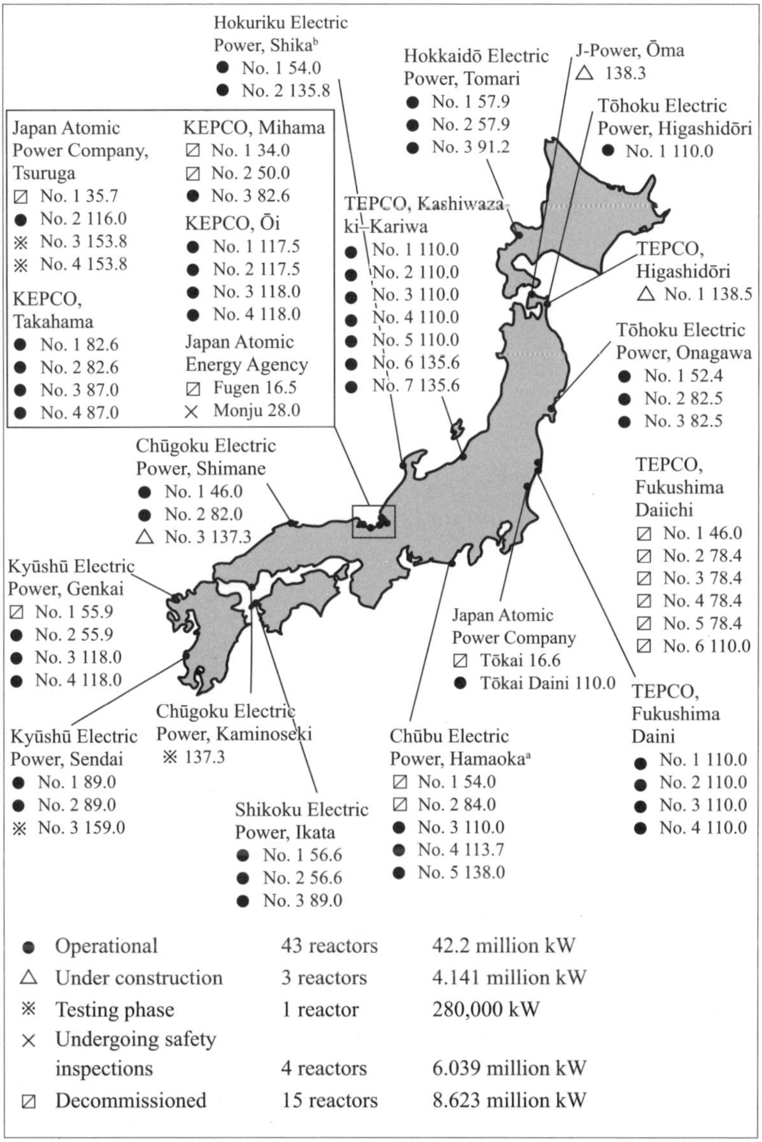

a Reactors 3–5 at Hamaoka are suspended pending completion of tsunami protection works.

b Reactor 2 at Shika is operating temporarily at a different output level but here its original output is given.

Source: Genshiryoku Shiryō Jōhō Shitsu (ed.), *Genshiryoku shimin nenkan 2015* (*Citizen's nuclear energy annual 2015*), Nanatsumori Shokan.

company in Japan in terms of assets. Its free cash flow was 389 billion yen (eleventh largest) and its planned capital investment for the 2010 financial year was 789.9 billion yen (second largest) (*Shūkan Tōyō Keizai*, April 23, 2011: 37). At the end of March 2011 the company had paid-up capital of 900 billion yen, 36,700 employees and sales of 5.15 trillion yen. If TEPCO's subsidiaries are included in the total, these figures rise to sales of 5.37 trillion yen and a workforce of 53,000.[4]

In order to procure materials and equipment for the construction of power stations, the utilities have extensive dealings with other companies. In 2010 TEPCO's total expenditure on procurement for materials and services was 1.2527 trillion yen. Furthermore, the utilities are not overly concerned with cost and frequently enter into no-bid contracts with suppliers. Because the utilities do not press their suppliers to lower costs, suppliers are able to make large profits from their dealings with the utilities.[5] This gives the utilities great influence over other companies.

With their top-heavy management structures, the utilities have also been active in the business world. Top TEPCO executives have held the posts of vice-chairperson of the Japan Business Federation (Keidanren), vice-chairperson and president of the Japan Association of Corporate Executives, and chairperson of the Divisional Group of Natural Resources and Energy of the Tokyo Chamber of Commerce and Industry.[6]

The top-heavy structures at the utilities are the direct result of government policy. The government guarantees a monopoly on both generation and supply to each utility in its designated region. Thus far, attempts to liberalize this highly regulated market have been slow. The profitability of the utilities is also guaranteed by the comprehensive cost-accounting method enshrined in Article 19 of the *Electricity Business Act 1964*.[7] This results in the utilities having little cost awareness. After the disaster, they received criticism for their use of no-bid contracts in their dealings with associated companies and for the generous salaries and conditions enjoyed by their employees and executives, all of which contributed to increasing their operational costs.

A further consequence of the comprehensive cost-accounting method is that the utilities' profits are guaranteed to increase in direct proportion to the amount they invest in infrastructure, such as additional generating capacity and extending the transmission network. As a consequence of this, the utilities have been proactive about investing in infrastructure.[8] They have done so with the full knowledge of the government. In January 1978 the government called on the industry to help stimulate the economy by matching its investment in electric power infrastructure to the government's own investment in public works (Saitō 2012: 218–9). Since then, TEPCO has increased its capital investment or placed orders early so as to keep up with government spending on public works. That is why TEPCO's level of investment in infrastructure is second only to that of the Japanese telecommunications giant NTT. This is despite the fact that the utility only provides power to the Kantō area, while NTT covers the entire country (Ōshika 2013: 267–8).

By the end of December 2010, TEPCO's government-approved approach to infrastructure spending had produced interest-bearing liabilities of 7.4641 trillion yen (*Shūkan Tōyō Keizai*, April 23, 2011: 34). Yet TEPCO has had no difficulty in accessing credit. Article 37 of the Electricity Business Act empowers utilities to issue special corporate bonds known as electric power bonds. These bonds are a particularly secure form of investment for creditors, who are guaranteed repayment ahead of any other creditors. As a result, electric power bonds enjoy the same high level of confidence as government bonds and the electric utilities have cheap and reliable access to the large amounts of credit necessary for infrastructure investment (Ōshika 2013: 262).[9]

Electric power bonds make up 20% of Japan's 15 trillion yen domestic market for corporate bonds. Of this, TEPCO-issued electric power bonds make up 35%, making the company the most important single player in the market. Although the banks would like to loan money to TEPCO, the company does not take out significant bank loans (Ōshika 2013: 210, 246). This makes TEPCO a formidable player in the business world.

Even in terms of the entire electric power industry, the nuclear energy sector is an enormous industry. In 2009 nuclear energy contributed 30.2% of Japan's total electricity generation and the utilities spent 2.1353 trillion yen on nuclear technology. In 2010 the government's annual budget for nuclear energy reached 432.3 billion yen. The nuclear energy sector employs 11,668 people in the electricity business and a further 33,714 people in mining and manufacturing (*Shūkan Daiyamondo*, May 21, 2011: 28–32). Because nuclear energy is such a large industry, any suggestion of abolishing nuclear power is met with a strong backlash in defense of budgets and jobs.

Government relationship

The relationship between the electric utilities and METI is complex. METI plays a supervisory role in the electric power industry. It regulates both electricity prices and the construction of electric power station. Historically, TEPCO has taken on former METI employees, either directly or through one of the industry bodies it dominates, in order to construct a cooperative relationship with the ministry. This practice is known as *amakudari* (descent from heaven).

Nevertheless, the electric utilities have at times clashed with METI. Prior to the Second World War, the Japanese electricity market was crowded with numerous private companies. In April 1939 all of the country's electricity generators and distributors were nationalized as part of the war economy. With nationalization came the creation of a single umbrella company, Japan Electric Generation and Transmission Company, and nine regional companies responsible for managing the distribution network. When the war ended, the state continued to administer the generation and distribution of electricity for some years. Then, in May 1951, the General Headquarters of the Supreme Commander for the Allied Powers ordered the establishment of nine private electric utilities. Each of these utilities would be responsible for both generation and distribution within a defined geographical region. (Following the restoration of Okinawa to Japanese sovereignty, a tenth company, Okinawa Electric Power, was created.) When the occupation ended in 1952, MITI wanted to renationalize the electric power industry and planned a number of interventions into the industry. This led to strained relations with the utilities, which wished to remain as private companies. When nuclear energy was first being developed, a fresh struggle also broke out over who would take the lead in implementing the new technology – a national corporation or the private electric utilities (NHK ETV Tokushū Shuzaihan 2013: 119–23; Kikkawa 2011: 123–40; Tahara 2011: 48–70, 83–100; Yoshioka 2011: 25–8, 83–91).

The three laws for electric power development

From the 1950s until the mid-1960s, Fukui and Fukushima prefectures tried enthusiastically to lure nuclear power station developments.

From the latter half of the 1960s, however, large-scale movements against the building of nuclear power stations appeared in some regions. In the 1970s, when the stations were beginning to come online, a series of accidents and faults provoked significant opposition to plans to build nuclear power stations. The electric utilities and MITI began to work together even more closely to overcome this opposition and promote the technology.

The relationship between TEPCO and the ministry is perhaps best symbolized by the enactment of a package of legislation known as *Dengen Sanpō* (the 'three laws for electric power development'). The three laws included the *Act on the Development of Areas Adjacent to Electric Power Generating Facilities*, the *Tax Act for Promotion of Power Source Development* and the *Act on Special Accounts for Electric Power Development Acceleration Measures*. When the oil shock crisis struck in October 1973, then Prime Minister Tanaka Kakuei designated the further development of nuclear power as an issue of national importance. He took personal responsibility for supervising the passage of the three laws through the Diet in a bid to provide an economic stimulus to host communities. The *Dengen Sanpō* passed into law in June 1974. It was MITI that drafted the legislation.

Under the *Dengen Sanpō* the utilities pay an electric power development levy on a designated amount of electricity sold (85 yen for every 1,000 kilowatt hours). The levy is paid into a special account designated for 'power resource promotion'. Municipal authorities in host communities (both those that host power stations on their own land as well as neighboring municipalities) receive subsidies from this account. The main subsidy is the 'grant-in-aid for host community promotion', but there are numerous other categories of grants and subsidies. These funds can be used for a range of designated purposes such as roads, social welfare, education and the construction of cultural facilities.[10]

The *Dengen Sanpō* facilitated cooperation between the electric utilities and MITI in pursuit of the state's nuclear agenda and helped to ease growing tensions between MITI and the electric utilities over the new technology. This also had the effect of making the electric utilities become more conservative, less like private companies and more like government agencies (Kikkawa 2011: 143–4).

Energy diversification

The oil shock crisis prompted MITI to place energy security at the center of its energy policy. From the 1980s the ministry began to advocate a policy of energy diversification. MITI's policy saw oil's share in the energy mix decline, while coal, natural gas and nuclear energy increased. The ministry also promoted the development of new energy sources, including renewables. Nuclear energy, in particular, was held up as an extremely stable and economical energy source that was 'as good as home-grown'. The ministry promoted the idea that nuclear power could provide the basis for energy self-sufficiency by devising a nuclear fuel cycle project. The new policy direction saw even more attention paid to measures for host communities. The greater part of the government's energy budget, as well as its public relations campaigns and its efforts to mollify local residents, went into nuclear projects (Honda 2005: 184–6; Yoshioka 2011: 184–5).

The nuclear fuel cycle is based on the idea that the leftover uranium in the spent fuel after it has been used to power a nuclear reactor and the plutonium that is produced through the nuclear reaction can be reclaimed and reused as a fuel source. The nuclear fuel cycle project envisioned the extraction of plutonium through the reprocessing of spent fuel and its combination with depleted uranium to produce a mixed oxide fuel (MOX). This would then be used to power a new type of nuclear reactor, the fast breeder reactor (FBR). FBRs were expected to produce both electricity and more plutonium, provoking high expectations that the 'nuclear dream' of energy self-sufficiency might be realized. However, although the Monju research FBR (located in Tsuruga City, Fukui Prefecture) was opened in 1995, it was shut down a little over three months later due to a sodium leak. It has remained idle ever since and commercial-scale FBRs remain a far-off dream. However, under the terms of the nuclear non-proliferation treaty, the United States government put pressure on Japan not to produce surplus plutonium. The Japanese government responded by asking the electric power industry to use the MOX fuel originally meant for FBRs in its existing light water reactors in an initiative known as the plutonium-thermal project.

Another component of the nuclear fuel cycle project is the spent fuel reprocessing facility being built in Rokkasho, Aomori

Prefecture. Like the FBR, the facility has been plagued with problems and remains in a testing stage. Japan continues to rely on Britain and France to reprocess its spent fuel. The highly radioactive waste liquid that is produced during reprocessing is another problem facing the project. Currently, this liquid is enclosed in borosilicate glass and buried but there is as yet no final disposal site for these vessels. In reality, hardly any aspect of the nuclear fuel cycle is functioning as anticipated.

Energy diversification also resulted in the further expansion of the subsidies provided under the *Dengen Sanpō*. From July 1980 the electric power development levy was increased to include an energy diversification allowance of 215 yen (designated for research and development) in addition to the host community allowance of 85 yen. The levy was increased in 1983 to include a 160-yen community allowance and a 285-yen energy diversification allowance (per 1,000 kilowatts).[11] The period during which host communities can apply for these subsidies was extended. The eligibility criteria were also relaxed so that, from October 1981, prefectural governments could also apply to receive the subsidy (Yoshioka 2011: 151–2; Honda 2005: 186–8).

The new component, the energy diversification allowance, created a budget for research and development that was divided equally between MITI and the Science and Technology Agency. During the 1980s, two-thirds of these funds were utilized for the development of nuclear energy. Half of this went to the Power Reactor and Nuclear Fuel Development Corporation (PNC) for the development of a fast breeder reactor.

Exchange of personnel

The partnership between METI and the electric utilities is also visible in their personnel. According to figures released on May 2, 2011, by METI, 68 former ministry employees have 'descended' to posts in one of the ten regional power companies or the two distributors (Electric Power Development Company (J-Power) and Japan Atomic Power Company) in the past 50 years. A further 108 METI bureaucrats are known to have 'descended' to energy-related corporations or incorporated foundations (Ōshika 2013: 397).

There is a long history of *amakudari* from METI to TEPCO, beginning in 1959 when MITI administrative vice-minister

Ishihara Takeo became a director of TEPCO, eventually rising to become vice-president. The post of TEPCO vice-president has since become 'the *amakudari* seat'. Three commissioners and deputy commissioners of MITI's Agency for Natural Resources and Energy have been employed as advisers or vice-presidents at TEPCO (Komatsu 2012: 71–3; Miyake 2011: 104–12; Gurūpu K21 2011c).[12] There are also numerous cases of children of senior officials of the Agency for Natural Resources and Energy using their connections to find employment with the utility.[13]

At other times, electric utility officials 'ascend to heaven' and join the bureaucracy. Between 2001 and 2011, 36 electric utility employees were employed as temporary public servants in various posts (12 in the Cabinet Secretariat, 15 in the Cabinet Office and nine at the Ministry of Education, Culture, Sports, Science and Technology (MEXT)). Another 102 individuals who are still employed at seven electric utilities, the Japan Atomic Power Company, Japan Nuclear Fuel Limited, CRIEPI, nuclear power station equipment manufacturer IHI and nuclear power promotional organization Japan Atomic Industrial Forum (JAIF)[14] are employed casually as public servants in the Cabinet Office, the Cabinet Secretariat, METI and MEXT (34 of these are TEPCO employees). None of these have been hired according to the rules of the National Personnel Authority. Instead, they have been retained without the usual public advertisement of the positions under special exemptions intended for technical specialists (Komatsu 2012: 71–4; Gurūpu K21 2011c).

Collusion with nuclear safety agencies

Utility staff, as well as employees of nuclear reactor manufacturers, have also been seconded to the Nuclear and Industrial Safety Agency (NISA), Japan's nuclear energy safety regulator. This has the potential to undermine the effectiveness and impartiality of the safety regime. Between 2001 and 2011, NISA employed 83 staff from the nuclear energy industry, including 22 from Toshiba, six from IHI and six from KEPCO. Of these, more than 50 were employed as nuclear safety inspectors. Some were responsible for inspecting the reactors owned by their former employers. In other cases, former inspectors have been re-employed by these same companies after leaving NISA (Doi 2011: 82–3).

Former employees of the electric utilities, nuclear technology manufacturers, CRIEPI and JAIF have also been employed by the Nuclear Safety Commission (NSC) in the Cabinet Office, which is also responsible for nuclear safety (Komatsu 2012: 71–4; Gurūpu K21 2011c).

Commenting on the collusion between NISA and the electric utilities, a TEPCO insider said:

> Public servants who have no nuclear energy experience get transferred to NISA and become safety inspectors. When they arrive, the first thing that happens is they get a lecture from TEPCO about nuclear energy. TEPCO also spends a considerable amount of money wining and dining these new officials. (*Shūkan Bunshun* special issue 'Tokyo Denryoku no daizai', July 27, 2011: 90)

When problems crop up during inspections, TEPCO officials take the relevant people to expensive nightclubs, where they try to negotiate with them, saying, 'can't you help us do something about the problems we were discussing earlier?'

The insider further explained the consequences when bureaucrats with no experience of nuclear issues are responsible for safety evaluations and become enmeshed with TEPCO:

> Both MITI and the [NSC] double-check the results of safety evaluations. However, the inspection documents are prepared by TEPCO. The utility prepares a Safety Inspection Report in collaboration with MITI. The reports are printed at TEPCO's expense, but are presented as the work of [MITI]...[Then] when the [NSC] meets to evaluate the report, MITI has to explain it to them. TEPCO prepares the papers for these meetings too. While such reports ultimately appear as Safety Inspection Reports of the NSC, they are actually produced by TEPCO. It is a case of 'industry self-regulation to suit the industry'. (*Shūkan Bunshun* special issue 'Tokyo Denryoku no daizai', July 27, 2011: 91)

This collusion also includes holding 'prearranged' public forums. On September 30, 2011, a government independent investigation committee found that, on at least seven occasions in the past five years, the head of public relations for the NSC and the public relations bureau in the Agency for Natural Resources and Energy recommended that electric utilities send their own staff to pose as members of the public at forums on nuclear energy issues (such as the plutonium-thermal project and earthquake resistance) or recruit pronuclear residents to attend the meetings and voice their opinions (*Asahi Shimbun*, October 1, 2011, morning edn). These reports are just the tip of the iceberg.

Liberalization

Thus far I have argued that the relationship between MITI and TEPCO is a cooperative one. However, when pro-liberalization bureaucrats within MITI tried to advocate unbundling the integrated service of power generation and transmission in the 1990s, it led to major conflict between the ministry and the company. When political actors cooperate with one another, it is difficult to determine where the balance of power lies. However, when they come into conflict, these power relationships finally become clear. In the end it was TEPCO, in alliance with elected politicians, who won the battle over electricity market liberalization.[15]

In the 1990s electricity prices in Japan were high in comparison with other developed countries (double those in Britain, France and the United States and 1.3 times those in Germany and Italy). The bureaucrats within METI who wanted to liberalize the electric power market believed that unless Japan could reduce its electricity prices, then Japanese industry would lose its competitive advantage. Murata Seiji was the leading figure in the pro-liberalization faction. In 1994 Murata became director of the public utilities department in the Agency for Natural Resources and Energy and in 1995 he led the charge to amend the Electricity Business Act. The amendment, known as the First Electricity Sector Reform, allowed new generators, known as Independent Power Producers (IPP), to enter the market. It created a bidding system whereby the existing utilities would set up a quota to purchase electricity from IPPs. Large companies that possessed their own in-house power stations became IPPs. Now they could compete to sell their surplus electricity to the utilities by offering the most attractive price.

Satō Shinji, Minister for International Trade and Industry, signaled the possibility of further reform on January 7, 1997, when he suggested during a press conference that the hitherto taboo subject of unbundling the integrated service of power generation and transmission ought to be explored. Satō's announcement was widely seen as having been coordinated by Murata, who by now had become deputy vice-minister. In July 1997 the ministry's Electricity Industry Council launched an inquiry to consider lowering the price of electricity by 20%, unbundling the integrated service of power generation and transmission, and opening the power generation market to new players.

This Second Electricity Sector Reform of March 2000 led to the establishment of a system whereby IPPs could transmit electricity over the utilities' existing network by leasing the transmission lines. However, the reforms only allowed new retailers to enter the market and to provide electricity only to large-scale customers requiring more than 2,000 kilowatts who were extra-high voltage (20,000 volts and above) supplied customers.

Nuclear cover-ups and the Third Electricity Sector Reform

The second reform was approved on the condition that its results would be evaluated after three years (March 2003), when decisions would be made about the future of the policy. In November 2001 the Electricity Industry Working Group of the Advisory Committee for Natural Resources and Energy began debating a third set of reforms. Pro-liberalization bureaucrats were concerned that, despite the provisions of the second reform, the high cost of transmission was keeping most potential retailers out of the market. In order to increase the transparency of the consignment costing system, they felt it would be necessary to introduce accounting separation between the generation, transmission and distribution parts of the utilities' businesses. The working group nominated the following issues to be dealt with as part of the Third Electricity Sector Reform: total liberalization of the retail sector, unbundling of power generation and transmission, opening up the electricity grid and establishing a facility to enable trading in wholesale power.[16] The electric power industry, together with the Federation of Electric Power Related Industry Workers' Unions of Japan (Denryoku Sōren), the union covering electricity industry workers, conducted an extensive public relations campaign against liberalization and increased its lobbying of politicians.

In July 2000, while the rumblings over liberalization continued, a General Electric technician turned whistleblower. He told MITI that he had discovered six cracks in the steam dryer of reactor

No. 1 at the Fukushima Daiichi Nuclear Power Station during a maintenance inspection. He further claimed that, following a request from TEPCO, his superiors at General Electric had instructed him to edit the videotape in such a way that the cracks would not be visible when he submitted it to MITI. NISA's response to the allegations was not to conduct its own inspections, but to inform TEPCO of the whistleblower's testimony and hand over documents about him to the company. Although NISA conducted a number of inquiries into the issue, in the end it did not demand the company take any urgent action.

In July 2002, when Murata became administrative vice-minister of Economy, Trade and Industry, NISA did an about-face. TEPCO's then public relations officer describes the way NISA suddenly demanded a report on how the cracks had been dealt with. On August 29, 2002, NISA revealed that TEPCO had covered up 29 incidents at its nuclear power stations. The company admitted that it had been falsifying recordings of cracks in reactor core shrouds and falsifying safety inspection reports since the late 1980s. Following these revelations, four current and former TEPCO chiefs took responsibility for the cover-ups and resigned: Araki Hiroshi (chairperson), Minami Nobuya (president), Hiraiwa Gaishi (adviser) and Nasu Shō (adviser). Enomoto Toshiaki, TEPCO vice-president and head of the nuclear power division, also resigned. He was the most senior of a further 35 staff who were forced to resign, were demoted or received pay cuts.

TEPCO's grudge against METI only intensified. In TEPCO's view, Murata had used the cover-up to quell resistance to electricity market liberalization and singled out TEPCO as the 'bad guy' despite the company's cooperation with NISA in investigating the problems at its power stations. TEPCO lobbied the Liberal Democratic Party (LDP) and Denryoku Sōren lobbied the Democratic Party of Japan (DPJ) to oppose liberalization. They also put pressure on the bureaucrats who had been involved in the liberalization discussions, encouraging them to 'distance yourself from what you may have said back then'. Murata suddenly declared that it was 'every man for himself' and withdrew his protection from his troops and the younger bureaucrats.

Eventually, the LDP and TEPCO held the carbon tax (which had been hastily introduced by METI in order to meet Japan's obligations under the Kyoto Protocol) hostage in order to prevent the unbundling of power generation and transmission. Murata and his allies were forced to postpone the unbundling of generation and transmission in order to secure the passage of the carbon tax legislation by the LDP. In the end, the Third Electricity Sector Reform included the following measures: accounting separation for the utilities' generation and distribution divisions, establishment of a neutral body to manage the use and reconfiguration of interconnected transmission lines, a re-examination of the consignment system for electricity sales,[17] relaxation of the eligibility requirements for electricity retailers to sell directly to customers from a minimum of 500 kilowatts in April 2004 to 50 kilowatts in April 2005, and the establishment of a trading mechanism for electricity wholesaling. It was also decided that a further investigation into the total liberalization of the retail electricity market would begin in April 2007.

However, in 2007 METI decided that it was not the time to expand the retail electricity market and the opportunity for liberalization was closed. Since then, there has been no substantial progress on liberalization. In 2007 there were only 13 companies operating in the new retail market. With transmission costs remaining high, the integrated distribution network was largely unused. The reforms were also unsuccessful in creating any competition between the nine regional utilities.[18]

The nuclear fuel cycle project

The final showdown between the electric utilities and the pro-reform bureaucrats in METI took place over the nuclear fuel cycle project. In March 2004 six young officials in the Agency for Natural Resources and Energy created a document entitled 'An invoice for 19 trillion yen: the limitless expansion of the nuclear fuel cycle project'. They circulated it first within the agency, then to members of parliament and the mass media.

In the document, they argued that the nuclear fuel cycle project was not cost effective. They calculated, despite initial estimates of 690 billion yen, the cost of building the Rokkasho Reprocessing Plant in Aomori Prefecture had already cost 2.2 trillion yen. If the plant operated for 40 years, it ought to cost 18.8 trillion yen. However, if it operated at lower than expected capacity, the cost could grow to 50 trillion yen.

If the project was set to cost such an enormous sum of money, why continue with it at all? The document discussed the real reasons the parties involved were continuing to pursue the nuclear fuel cycle. If the government changed its policy, it would be subject to claims of up to 2 trillion yen compensation from the utilities to cover their contribution to the cost of construction thus far. In turn, the utilities were faced with the threat of having to reimburse their customers for the portion of their power bills that had been collected to pay for the reprocessing facility. Finally, the politicians, who received financial backing from energy-related companies and unions, were unable to change the policy.

The young bureaucrats had received the go-ahead from Terasaka Nobuaki, director-general of the Electricity and Gas Industry Department, who had told them to 'stir things up, but don't go too far'. However, once again Murata failed to protect the young bureaucrats. He was unable to persuade Minister for Economy, Trade and Industry Nakagawa Shōichi, who was a supporter of the nuclear fuel cycle project and an advocate for Japan possessing nuclear weapons, to accept the young bureaucrats' arguments (Ōshika 2013: 278–80; Takeuchi 2013: 85, 89–91).

The attempt to revise the nuclear fuel cycle project also aroused the ire of the electric utilities. Aomori Prefecture was accepting the spent fuel from their reactors, so the companies could hardly afford to damage their relationship with the prefectural authorities. In July 1998 Aomori's prefectural governor, the mayor of Rokkasho and the president of Japan Nuclear Fuel Limited (JNF), the operator of the reprocessing plant, had signed a memorandum in the presence of the chairperson of the FEPC. The memorandum stated that, in the event of major problems with reprocessing, JNF would immediately take the necessary measures to deal with the situation, including removing the spent fuel from the facility. If the electric utilities were to abandon the nuclear fuel cycle project, Aomori Prefecture might ask them to collect the spent fuel they had deposited at Rokkasho. If they were forced to do so, the spent fuel pools in the reactors would be full and it would be impossible to continue operating their nuclear reactors (Asahi Shimbun Aomori Sōkyoku 2005: 57–8).

A senior bureaucrat who was involved in producing 'An invoice for 19 trillion yen' has testified that one afternoon in the summer of 2004 he received a call from the FEPC. The caller told him that 'this evening there will be an announcement. You have been transferred.' For the caller to have this information he must have been in possession of the official transfer order, which only the administrative vice-minister or deputy vice-minister ought to have known about. The caller continued, 'Shall I send it to you?', and the order arrived through the fax machine on his desk. It was a clear demotion. This particular bureaucrat had previously been warned that 'the politicians are allied with the industry. The utilities buy huge numbers of tickets to their fundraising events. If you make too much of a fuss you'll run into trouble.' He explained, 'I didn't realize they had so much power. The order must have been given by the Minister at the behest of the FEPC.'[19]

In June 2004 Murata retired. The faction that favored maintaining the status quo took charge and the pro-reform elements were suppressed. Their actions were 'disregarded' and they were either seconded elsewhere or transferred to other departments. The bureaucrats who had tried to pursue liberalization found that their career progression was stymied (Ōshika 2013: 280–1; Takeuchi 2013: 91).[20] After that, the Atomic Energy Commission decided to continue with the nuclear fuel cycle project. In December 2004 test operations at the Rokkasho Reprocessing Plant commenced. Following Murata's retirement, support for electricity market liberalization within METI died down. The electric power industry had triumphed over the liberalization faction within METI.

The Liberal Democratic Party

Donations

Through their struggle with METI, the electric utilities moved closer to the LDP. Historically, the electric power industry was known as one of the LDP's 'big three donors', after the banks and the iron and steel industries. However, when electricity prices went up in 1974, allegedly due to the sudden increase in the price of oil, Ichikawa Fusae (member of the House of Councillors) and her supporters launched the 'one yen underpayment' movement to protest about the utilities giving donations to political parties. TEPCO chairperson Kigawada Kazutaka declared that it was unseemly for a public utility that had been granted a regional monopoly to make political donations. He issued a ban on further donations and the other electric utilities followed suit (Shimura 2011: 151–4; Asahi Shimbun Tokubetsu Hōdōbu 2014: 168).

However, the utilities found other ways to continue channeling money to the party. First, individual electric utility executives made donations. The utilities' official line was that political donations were a matter of individual conscience and the company did not interfere. In reality, however, the donations were coordinated at an organizational level by each utility's general affairs department. In the 15-year period between 1995 and 2009, 448 TEPCO executives donated a total of 59.57 million yen to the LDP's fundraising organization, the National Political Association. TEPCO executives were expected to donate a fixed amount per year depending on rank. The chairperson and president were to donate 300,000 yen, vice-presidents 240,000 yen and executive directors 120,000 yen (Asahi Shimbun Tokubetsu Hōdōbu 2014: 170; Shimura 2011: 157–9). In total, 912 individual executives of the nine electric utilities (excluding Okinawa Electric Power, which does not maintain any nuclear reactors) donated a total of 115.67 million yen to the party from 2007 to 2009 (Gurūpu K21 2011a). In 2009 the utilities' associated companies also donated more than 80 million yen to the National Political Association (Doi 2011: 132).

The second method the utilities found to donate to the LDP was by advertising in the party's internal publications. In 1993 it was revealed that the electric power industry had spent 2.5 billion yen on advertising in these publications over a three-year period. Critics dubbed the practice a 'new means of making political donations' and it has now been stopped (Asahi Shimbun Tokubetsu Hōdōbu 2014: 170).[21]

Third, the utilities purchase tickets to politicians' fundraising events en masse. TEPCO spent more than 50 million yen each year on tickets to fundraising events held by more than 50, mainly LDP, politicians. The TEPCO group has more than 100 associated companies. Collectively, these companies purchased a further 100 million yen worth of tickets. However, these purchases are hidden. The *Political Funds Control Law* specifies that donations above 200,000 yen must be declared. When purchasing these tickets, the utilities ensure that they stay below this threshold. The annual budget for purchasing fundraising tickets is controlled by TEPCO's general affairs department, which does not have to declare its operating expenses. The purchases are directed by the head of the department based on an assessment of each politician's importance to nuclear energy policy and their degree of cooperation with the utilities. Diet members who represent electoral districts in Aomori, Fukushima and Niigata prefectures (where TEPCO's nuclear power stations and related facilities are located) and the top three Diet members with responsibility for METI (minister, vice-minister and parliamentary secretary) receive the largest donations.

The politician who has received the highest assessment by the utilities in recent years is Amari Akira (LDP, House of Representatives), the leading representative of the nuclear policy tribe in the Diet. Since he became Minister for Economy, Trade and Industry in 2006, the nine utilities have purchased one million yen worth of tickets to each of Amari's fundraising events. In order to reach this amount, each utility purchased the designated amount of tickets two or three times per year. The number of tickets purchased by each utility depends on the size of its business (Asahi Shimbun Tokubetsu Hōdōbu 2014: 162–6).[22]

Finally, the utilities cultivate political relationships through frequent entertaining and through bribery. At breakfast meetings organized by the FEPC, government relations staff members of the nine electric utilities meet face to face with LDP politicians. Government relations staff members also divide responsibility for different politicians among themselves. They entertain the politicians and support them with campaign funds at election time (Asahi Shimbun Tokubetsu Hōdōbu 2014: 180–3).[23]

Relationships with the LDP

Through its conflicts with METI and its promotion of nuclear power, TEPCO has become integrated with the structures of the LDP. However, in the 1990s the company began to feel the need to distance itself from politics. Following the conclusion of the Plaza Accords in 1985, a strong yen, cheap oil and low interest rates saw the nine utilities record more than 1.5 trillion yen in operating income, leading to calls from the LDP for a reduction in their profit margin. The LDP pressured the utilities to use their profit margin to not only reduce prices, but to construct coastal roads and new bullet trains and to bury existing power lines (*Shūkan Tōyō Keizai*, April 23, 2011: 44).

Then TEPCO chairperson Hiraiwa saw the need to counter the negative effects the LDP's entrenched rule was having on the company. He took drastic action. In August 1993 a new non-

LDP coalition government was established under the leadership of Hosokawa Morihiro (Japan New Party, member of the House of Representatives). In September, Hiraiwa, in his capacity as chairperson of Keidanren, decided that the federation would stop soliciting political donations from its members and funneling them to the LDP. He also moved closer to the Hosokawa Cabinet.

This angered the LDP. At the end of the year, when Hiraiwa led a delegation to the party's offices to present a log of claims for the speedy implementation of political reforms, he was jeered by the LDP leadership (Saitō 2012: 213, 234). Minister for International Trade and Investment Satō's positive reception of the discussion around the Second Electricity Sector Reform can be seen as a reprisal for Hiraiwa's decision to end Keidanren's support for the party's fundraising efforts.

In the end, TEPCO was unable to break away from the LDP and eventually struck an agreement with the party. During the 1998 House of Councillors elections, TEPCO vice-president Kanō Tokio ran in a proportionally represented constituency on an LDP ticket (Saitō 2012: 234, 271).

In the LDP, candidates for proportional representation typically have to organize their own register of political party supporters known as a *kōenkai*. On December 15, 1997, LDP Secretary-General Katō Kōichi (House of Representatives) made the following observations during a speech at Kanō's campaign launch:

> We have a system for ranking candidates' supporters' register. Out of a possible one hundred points, fifty are allocated based on the accuracy of the register of 1 million supporters compiled by Mr. Kanō. Once he has submitted his supporters' register, we call a random sample to ask them whether they do in fact support him and the LDP. If they say no, then we put a mark next to their names. I know that we have some representatives of the nine utilities here today. Let me say, we are confident that if Mr. Kanō and our party have your help in constructing our supporters' registers then he will certainly be ranked highly in LDP's proportional representation list. (Lee 2012: 185)

According to one LDP insider:

> the electric utilities have the most accurate registers. Candidates from the major construction firms have a haphazard approach. They

> might just photocopy a list of people who graduated from the same university as them. The utilities on the other hand, they mobilize all of their associated companies and business partners to create the register so they can ensure that they turn out the vote. A condition of compiling these registers is that candidates assemble '20,000 party member and supporters'. This means gathering membership fees from 20,000 members. At 10,000 yen each, that makes a total of two hundred million yen. (Lee 2012: 185–6)

At the time, the LDP was trying to compensate for a decline in political donations. Not only had Keidanren stopped helping to channel donations to the party, the injection of government funds into the banks to help them deal with masses of bad debts meant that they, too, could no longer make large donations to the party (Lee 2012: 184–8).

By temporarily lending its support to pro-liberalization bureaucrats to pursue their electricity market reform proposals, the LDP succeeded in pulling the utilities back into their orbit to provide 'votes and money'.

The nuclear energy policy tribe

However, Kanō was not the type of person to be held hostage by the LDP. As a representative of TEPCO's interests inside the LDP, he was a powerful force. In November 1999 Kajiyama Seiroku (House of Representatives) and other powerful LDP representatives founded the bipartisan Renewable Energy Caucus with 250 members. In April 2000 the caucus put together a bill that would compel the utilities to formulate renewable energy plans based on targets to be determined by the government. Under the provisions of the bill, the utilities would be required to lodge their purchase agreements for renewables with the government.

In response, one TEPCO executive stated publicly that 'electricity produced through wind energy is unstable and dirty. We don't want even a single watt.' A senior MITI official also declared, 'we will not allow the politicians to pass such a bill' (Kyodo News 2013b). MITI responded by proposing its own legislation, the *Act on Special Measures Concerning New Energy Use by Operators of Electric Utilities* (known as the RPS Law). This legislation would not require the utilities to purchase all of the available renewable energy supply.

Instead, MITI would set a target for renewable energy use and require the utilities to purchase a proportion of what was available on the market.

In August 2000 Hashimoto Ryūtarō (LDP, House of Representatives) was appointed chair of the Renewable Energy Caucus, replacing Aichi Kazuo (LDP), who had lost his seat in the June House of Representatives elections that year. Hashimoto changed the caucus's position, backing the government's proposal of the RPS Law over the bill originally put forward by the caucus. When caucus secretary Katō Shūichi (Komeito, House of Councillors) asked to meet with Hashimoto to request a general meeting of the caucus, he was rebuffed. Despite being chair of the caucus, Hashimoto refused to meet with Katō, claiming that he was too busy. These maneuvers stymied the caucus.

In April 2000 an energy policy working group was established within the LDP by the Oil and Energy Resources Policy Committee to prepare for the enactment of the *Basic Act on Energy Policy*. The head of this working group was Amari Akira. Kanō Tokio (LDP, House of Councillors) was its secretary. This bill proposed that energy policy be based on the principles of 'stability of supply' and 'environmental compliance', and 'activating market principles' after paying due attention to the first two. In reality, the Basic Act was seen as an attempt to prevent the unbundling of the utilities' generation and transmission functions and to continue the spent fuel reprocessing project.[24] While the Basic Act did not explicitly mention promoting nuclear power, out of respect for Komeito's objections, by advocating stability of supply and the prevention of global warming its basic message was that nuclear power was necessary.

In the spring of 2001 the Renewable Energy Project Team held its first meeting. Kanō opened proceedings by speaking about the RPS Law. Katō Shūichi interrupted him, arguing that it was necessary to force the utilities to buy all of the renewable energy available on the market. Amari, the chairperson of the Project Team, shouted back, 'we don't need an idiot like you who doesn't understand the issues'. From that point on, the LDP led the discussion within the Project Team. In August 2001 the Project Team's proposal to make the RPS Law the basis of renewable policy was put forward at a meeting of the Renewable Energy Caucus. Ultimately, the RPS Law was introduced into the Diet with the backing of the Cabinet and passed through the

Diet in May 2002. The following month, the *Basic Act on Energy Policy* also passed through the Diet.

The period between April 2003, when the system for purchasing electricity under the RPS Law commenced, and August 2011, when the *Act on Special Measures Concerning Procurement of Electricity from Renewable Energy Sources by Electricity Utilities* passed through the Diet following the nuclear disaster, is known as the 'dark age of renewable energy'. METI set such a low target for renewable energy use that generation ground to a halt.[25] Once again, TEPCO had united with the LDP to prevent the unbundling of power generation and transmission, smash the renewable sector and maintain support for the nuclear fuel cycle project.

The Diet members in the nuclear energy policy tribe also influenced the formulation of the Basic Energy Plan by ensuring it contained the phrases 'nuclear energy is designated as a key source of energy' and 'integration of electricity generation and distribution' (*Asahi Shimbun*, May 25, 2011, morning edn). In 2006 the Nuclear Energy Division of the Electricity Industry Working Group in METI's Advisory Committee for Natural Resources and Energy released the Nuclear Energy National Plan. This report advocated the continued development of nuclear energy and called on the government to take a strong lead in promoting this development. A Basic Energy Plan containing the main elements of this report received Cabinet approval in March 2007 (Yoshioka 2011: 41, 340–2).

The academy

Academic teaching and research provide the basis for the nuclear industry. Nuclear engineering researchers based in the universities educate the technicians who work for nuclear reactor manufacturers and electric utilities. They also sit on committees such as the Atomic Energy Commission, which sets nuclear energy policy, and the Nuclear Safety Commission, which is responsible for determining and monitoring safety standards.

These researchers have close connections with the nuclear industry. They receive large amounts of funding from nuclear energy companies, industry bodies, government agencies and the numerous quasi-independent institutions that fall under their umbrella. These funds include grants-in-aid, project-based funding and collaborative research funding (Sasaki 2011b: 102–4). The industry also sponsors university courses in nuclear engineering.[26]

The result of this cozy relationship was that it became difficult to criticize nuclear power within academia. For example, Anzai Ikurō (Emeritus Professor, Ritsumeikan University) was a member of the first cohort of students in the Department of Nuclear Engineering at the University of Tokyo. A specialist in radiation protection, he raised a number of problems with Japan's nuclear policy and became an influential figure within the social movements. After being hounded from his position in the Department of Nuclear Engineering, he was employed as an assistant lecturer in the Faculty of Medicine. Nevertheless, he continued to experience harassment for his views.[27]

The labor unions

While the electric utilities management were providing 'votes and money' to the LDP to promote nuclear power, the electric power unions did the same for the opposition DPJ.

Denryoku Sōren is the union federation covering the electric power industry. Its member unions represent employees of the ten regional companies, as well as Japan Atomic Power Company, Electric Power Development Company (J-Power) and all associated companies. Denryoku Sōren (219,153 members as at May 27, 2014) is the eleventh-largest industry federation within Japan's largest national union federation, the Japanese Trade Union Confederation (JTUC – Rengō).[28] However, Denryoku Sōren's political muscle is much larger than its membership alone would suggest.

Denryoku Sōren's importance within Rengō is demonstrated by the fact that Sasamori Kiyoshi, fourth president of Rengō (2001–05) and special Cabinet advisor to Kan Naoto, came out of the Tokyo Electric Power Workers' Union, one of Denryoku Sōren's member unions. Candidates in proportionally represented constituencies in the House of Councillors elections who have a background as Denryoku Sōren officials tend to receive a higher number of votes than other DPJ candidates. In the 2001 House of Councillors elections, Fujiwara Masashi (Kansai Electric Power Workers' Union) received 259,576 votes (the second-highest number in the party). In the 2004 House of Councillors elections, Kobayashi Masao (Tokyo Electric Power Workers' Union) received 301,322 votes (the highest). In the 2007 House of Councillors elections, Fujiwara received 194,074 votes (ninth highest) and Kobayashi received 207,227 votes (fourth highest). In the 2013 House of Councillors elections, Hamano Yoshifumi (Kansai Electric Power Workers' Union) received 235,917 votes (second highest) and in the 2016 election, Kobayashi received 270,305 votes (highest). In each case, the Denryoku Sōren candidates received almost the same number of votes as there are members of it. This demonstrates its superior ability to mobilize its members to vote when compared with other unions.

Other DPJ candidates have also made use of this mobilizing power. Of 48 candidates who received Denryoku Sōren endorsement in the 2010 House of Councillors elections, 25 were elected (Komatsu 2012: 61–6).[29] The former secretary of one DPJ representative said of Denryoku Sōren, 'we all want their endorsement at any cost. Their vote is that reliable. Once the union decides to endorse a candidate, an official will come around with a list of union members and explain "there are ninety workers in your district"'. However, their support comes with its own conditions. The former secretary explained that, when he went to the offices of the Kansai union, he was given a policy agreement written on a single A4 sheet and some booklets on nuclear energy and told to 'read them carefully': if the candidate did not sign the agreement, which stated that he or she approved of nuclear power, then he or she did not get the endorsement (Kyodo News 2013d).

Denryoku Sōren's influence within the DPJ also derives from the federation's donations to the party. The union's political wing, the Denryoku Sōren Political Action Committee, had a total income of approximately 2.5 hundred million yen in 2009. Of this, approximately 70 million was spent on political campaigning. Kobayashi Masao (DPJ, House of Councillors), who was standing for re-election in 2010, received approximately 30 million yen for election expenses. The remainder was distributed to each member union's political organization. Furthermore, the Denryoku Sōren Political Action Committee contributed 107.65 million yen to the DPJ and its representatives in the three-year period between 2007 and 2009. Of this, Kobayashi received 40 million and Fujiwara (DPJ, House of Councillors) received 33 million. The committee also purchased 5.3 million yen worth of tickets to fundraising events hosted by DPJ politicians.

According to the Ministry of Internal Affairs and Communications' 2010 report on political donations, a total of 750 million yen was collected in political contributions by Denryoku Sōren's political committees from 127,000 union members working at the ten electric utilities and three associated companies through memberships and other payments. Of this, 64 million yen went to the Denryoku Sōren Political Action Committee. The political action committees in each of Denryoku Sōren's member unions also purchase tickets to DPJ representatives' fundraising events and make political contributions to the 150 municipal and prefectural government representatives who are also union members. These contributions are used for entertaining at clubs and restaurants and for regional representatives' travel expenses (Gurūpu K21 2011b; Doi 2011: 130–4; Komatsu 2012: 64–5).

In November 2012 it was revealed that of 99 current employees of the nine electric utilities serving in local government (TEPCO had the largest number, 23), 91 were still receiving salaries from the utilities: 52 of those were still receiving salaries even though they were on leave to pursue their political activities, thanks to a special exemption (prior to raising the price of electricity in August 2012, TEPCO abolished these exemptions). In addition, the fundraising bodies and supporter groups of 90 of these employee political representatives had received 600 million yen in contributions from union political committees in the three years up to 2010.

After the 2011 disaster, most of these employee-politicians continued to oppose antinuclear initiatives in regional assemblies and to promote nuclear energy in their constituencies.[30]

Local government

As discussed above, protests by local residents against planned nuclear developments had become commonplace by the 1970s. TEPCO and the government needed to develop plans to counter these movements at the local level. While safety concerns tended to loom large in local antinuclear movements in Europe and the United States, in Japan money was a bigger issue.

Monetary compensation

Property and fishing rights have tended to be more jealously protected in Japan than in Europe or the United States. As a result, opposition from landowners and fishers presented the greatest problem for siting nuclear reactors. The utilities' main priority was negotiating with landowners to purchase land and with fishery cooperatives to relinquish their fishing rights (Yoshioka 2011: 198–9; Honda 2005: 24). In Japan fishing rights are managed by the fishing cooperatives, of which individual fishers are members. Therefore, the utilities and the government tried to influence local movements using the power of money.

Up until the mid-1970s a number of intense battles were fought in which fishing cooperatives used their fishing rights to prevent the construction of nuclear power stations. However, the cooperatives are tied to conservative local powerbrokers through personal networks and they also tend to be LDP supporters. They were unable to form sustainable joint struggles with progressive antinuclear groups and grassroots movements. The utilities used local powerbrokers to break the resistance of the fishing cooperatives, leading many cooperatives to moderate their demands and trade outright opposition for compensation claims (Honda 2005: 114).

However, paying compensation for lost fishing rights began costing the utilities more and more each year. In 1966 the agreement TEPCO forged with a fishing cooperative for the construction of the Fukushima Daiichi Nuclear Power Station provided for 100 million yen in compensation for lost fishing rights, while the 1973 agreement for the construction of Fukushima Daini Nuclear Power Station (located in Naraha Town and Tomioka Town in Futaba District, Fukushima Prefecture) included 3.5 billion yen in compensation. The figure increased again in the 1979 agreement for Tōhoku Electric Power's Onagawa Nuclear Power Station, which included 9.83 billion yen in compensation for the cooperatives (Honda 2005: 142).

As discussed above, the *Dengen Sanpō* was enacted to address local opposition to the construction of nuclear power stations. In 2009 the total amount of compensation provided to municipal governments under the *Dengen Sanpō* stood at 122.2 billion yen (*Shūkan Daiyamondo*, May 21, 2011: 34–6, 42). However, from the late 1970s few reactors have been constructed at new sites. Since then, those new power stations that have been built were already in the planning stage at the end of the 1960s. Most of the new reactors that have been built have been added to existing sites.[31]

Why are most new reactors built inside existing power stations? Under the *Dengen Sanpō*, the first compensation payments are made at the beginning of the environmental impact assessment process and they continue for ten years after the reactors have commenced operations. After that, the amount paid out slowly decreases. For example, if it takes three years for a hypothetical nuclear development to proceed from the initial environmental impact assessment process until construction starts and another seven years before it begins operating and the operational life of the reactor is 35 years, then the total compensation payable over the entire 45-year period will be 245.5 billion yen. However, of this total, 44.9 billion yen will be paid out in the ten years before the reactor starts operating. The utilities also pay a significant amount of property tax to the municipal government. However, there is a statutory depreciation period of 16 years, after which the utilities have to pay much less in property tax. Furthermore, as municipalities use the compensation provided under the *Dengen Sanpō* to construct new infrastructure, maintenance costs put pressure on local budgets. This is when municipalities start looking for the next injection of compensation that will accompany new reactor construction.

For example, the municipality of Futaba, where the Fukushima Daiichi Nuclear Power Station is located, had a ratio of net cost of bonds of 29.4% in 2009. This meant that the municipality was designated as requiring intervention to achieve early financial soundness. In this context, the municipal assembly voted to approve the construction of reactors No. 7 and No. 8. Between 2007 and 2009 Futaba Town received 980 million yen as part of the first-stage of compensation funding for these new reactors. The utilities are well aware of these issues. By encouraging municipalities to build nuclear reactors, they created a base of support for the further development of the technology.[32]

In addition to the compensation payments made under the *Dengen Sanpō* and municipal property taxes, the utilities donate to municipal governments that host nuclear power stations. TEPCO has been making these hidden donations to municipal governments since the 1960s. Between 1990 and 2009 the company donated 1.8 billion yen to Mutsu City in Aomori Prefecture, where TEPCO and the Japan Atomic Power Company are building interim storage facilities for radioactively contaminated waste. It also donated 13 billion yen to Kashiwazaki City and Kariwa Village (Kariwa District) and Niigata Prefecture, where the company's Kashiwazaki-Kariwa Nuclear Power Station is located, and 19.9 billion yen to the four municipalities and Fukushima Prefecture that host the Fukushima Daiichi Nuclear Power Station and the Fukushima Daini Nuclear Power Station. The FEPC also donated 17 billion yen, of which 5 billion was provided by TEPCO, to various entities in Aomori Prefecture in order to promote the nuclear fuel cycle project.[33] Other donations, whose purpose is unclear, total more than approximately 45 billion yen.

TEPCO budgeted 1–2 billion yen for donations at the beginning of each financial year. It frequently exceeded this amount and the yearly average was more than 2 billion yen. The funds were distributed among the various prefectures based on a rough approximation of the amount of electricity generated by each. The mechanism for accessing these donations was via a request from the local mayor and they were approved at the executive level within TEPCO.[34]

Prefectural governments also imposed a non-statutory nuclear fuel tax. In August 1976 Fukui Prefecture asked the Ministry of Home Affairs to approve the creation of the tax due to uncertainty over tax revenues from nuclear power stations that had already been built under the *Dengen Sanpō*. The tax would apply from the time nuclear fuel was loaded into the reactor. It was set at 5% of the purchase price of the fuel, payable to the prefecture yearly for ten years. Following the introduction of the nuclear fuel tax in Fukui, it was introduced in all prefectures hosting nuclear reactors, including Aomori, where the Rokkasho Reprocessing Plant is located.

The nuclear fuel tax, unlike the compensation provided under the *Dengen Sanpō*, was not tied to the construction of specified public buildings and was not time-limited or related to whether the reactors were operating or not. Furthermore, unlike municipal property tax, nuclear fuel tax was not subject to depreciation. The construction of new nuclear reactors and the consequent increase in the use of fuel would produce increased taxation revenues. Honda (2005: 142–3) describes the tax as a form of sponsorship of those prefectures that have already accepted nuclear power stations. In 2010 the nine utilities paid a total of 24.2 billion yen in nuclear fuel tax (*Asahi Shimbun Digital*, March 26, 2014).

When the utilities pay construction contractors to build a nuclear power station, the price includes the cost of placating host communities. These contractors hire a further layer of subcontractors to carry out this work. The subcontractors take local powerbrokers and politicians out to drink at bars and organize sightseeing trips, ostensibly to inspect other nuclear power stations. They also assist local politicians during elections. In return, local powerbrokers and politicians help the contractors to acquire land and deal with local opposition. The utilities also provide electoral assistance, give jobs to the children of local powerbrokers and politicians, and build friendly relations with them. They donate tea for events put on by the local shopkeepers' association or the neighborhood committee. They also source all the materials used in the nuclear power station from local suppliers (*Shūkan Daiyamondo*, May 21, 2011: 36–7; *Shūkan Tōyō Keizai*, April 23, 2011: 43).[35]

Through the various mechanisms outlined thus far, large amounts of money flow from the utilities and the government to host communities. According to a survey of 44 nuclear host communities conducted by NHK, the municipalities have received 3.11 trillion yen thus far in declared compensation alone through the *Dengen Sanpō*, donations and nuclear fuel tax (of which 164.12 billion yen came from donations). This makes up 35% of their annual revenue. Kashiwazaki City was the largest recipient, at 300 billion yen (NHK 2012). The utilities also provide jobs to host communities. Taken together, these measures give them extraordinary power in host communities. Their influence is equivalent to that of a major corporation in a company town. Perhaps it is even greater.

Onda Katsunobu, who interviewed people in the towns of Ōkuma and Futaba, host communities of the Fukushima Daiichi Nuclear Power Station, describes the stifling atmosphere that exists in such company towns. Local residents who worked for subcontractors at the station told him, 'if the company finds out I talked to a magazine or newspaper reporter like you, they'll tell me "don't bother coming to work tomorrow"'. Since the 1970s a secret group known as TCIA (TEPCO's Central Intelligence Agency) has also been operating in host communities.[36] TCIA monitors the registration numbers of cars at antinuclear rallies. During elections it checks which local residents are frequenting the antinuclear candidate's home or office. TCIA even went so far as to buy every copy of weekly magazines and other publications that contained articles about TEPCO in the local newsagents and bookshops, making it impossible for locals to obtain one in town. As a result, people who live in reactor host communities hesitate to talk about nuclear power or TEPCO, whether at work or at home (Onda 2011: 158–60).

From the end of the 1980s until the 2000s, organized harassment of antinuclear groups took place not only in host communities but across the country (Kaido 2014). Those responsible for this harassment have never been identified.

Resistance

So far I have described how the utilities gained approval from municipal governments for plant construction using the power of money. While this is an example of the utilities' power, it is also a sign of the municipal governments' power of resistance. This is what forces the utilities and the government to expend such vast amounts of money (although ultimately these funds come from higher electricity prices).

So how do the municipalities resist plant construction? The first form of resistance is that wielded by the fishing cooperatives and landowners. If the cooperatives were unwilling to give up their fishing rights and the landowners to sell their land, it would be impossible to build a nuclear power station. However, this kind of opposition is overcome relatively easily through financial incentives.

The second form of resistance developed out of a local referendum on the construction of a nuclear power station in Maki Town (Nishikanbara District, Niigata Prefecture; now merged into Niigata City) on August 4, 1996. The referendum movement quickly spread across the country, giving expression to the opposition of local residents to plant construction. However, pronuclear mayors and municipal councils are usually able to block the passage of such referendum proposals.

A third form of resistance is the power of the municipal mayors and prefectural governors to withhold their approval for nuclear power station construction. Legally, only the central government has the power to approve nuclear energy-related projects. Municipal and prefectural governments have no such power. Nevertheless, without their approval the process becomes much more difficult. Since the 1970s it has become customary for MITI to seek the cooperation of municipal mayors and approval from local governors as part of its internal approvals process for power station siting. It has also become standard for the utilities to make safety plans and other arrangements in cooperation with local government bodies. It has therefore become possible for prefectural governors to interrupt the construction and operation of nuclear power stations by withholding their consent or refusing to sign off on these safety plans.

The utilities and the government have had little difficulty in obtaining the consent of municipal mayors and a majority of municipal councillors for nuclear power station construction. However, some prefectural governors have demonstrated strong opposition to nuclear projects.[37] Since an accident at the Monju fast breeder reactor, it has become increasingly common for governors to exercise their de facto jurisdiction (Honda 2005: 24–6; Yoshioka 2011: 264–8).[38]

The legal system

One mechanism whereby citizen groups can potentially prevent the construction or operation of a nuclear power station is by taking the government or utility to court. Before the 2011 disaster, there were about 20 such antinuclear lawsuits,[39] some of which highlighted flaws in the safety inspection process.[40] Plaintiffs also highlighted a number of other issues with the nuclear power stations and questioned what would happen if an earthquake was to occur that exceeded the level anticipated by the utilities. They asked whether a nuclear power station could survive such an event, whether blackouts would occur and whether a tsunami might cause a complete loss of power to the plants. It is as if they had anticipated the 2011 disaster (Kaido 2011: 58–79). However, with two exceptions, these lawsuits all ended in defeat for the plaintiffs.[41] In both these cases, where the lower courts found in favor of the plaintiffs, the decision was overturned in higher courts. Why have judges consistently ruled in favor of the utilities and the government?

In general, large companies tend to predominate in court proceedings due to their economic power and social position. In antinuclear lawsuits, in particular, top-class researchers and technicians act as witnesses for the defense and make favorable statements about nuclear safety (Sakurai 2011: 156–7). Furthermore, in civil cases, where the burden of proof lies with the plaintiff, the documents and information they need are held by the defendants – that is, the utilities. The utilities are very reluctant to release this information, citing commercial-in-confidence and security concerns. The information differential between citizens and corporations makes it difficult for residents to win their case (Isomura and Yamaguchi 2013: 40–7).

However, in antinuclear lawsuits the plaintiffs typically receive help from antinuclear lawyers and researchers. In a number of such cases, they have competed strongly against the state and the utilities. These antinuclear experts help to make the competition fairer, somewhat offsetting the superior position occupied by the government and the utilities.

The Supreme Court

Another issue is that the courts take into account the political context of antinuclear lawsuits and are influenced by the Supreme Court's tendency to avoid making judgments on matters of public policy. The jurisprudential principle known as the 'theory of legislative discretion' is deeply rooted in the judiciary, especially the Supreme Court. According to this principle, lawsuits are only meant to test whether a particular law has been broken and not to decide on matters of public policy. While this principle is based on the democratic idea that unelected judges ought not to interfere with elected representatives on matters of public policy, it is also a defensive position that has been adopted by the courts to stave off political interference.

From the late 1960s and into the 1970s, the LDP and sections of the media conducted a campaign in which they accused the legal profession of 'leftwing bias'. Judges who gave activist rulings on constitutional matters and lawyers who participated in the Japan Young Lawyers Association were accused of having 'communist sympathies'. From that time, the Supreme Court's conservativism has increased and the doctrine of judicial restraint has held sway. The courts have taken the position that if they hand down judgments that oppose government policy, the government might play a more interventionist role in their internal affairs. As a result, the courts avoid making such decisions.

Individual judges, too, are mindful of the opinions of the Supreme Court, which has authority over judicial appointment and they, too, are reluctant to issue judgments that go against public policy. In the district courts, antinuclear lawsuits are classified as 'notifiable cases', which must be individually reported to the Supreme Court. Notifiable cases include all cases that may be of interest to the government. The existence of such a notification system is evidence of the extreme caution with which the Supreme Court treats its relationship with the government (Isomura and Yamaguchi 2013: 76–89).[42]

Furthermore, the Supreme Court has certainly tried to influence the decisions of the lower courts in antinuclear lawsuits. For example, in one case, the presiding judge in a lower court hearing to revoke the development approval for the No. 1 reactor at the Ikata Nuclear Power Station was suddenly replaced in the midst of the proceedings, just after the cross-examination of witnesses had taken place. In another case, the Supreme Court's General Secretariat organized a meeting to discuss legal interpretation during an antinuclear lawsuit at the district court level in October 1979. Judges were summoned to attend the meeting, where they discussed the question of whether or not local residents in reactor host communities have legal standing to sue for an administrative cancellation. In summing up the discussion, a representative of the Administrative Affairs Bureau in the Supreme Court's General Secretariat observed that, as there was very little chance that a nuclear accident would occur even if one were to take a negative view of the matter, there was no cause for concern. A third example of the Supreme Court's effort to shape the outcome of antinuclear lawsuits was a similar legal conference held in October 1988 that considered the format for trying antinuclear lawsuits. During the conference, Supreme Court officials gave their opinion that, due to the high level of technical knowledge necessary to discuss nuclear power, when trying such cases the courts ought to defer to the administrative agencies that employ suitable staff. It would be enough for the courts to rule on whether the decisions of these agencies were made rationally and proportionally. The opinion given during this meeting closely resembled a decision reached by the Supreme Court in October 1992, which I discuss below. It seems that the court had already made its opinion on antinuclear lawsuits clear at this much earlier stage (Kaido 2011: 6–8, 222–6; Isomura and Yamaguchi 2013: 163–8; Asahi Shimbun 'Genpatsu to Media' Shuzaihan 2013: 106–9).

Nevertheless, there are cases where lower court judges have issued injunctions stopping the operation of nuclear reactors. It seems that they did not experience any ill effects in terms of their career advancement because of these decisions.

Individual judges

Quite apart from the power of the utilities themselves, judges may be reluctant to issue injunctions against the construction or operation of nuclear power stations for other reasons. First, the debate on nuclear safety is highly technical. Judges, who are amateurs when it comes to technical issues, find it difficult to make decisions about them. They are also reluctant to make decisions that might have significant social consequences, such as ordering a halt to plant construction or operations. In its first ruling on an antinuclear lawsuit that sought to revoke the development approval for the No. 1 reactor at the Ikata Nuclear Power Station on October 29, 1992, the Supreme Court indicated its preference for leaving such matters in the hands of the relevant administrative agencies. It stated that, as the system for evaluating the proposal set down by the agencies had been made with the input of scientific and technical experts, the courts ought to give precedence to the decisions made by these agencies, except in the case of a glaring error that could not be easily overlooked. This 'principle of expert discretion' has continued to influence subsequent judgments (Isomura and Yamaguchi 2013: 28–30; Shindō 2012: 50–6).

Second, prior to the Fukushima nuclear disaster, the majority of Japanese people thought that, despite their reservations about nuclear power, they had little choice but to rely on the technology or face electricity shortages. It may be that most judges felt the same way (Isomura and Yamaguchi 2013: 47–51). Most judges, like the populace as a whole, probably believed in the 'myth of nuclear safety'. Furthermore, judges tend to be highly sensitive to public opinion.

The mass media

There is not enough space here to list all the ways in which the utilities have used public relations and their influence over the media to shape public opinion in favor of nuclear power.[43] There are also numerous reports of the utilities applying pressure to television stations to stop them from broadcasting programs critical of nuclear power (Komatsu 2012: 115–9; Asahi Shimbun 'Genpatsu to Media' Shuzaihan 2013: 310–3; Katō 2012: 138–83; Gurūpu K21 2011d; Kyodo News 2013e) and the government has been very active in carrying out public relations campaigns in support of nuclear power.[44]

It was the extensive propaganda campaign carried out by Shōriki Matsutarō, proprietor of the mass-circulation *Yomiuri Shimbun* and the first chairperson of the Japan Atomic Energy Commission, that saw nuclear power initially gain acceptance in Japan, despite the history of the atomic bombing of Hiroshima and Nagasaki. In 1954 Shōriki launched his campaign for the peaceful use of nuclear energy in the pages of the *Yomiuri*. In 1955 all Japan's major newspapers held expositions and fairs promoting the peaceful use of nuclear energy across the country, in collaboration with American government agencies. These campaigns raised peoples' expectations of the potential of the technology (Jōmaru 2012: 74–127; Kanbayashi 2012: 90–3). Then, in the 1970s, as opposition to nuclear power station construction in host communities grew, the FEPC bolstered its public relations campaign and began running advertisements in the newspapers promoting nuclear energy (Suzuki 1983: 38–44; Asahi Shimbun 'Genpatsu to Media' Shuzaihan 2013: 288–91; *Shūkan Tōyō Keizai*, June 11, 2011: 57). As a result, it became difficult for newspapers to report on the dangers of nuclear power. Nuclear advocates used the mass media to shape public opinion in favor of the technology.

The first aspect of this media campaign was the massive investment by the utilities in mass media advertising. In March 2011 the ten utilities' combined budget for expansion and development (which includes advertising, as well as expenses such as the running of nuclear energy-related museums etc.) stood at 86.6 billion yen. TEPCO's share of that total was 26.9 billion yen, including 7 billion for television and radio broadcasting and 4.6 billion for advertising and publicity.[45] The FEPC has an advertising budget of approximately 2 billion yen (Komatsu 2012: 98–9).

There are also close ties at the management level between the utilities and the television stations. At the time of the 2011 earthquake, former TEPCO president Minami Nobuya was an external director of Fuji Media Holdings, while former TEPCO chairperson Araki Hiroshi was an external director at TV TOKYO Holdings Corporation. Until May 2011, TV Asahi also had TEPCO chairperson Katsumata Tsunehisa on its programming commission.[46]

The utilities also spend time entertaining individual reporters.[47] In the 1990s one science writer at a major newspaper described how the utilities wooed reporters in the science and economics bureaus by taking them on tours to visit nuclear power stations, entertaining them at high-class clubs or taking them golfing. Having earned their trust, the utilities would then give the reporters access to highly confidential documents or organize interviews for them with senior executives. If they wanted to go overseas, the utilities would give them money and if they were transferred to another office, then the utilities would host a farewell party. At times, TEPCO helped these reporters to gain appointments to inspection committees within METI or sent them on Japan Electric Power Information Center, Inc. overseas fact-finding missions that were organized and paid for by the utilities to investigate nuclear energy in other countries. Others were given positions at CRIEPI or other industry bodies (Kanbayashi 2011: 52–3).

The utilities also found positions for retired journalists, just as they did for retired bureaucrats who 'descended from heaven'. Four former newspaper reporters, one former broadcaster and one former news agency employee were employed at CRIEPI as research advisers. They received 200,000 yen per month for working two to three days per week rewriting manuscripts prepared by experts to make them accessible to a general audience. The average length of contract for these positions was six years (Jōmaru 2012: 289–90, 331–4, 354–6; Asahi Shimbun 'Genpatsu to Media' Shuzaihan 2013: 279–80; Takahashi Atsushi 2012: 95–102; Doi 2011: 101–4).[48]

The education system

As part of their strategy to influence public opinion regarding nuclear power, the utilities even intervened in the school system to promote teaching about nuclear energy. The utilities also paid for textbook editors to visit nuclear power stations (Kyodo News 2013c).

In 1996 the Atomic Energy Society of Japan, an organization whose staff includes TEPCO executives, began conducting regular studies of how nuclear power was being portrayed in primary and secondary school textbooks. It compiled the results of this study, along with suggestions for change, into a report that was given to the MEXT textbook department and to the textbook publishers (Asahi Shimbun 'Genpatsu to Media' Shuzaihan 2013: 359–62). In January 2009 the society recommended that references to 'concerns' and 'doubts' about nuclear safety in certain textbooks be replaced with 'the more appropriate phrase "challenges remain"'. The society also criticized the presentation of renewable energy in middle school science and social science textbooks for being overly optimistic.[49]

It seems that MEXT accepted these suggestions and included them in its evaluation criteria for school textbooks. While conducting routine evaluation of school textbooks in April 2005, MEXT issued eight publishers with notices asking them to revise their presentation of nuclear energy. Five of these notices advised that 'the problems of nuclear energy have been over-emphasized' and instructed the publishers to downplay the possible risks associated with nuclear energy and to add the benefits of nuclear energy. A further six notices instructed publishers to be more circumspect in their coverage of renewables and to emphasize the problems remaining with these technologies.[50]

In the spring of 2010 MEXT distributed books on nuclear energy to primary and secondary schools that it jointly produced with METI's Agency for Natural Resources and Energy. The book produced for middle school students contained a statement that nuclear power stations were designed to withstand large earthquakes and tsunamis. After the Fukushima disaster, this book created an uproar and was recalled.[51]

Conclusion

In this chapter I have investigated the ways in which TEPCO exercised its political and economic power prior to the earthquake disaster. The roots of TEPCO's power lay in the comprehensive cost-accounting method specified in the Electricity Business Act. The financial resources the company was able to develop thanks to this law enabled TEPCO to boast of its unparalleled dominance of the economy in its role as a large procurer. The utilities gained the support of LDP politicians by providing them with votes and money and used that support to smash attempts by the bureaucracy to contain their power. By providing votes and money to the main opposition party, the DPJ, the electric power unions made sure it, too, would support a pronuclear position. By colluding with the nuclear safety regulators and academic researchers, the utilities watered down the regulations. They also used the power of money (more than 3 trillion yen) to gain approval for power station siting from municipal governments. By winning over the mass media, TEPCO contained any criticism of nuclear power from that quarter. In its arrogance, it even intervened in the education system.

These examples demonstrate TEPCO's enormous power, far exceeding that of most private companies, which it used to promote nuclear power. This structure of power enabled TEPCO to create the 'myth of safety' and to continually disregard the numerous warnings it received about the potential for a severe accident.

3 Nuclear Policy after 3.11

Kamikawa Ryūnoshin

In this chapter I analyze changes in Japan's nuclear energy policy from the time of the Fukushima accident of March 2011 to August 2016. After the Fukushima accident, the Democratic Party of Japan (DPJ)-led coalition government under then Prime Minster Kan Naoto moved to phase out nuclear power. However, the DPJ government's attempt to turn around Japan's existing nuclear energy policy proved unsuccessful. When the Liberal Democratic Party (LDP)–Komeito coalition returned to power in December 2012, Prime Minister Abe Shinzō tried to reverse the DPJ's changes to nuclear energy policy. However, his plan to restart Japan's nuclear reactor fleet has stalled and the government has deferred decisions about the construction of new nuclear power stations. In this chapter I discuss the complex network of interests embedded in Japan's nuclear energy policy. I argue that the failure of both the DPJ and LDP–Komeito governments to demonstrate any leadership on this issue and their continuing deferral of important decisions on nuclear energy are a result of the complex network of interests embedded in this policy area. The result has been continuing policy stagnation.

According to the *Act on Compensation for Nuclear Damage 1961* (Compensation Act), nuclear power station operators are liable for nuclear damage regardless of whether or not they can be proved to have been at fault (no-fault liability). The Compensation Act also places no limit on the extent of an operator's liability for nuclear damage (unlimited liability). The Tokyo Electric Power Company (TEPCO) is therefore responsible for the entire cost of the Fukushima nuclear accident, including nuclear damage compensation, decontamination costs and the cost of decommissioning the damaged reactors. The total cost is likely to exceed ten trillion yen. This means that TEPCO is, in effect, insolvent. However, the government at the time of the accident (DPJ) decided to protect TEPCO from insolvency and allow it to continue to operate. There were five reasons for the government's decision. First, the government wanted to make TEPCO pay for the cost of nuclear damage compensation. Second, it wanted TEPCO to clean up after the nuclear accident. The third consideration was enabling TEPCO to continue to be able to provide electricity in its supply area. Fourth, the government was anxious to stabilize the bond market by guaranteeing TEPCO's electric power bonds. Finally, the government wanted to maintain the stability of the financial system by insulating those financial institutions that have loaned large sums of money to TEPCO against heavy losses (Endō 2013: 145).

On September 12, 2011, the government set up the Nuclear Damage Compensation Facilitation Corporation (NDF) to prevent TEPCO from becoming insolvent. All of the 12 nuclear energy companies, including TEPCO, and the government are shareholders in the corporation and each has a 50% stake. The system works as follows. In each accounting period, TEPCO accounts for the expenses it incurs in paying nuclear damage compensation as an extraordinary loss. After auditing TEPCO's accounts, the government transfers an amount of money sufficient to counterbalance this loss from the NDF to TEPCO. TEPCO then accounts for this income as an extraordinary profit. The NDF is financed by issuing a type of government bond known as a grant bond. Grant bonds can be redeemed for cash at any time but the government does not need to include the cost of paying out on these bonds at the time of issuance. This setup is designed to enable the NDF to repay the government's investment over a long period of time. This will be achieved by placing a special levy on TEPCO, the operator responsible for the accident, representing a portion of its liability, and a general levy on the 12 electric power companies (including TEPCO). The structure of the scheme reflects the Ministry of Finance's desire to avoid investing public finance in nuclear damage compensation.

Considering the cost of decommissioning Fukushima Daiichi and decontaminating the disaster zone, TEPCO would inevitably have been insolvent without an injection of capital by the government. In order to prevent this from happening, the NDF invested one trillion yen in TEPCO on July 31, 2012. This investment included the purchase of 320 billion yen of Class A preferred shares (voting shares) at 200 yen per share and 680 billion yen of Class B preferred shares at 2,000 yen per share. The latter are not voting shares, but had the potential to be converted into voting shares in the future. This gave the NDF a 50.11% stake of TEPCO's Class A shares, giving it a controlling interest in the company. The NDF can also convert its

Class B shares into preferred shares with voting rights, increasing its voting share to 75.84% (*Nihon Keizai Shimbun*, May 21, 2012, digital edn). This means that TEPCO has actually been nationalized. Sengoku Yoshito, Executive Acting Chairman of the DPJ's Policy Research Council, together with the Ministry of Economy, Trade and Industry (METI), selected a new chairperson and directors for TEPCO. They appointed METI's Shimada Takashi as a director and operating officer at the company to act as the government's eyes and ears. In practice, TEPCO has been placed under METI control.

In the previous chapter in this collection, I argued that TEPCO had utilized its enormous power to advocate for the expansion of nuclear energy. Since the Fukushima accident, TEPCO's effective nationalization has led to a considerable decline in its power and influence. The clearest manifestation of this is the government's implementation of electricity sector reform. Under the DPJ government, a METI committee thrashed out a plan for the complete liberalization of the retail electricity market and the unbundling of the integrated service of power generation and transmission. Since the change of government, the LDP–Komeito coalition has continued with the electricity sector reform. The retail electricity market was fully liberalized in April 2016 and the unbundling of power generation and transmission is planned for April 2020. While the electric power industry has protested these changes, with TEPCO under the control of METI, the ministry responsible for leading the reform, these objections have not hindered the steady implementation of the reform.

However, despite the decline in TEPCO's influence, there has not been a major shift in nuclear energy policy. Japan, unlike Germany, which decided to give up nuclear power after the Fukushima nuclear accident, has decided to stick with the technology. The DPJ Cabinet did endorse the Innovative Energy and Environmental Strategy on September 19, 2012, which included the goal of 'developing policy resources to enable a complete phase out of nuclear power in the 2030s'. This led many people to assume that the government would try to phase out nuclear power. However, in July 2012 the DPJ government approved the restart of Kansai Electric Power Company's (KEPCO) Ōi Nuclear Power Station (Ōi Town, Ōi District, Fukui Prefecture). The Innovative Energy and Environmental Strategy, which contained this call for a phase out of nuclear power, was a contradictory document that stood little chance of ever being implemented. In addition, when the LDP–Komeito coalition government returned to power in December 2012, it abandoned the DPJ's 'zero nukes' policy. On April 11, 2014, the LDP–Komeito Cabinet approved a new Basic Energy Plan that designated nuclear power as 'an important source of baseload power'.[1] On August 11, 2015, Kyūshū Electric Power (Kyūden) restarted reactor No. 1 at the Sendai Nuclear Power Station (Satsumasendai City, Kagoshima Prefecture), signaling the end of 'zero nuclear'. The LDP has reversed the DPJ government's policy on nuclear energy and returned to a pronuclear position.

Nevertheless, nuclear energy policy has not reverted to its pre-3.11 state. At the time of writing in August 2016, only reactors No. 1 and No. 2 at the Sendai Nuclear Power Station and reactor No. 3 at the Ikata Nuclear Power Station (Ikata Town, Nishiuwa District, Ehime Prefecture) are operating. Prime Minister Abe has stated on a number of occasions that his government has no plan to construct any new nuclear capacity. The Abe government has defied expectations by failing to fully restore the nuclear energy program.

Why was the DPJ government unable to substantively shift nuclear energy policy so as to phase out nuclear power? Why has the DPJ–Komeito government failed to fully restore Japan's nuclear energy program? These are the two questions I address in this chapter.

The DPJ's nuclear energy policy

Kan Naoto, who became Prime Minister on June 8, 2010, advocated expanding the renewable energy sector and once considered nuclear energy to be no more than a 'transitional' energy source. However, after becoming Prime Minister, he continued with the nuclear energy policy that was adopted by the DPJ in 2009. At that time, the party had decided to build more domestic nuclear power plants in order to combat climate change and to export nuclear technology overseas in order to promote economic growth. The Fukushima nuclear accident of March 11, 2011, proved to be a turning point in Kan's position on nuclear power. After the disaster, he ordered the Japan Atomic Energy Commission to come up with a simulation for a worst-case scenario. When the simulation was delivered on March 25, Kan was told that in a worst-case scenario (which included further hydrogen explosions at the Fukushima Daiichi Nuclear Power Station), the government would have to forcibly relocate everyone living within a 170-kilometer radius of the plant. Those living within a 250-kilometer radius, an area that includes most of Tokyo, would have to be evacuated. When presented with this plan to evacuate 30 million people, Kan came to believe that the risks posed by nuclear power were so significant as to constitute a threat to the continued existence of the Japanese state. He became convinced that the safest course of action would be to eliminate Japan's dependence on nuclear power (Ōshika 2013: 187–92; Kan 2012: 150–2; Kan 2014: 29–32; Institute for Renewable Energy Policies 2012: 69–70).

Suspension of operation of Hamaoka Nuclear Power Station

On April 18 Kan stated publicly that nuclear energy policy would have to be thoroughly re-examined in order to determine whether or not its safety could be guaranteed. He also stated that plans to build new nuclear power stations would have to be revised. On May 6 Kan asked Chūbu Electric Power (Chūden) to suspend operations at its Hamaoka Nuclear Power Station (Omaezaki City, Shizuoka Prefecture). Scientists have determined that a future earthquake along the Tōkai fault is inevitable. The Hamaoka Nuclear Power Station is located directly above the projected hypocenter of this earthquake and antinuclear activists have been trying to draw attention to this risk for some time. At a meeting on April 27, the government's Central Disaster Management Council found that there was an 87% chance that a magnitude 8.0 earthquake would occur along the Tōkai fault within the next 30 years. Nevertheless, METI still hoped that by suspending operations at the Hamaoka facility, the government would be able to placate opponents of nuclear power and thereby maintain the remainder of the nation's nuclear reactor fleet.

METI prepared a statement to be read by the Minister for Economy, Trade and Industry Kaieda Banri at a press conference held to announce the suspension of operations at the Hamaoka Nuclear Power Station. The statement explained that operations at Hamaoka were being suspended due to the exceptional risk posed by an earthquake in that region. However, with regard to the rest of Japan's reactor fleet, it aimed to reassure the public that the reactors were safe. After Fukushima, it continued, the Nuclear and Industrial Safety Agency

(NISA) had implemented the appropriate emergency measures needed to deal with an earthquake and tsunami. However, Kan addressed the press conference and said nothing about the remaining nuclear reactors, calling only for Chūden to suspend operations at Hamaoka. While the government lacks a legal mechanism to compel operators to shut down a nuclear power station, Chūden acceded to the request (Kan 2012: 155–8; Ōshika 2013: 308–26).

Kan, who recognized that METI was trying to maintain the nuclear reactor fleet, felt strongly that energy policy could not be left in the hands of the ministry. On May 18 Kan publicly aired his concern that both the Agency for Natural Resources and Energy, which is charged with promoting nuclear power, and NISA, which is supposed to regulate nuclear safety, were located within METI. He suggested that the regulator ought to be separated from METI and also indicated that he was considering unbundling the currently integrated services of power generation and transmission. On June 7 he even shifted the Energy and Environment Council, which was initially supposed to be attached to METI, to the Cabinet Secretariat's National Policy Unit. Kan decided to use the council as a forum to debate future changes to energy policy (Ōshika 2013: 333–4, 356–61, 401–4).

The stress test regime and Kan's resignation

While these events were unfolding, there were moves within the Diet to oust Kan. Kan's grip on the leadership was already tenuous before the disaster. At the July 2010 House of Councillors election, the DPJ lost its upper house majority. Former secretary-general Ozawa Ichirō and his supporters within the DPJ attacked Kan, whom they blamed for the party's poor performance in the election. They felt that the Prime Minister's earlier suggestion that the government might raise consumption tax had aroused the ire of voters, costing the party its majority. In the House of Councillors, the opposition launched an offensive against the government, using its new-found majority to censure a number of ministers. When a Chinese fishing boat collided with a Japan Coast Guard vessel on patrol near the Senkaku Islands in September 2010, Kan's awkward handling of the incident gave the opposition a further excuse to attack the Prime Minister. The government began to struggle in the polls. After the Great East Japan Earthquake a temporary truce had existed between the government and opposition. However, from the end of May, as the initial phase of the disaster response drew to a close, the opposition placed increasing pressure on the Prime Minister for his alleged mishandling of the disaster.

On June 2 the LDP, Komeito and other opposition parties submitted a motion of no-confidence in Kan's government to the Diet. Behind the attack lurked DPJ powerbroker Ozawa Ichirō. Ozawa had made it clear to the LDP that were they to introduce a no-confidence motion, it would have the support of his parliamentary group within the DPJ. A growing rebellion against Kan within the party meant that the motion looked likely to pass. The Prime Minister tried to rally support. He told his colleagues, 'once I have discharged my responsibilities, then it will be time for the younger generation to take the reigns' (*Asahi Shimbun*, June 3, 2011, morning edn). The rebels interpreted the statement as a sign that Kan was willing to step down. With the rebellion appeased, the motion failed. At a press conference on June 26, Kan specified the conditions for his resignation. The Diet must pass three bills: the second supplementary budget for 2011, a bill authorizing the government to issue special deficit-financing bonds and a bill to establish a feed-in tariff to encourage renewable energy. These became known as the 'three conditions' (Ōshika 2013, 358–90; Kan 2012: 161–2, 170; Kan 2014: 111–2; Anzai 2012: 188–90).

METI was planning to restart reactors No. 2 and No. 3 at Kyūden's Genkai Nuclear Power Station (Genkai Town, Higashimatsuura District, Saga Prefecture), which were offline while undergoing routine maintenance checks. After that, METI intended to restart the remainder of the reactors that were also undergoing routine inspections. Without consulting Kan, METI entered into secret negotiations with Saga Prefecture and Genkai municipality, where the nuclear power station is located, and obtained approval from both tiers of local government to restart the Genkai reactors. On July 6 Kan announced the introduction of a new 'stress test' regime that would apply to all nuclear reactors. He justified the new safety procedures by claiming that people would never accept the restart of Japan's nuclear fleet based solely on the reassurance of NISA, which had failed in its duty to prevent the Fukushima accident.[2] The stress test regime involved electric utilities evaluating the safety of their own reactors and then submitting the results of these evaluations to NISA for review. Once NISA had approved the results, the Cabinet's Nuclear Safety Commission would give the final approval for reactor restarts.

The European Union had already begun carrying out stress tests after the Fukushima accident. On June 21 an International Atomic Energy Agency (IAEA) Ministerial Conference on Nuclear Safety issued a declaration calling on member states to carry out comprehensive risk and safety assessments of their nuclear power plants.

At this time, evidence that Kyūden was secretly intervening in support of the restart of the Genkai Nuclear Power Station came to light. A video was circulated on the internet that showed NISA explaining the results of its evaluation of the emergency safety measures at Genkai to the executive of Saga Prefecture. Responding to the video, the Saga prefectural government office sent a request to Kyūden, asking the company at the governor's request to go online and submit comments supporting the restart. Kyūden then emailed its employees, asking them to voice their opinions (on the internet) in favor of restarting the power station. On June 26 METI organized an information session for prefectural residents that was to be broadcast over cable television and the internet. On June 21, just before the scheduled information session, Saga Governor Furukawa Yasushi asked Kyūden's vice-president to 'make sure that comments and questions supporting the restart are submitted to the session via the internet' (Ōshika 2013: 416). Kyūden emailed managers in its nuclear division, as well as former Kyūden staff serving in four other companies with links to its nuclear energy operations. The utility asked them to submit comments to the program supporting the restart of Genkai. The broadcaster ultimately received 473 emails, of which 141 were 'fake' emails from people connected to the company that were intended to sway the session in favor of restarting the nuclear power station.

When the evidence of what had happened was revealed, Kyūden established an independent committee to carry out an internal review. However, when the committee produced a report acknowledging the intervention by the governor, the company attacked it. Kyūden's president initially suggested that he would resign over the fake email scandal, but then announced that he would remain in his post. The strong public opposition that resulted from this scandal made it difficult for Kyūden and the government to restart the Genkai Nuclear Power Station (Ōshika 2013: 410–25, 530; Kan 2012: 170–3).

On July 13 Kan held a press conference in which he announced his desire to phase out nuclear power completely. At the press conference, he explained how, after reflecting on the Fukushima accident:

> these thoughts led me to conclude that with regard to Japan's future nuclear power policy, we should aim to achieve a society that is not dependent on nuclear power. In other words, we should reduce our dependence on nuclear power in a planned and gradual manner and aim to realize a society in the future where we can do without nuclear power stations. (Kan 2012: 174)

However, at this point the idea of phasing out nuclear power had not been endorsed by the Cabinet, nor had it been discussed with the relevant ministries. The opposition, as well as business leaders, criticized Kan over the statement, as did members of his own Cabinet. On July 15 the Prime Minister issued a clarification in which he insisted that the statement had simply been his personal opinion.

However, on July 29 the Energy and Environment Council endorsed a plan to reduce Japan's dependence on nuclear power. This plan, the Innovative Energy and Environmental Strategy, was the result of a year of extensive public consultation and was intended to lay out the steps involved in realizing a nuclear-free society. On August 15 the Cabinet decided to remove NISA's nuclear regulation division from METI's remit and establish a new external agency, referred to provisionally as the Nuclear Regulatory Agency, under the Ministry of the Environment (Kan 2012: 173–7, 179–80; Ōshika 2013: 431–2; Institute for Renewable Energy Policies 2012: 69–72). These steps were part of Kan's plan to prevent the new antinuclear policy direction from being reversed after he resigned. On August 26 the Diet passed Kan's bill authorizing the issuing of special deficit-financing bonds and also passed the *Act on Special Measures Concerning Procurement of Electricity from Renewable Energy Sources by Electricity Utilities*. This Act established a new feed-in tariff that would commence in July 2012 to force the electric utilities to purchase all available supplies of renewable energy. Having fulfilled his three conditions, Kan announced his resignation.

Restarting the Ōi Nuclear Power Station

On August 30 the DPJ chose Kan's replacement for Prime Minister and party leader, Noda Yoshihiko. Noda indicated that it was unlikely construction of new nuclear power stations would go ahead and that those facilities that had come to the end of their lives should be decommissioned rather than having their operating license extended. Without either new construction or the extension of existing plant, Japan will eventually be left without any source of nuclear energy. At the same time, Noda also stated that he would try to restart Japan's reactor fleet beginning in the summer of 2012.

After the Fukushima accident, the remainder of Japan's nuclear reactor fleet continued to operate as usual. However, when reactors were taken offline for routine safety inspections, they were not immediately restarted. Under normal circumstances, nuclear reactors are taken offline for routine safety inspections after every 13 months of continuous operation. Unless the idle reactors were restarted, by May 2012 the nation's entire fleet would be idle. This prompted concerns that the reduced electricity supply would be insufficient to meet the increased summer demand. Noda's announcement that he intended to restart those reactors that had passed the new safety checks and for which local governments had given their consent was a response to these concerns. On September 22, during an address to the United Nations, Noda revealed that his government was also planning to proceed with the export of nuclear technology (Ōshika 2013: 441–2; *Asahi Shimbun*, September 26, 2011, morning edn).

On February 13, 2012, NISA determined that reactors No. 3 and No. 4 at KEPCO's Ōi Nuclear Power Station, which had been taken offline for routine safety inspections, could be restarted safely. This was followed by an announcement on March 23, by the Nuclear Safety Commission, ratifying the results of the NISA inspection. Having completed the two-stage process that was introduced following the controversy over the restart of the Genkai Nuclear Power Station, all that was now needed to proceed with the restart of Ōi was the approval of local government leaders in host communities (Ōshika 2013: 556–7, 561). Anticipating the results of the stress tests, the Noda government had already opened negotiations with local government leaders. At the end of 2011 the government approved the extension of the Hokuriku Bullet Train line to Fukui, a longstanding wish of Fukui Governor Nishikawa Issei. The government and the plant operator also promised to pay for the entire 42.2 billion yen cost of construction for roads in the vicinity of the Ōi plant, a condition that Ōi's mayor had placed on his approval (*Asahi Shimbun*, June 17, 2012, morning edn).

On January 31, 2012, the government submitted a bill to establish a new nuclear regulation agency on April 1, 2012. The LDP delayed the bill's referral to committee, calling for the new regulator to be given greater independence. On April 20 the LDP and Komeito submitted their own version of the bill to the Diet. Finally, on May 29, the bill was referred to committee. As a result of these protracted negotiations in the Diet, the Nuclear Safety Commission was only effectively functioning until March. This meant that only reactors No. 2 and No. 3 at Ōi, which had passed the two-stage process defined earlier by Kan, were able to be restarted. From April the final decision about reactor restarts was to be referred to a committee of four Cabinet ministers – Prime Minister Noda, Chief Cabinet Secretary Fujimura Osamu, Minister for Economy, Trade and Industry Edano Yukio, and Minister for the Environment and Minister Responsible for the Nuclear Accident Hosono Gōshi. Sengoku Yoshito, Executive Acting Chairman of the DPJ's Policy Research Council, and Saitō Tsuyoshi, Deputy Chief Cabinet Secretary, representing the Prime Minister's Office, attended the meeting as observers. From the time that Noda became Prime Minister, responsibility for important matters related to electricity sector reform were discussed in the 'three plus two' committee of Edano, Hosono and Furukawa Motohisa, Minister Responsible for the National Policy Unit, plus Sengoku and Saitō. In February the 'three plus two' committee began working towards the restart of the Ōi Nuclear Power Station (Sengoku 2013: 158–63).

In reality, Fukui Governor Nishikawa wanted to see Ōi restarted quickly. Thirteen commercial nuclear reactors and one research reactor (the Monju fast breeder reactor in Tsuruga City) are located in Fukui Prefecture. The electric power industry makes up some 12.6% of the prefectural economy, and the economy and public finances are structurally dependent on nuclear power. However, during the April 2011 gubernatorial elections, Nishikawa had promised to strengthen nuclear safety measures and so he placed two conditions on his support for restarting Ōi. He had labelled the stress test regime as inadequate and so his first condition was that the interim findings of an inquiry into the Fukushima accident be made public so that additional provisional safety standards based on these findings could be imposed on future reactor restarts. His second condition was that Prime Minister Noda and Minister for Economy, Trade and Industry Edano make a statement to the nation explaining the significance of nuclear power and the need to restart the nuclear reactor fleet. Nishikawa was concerned that if he took a leading role in the push to restart the reactors, he would be exposing himself to criticism from nuclear opponents. He therefore wanted the national government to lay the groundwork for the restart. Nishikawa was particularly unhappy that Edano had failed to take a clear stand in favor of restarting Ōi.

While Edano had vacillated over the decision, he was not necessarily opposed to reactor restarts. As much as 44% of KEPCO's total generating capacity is provided by nuclear power. Edano was concerned that unless Ōi was restarted, the electricity supply in the Kansai area would be unable to meet the increased demand during the summer months. As Chief Cabinet Secretary in the Kan government, Edano had been involved in implementing the stress test rules. Considering Ōi had now cleared those hurdles, he could not really oppose the restart.

Nishikawa was also irked by the attitude of other regional government leaders. Osaka Mayor Hashimoto Tōru, Shiga Governor Kada Yukiko and Kyoto Governor Yamada Keiji were all cautious about restarting Ōi. They wanted all local governments within 100 kilometers of the power station to be given a say in whether or not to restart the reactors, rather than just the local governments in the host community.

After much hesitation, the Noda government's ministerial committee finally approved the restart on April 13. However, Nishikawa withheld his consent, calling on the government to issue a clear statement to the nation that nuclear power was a key source of electric power and to pay more attention to the local communities that host nuclear power stations. Hashimoto, as the leader of a regional political party, the Japan Restoration Party, had his eyes set on entering the national political arena. He opposed the restart, claiming that his party would beat the DPJ at the next lower house election (Sengoku 2013: 152, 167–75; Ōshika 2013: 556–67). Opponents of the restart inside the DPJ also let their feelings be known. Public opinion was strongly against the move.

On April 23 KEPCO put out a public statement claiming that unless Ōi was restarted there would be a 19.3% deficit in the electricity needed to meet anticipated peak demand of 30.3 million kilowatts in its service area. The company later revised its initial estimate of the deficit to 15.7% (Anzai 2012: 35). With its nuclear reactors offline after the Fukushima nuclear accident, TEPCO also anticipated a 10.3% deficit in its service area during the 2011 summer peak demand period. KEPCO warned that the situation in Kansai would be even more severe and that restarting Ōi was essential.

On May 19 at a meeting of the Union of Kansai Governments, which brings together local government leaders from the Kansai region, Nuclear Accident Minister Hosono and Deputy Chief Cabinet Secretary Saitō put forward the government's position. At this meeting, Hashimoto suddenly announced that, at least until the new Nuclear Regulation Authority was established and the new safety guidelines were implemented, it would be reasonable to restart the Ōi reactors temporarily in order to meet peak demand. In the end, on May 30, Hashimoto and eight other regional government leaders released a joint statement giving their assent to the restart. In the statement, they 'urge[d] the government to take appropriate action with regard to a limited restart of the Ōi Nuclear Power Station and provided that the government's provisional safety standards have been met'.

The leaders' conversion was partly the result of growing pressure from local business groups that wanted to see the reactors restarted. Managers of small to medium businesses were anxious about talk of a 15% deficit in supply. Local business leaders targeted individual prefectural and municipal councillors from Hashimoto's Japan Restoration Party and pressed them to accept the restart. Shiga Governor Kada, too, received urgent demands from local business groups to save them from potential scheduled interruptions to the power supply. Executives from major manufacturing firms in the prefecture stated that unless the governor gave her approval to the restart, they would move their factories elsewhere.

On June 8 Noda declared that he was authorizing the restart of the Ōi Nuclear Power Station in order to 'protect people's livelihoods'. Fukui Governor Nishikawa opposed a temporary restart, calling on the government to clearly state the importance of nuclear power as a 'key source of electric power'. Noda responded by labelling nuclear power an 'important source of electric power' and distancing himself from the idea of a 'temporary restart' (Ōshika 2013: 572–8; Sengoku 2013: 175–6; *Mainichi Shimbun*, June 23, 2012, morning edn; *Asahi Shimbun*, June 17, 2012, morning edn). On July 5 reactor No. 3 at the Ōi Nuclear Power Station was restarted, followed by reactor No. 4 on July 21. However, a growing public consciousness of the need to conserve electricity meant that peak demand in KEPCO's service area only reached 26.82 million kilowatts, 11.1% lower than previously anticipated. It was clear that even without the 2.37 million kilowatt output from the Ōi Nuclear Power Station, Kansai could have made it through the summer (Ōshika 2013: 578; Anzai 2012: 21).

Growing opposition to nuclear power

METI's Advisory Committee for Natural Resources and Energy is an umbrella group for a number of committees and subcommittees that provide advice to the government on technical matters related to energy policy. In order to develop the Innovative Energy and Environmental Strategy, the organization's Fundamental Policy Committee discussed the revisions to the Basic Energy Plan and the 2030 energy mix targets. Previously, pronuclear academics and business figures had dominated expert committees related to nuclear energy policy. With only two or three antinuclear representatives represented, pronuclear positions had always prevailed. However, the new committee contained equal numbers of pronuclear, neutral and antinuclear experts, leading to heated discussions. The committee could not reach an agreement. On June 19, 2012, METI's Agency for Natural Resources and Energy, which acts as the secretariat for the committee, stepped in. The agency proposed four options: '0% nuclear power', '15% nuclear power', '20% to 25% nuclear power' and a fourth option 'not to give a specific target'.

The energy mix targets for 2010 were nuclear, 26%; renewables, 10%; and fossil fuels, 63%. The dominant thinking in the Agency for Natural Resources and Energy favored maintaining the current level of 25%. However, agency officials were willing to countenance the possibility of reducing the nuclear role in the mix to 15%. Upon assuming office, Noda had declared that no new nuclear reactors ought to be built and that reactors that have reached the end of their working life should be retired. In January 2012 Hosono stated that the life of a nuclear reactor ought to be capped at 40 years and that if, in an exceptional case, the operating life of a nuclear reactor was to be extended, the maximum possible extension would be 20 years. Following these policies will result in 15% for nuclear in the 2030 energy mix. From this point on, support for the 15% target grew inside the government.

Having received the Fundamental Policy Committee's proposals, the Energy and Environment Council decided to reduce the range of options to three: 0%, 15%, and 20% to 25%. The council also decided to carry out public consultations before finalizing the

Innovative Energy and Environmental Strategy in August (Ōshika 2013: 444–59, 498–501, 579–83; Yamaoka 2015: 131–4; National Policy Unit website a). The committee identified four important factors to consider when deciding on the energy mix. First was the question of guaranteeing nuclear safety and reducing future risks. Second was the strengthening of Japan's energy security. Third was how to address climate change, and fourth was how to reduce costs and prevent deindustrialization. The second, third and fourth points would be difficult to address in the 0% scenario.

The National Policy Unit (NPU) decided to undertake public hearings, conduct a deliberative poll and seek public comment in order to forge a popular consensus for the new proposal. From July 14 the NPU organized public hearings on Japan's energy and environmental options in 11 cities across Japan. Staffers from the NPU were to explain each of the three options. Applications were sought from people who wanted to attend the public hearings and those who were chosen were asked to state their opinion during the hearing and then afterwards to complete a questionnaire. The information materials provided to participants by the NPU stated that the 0% option would have 'major economic consequences', such as increasing the cost of generation and would therefore lead to higher prices. For the 15% option, the information stated that 'this option would enable us to deal flexibly with environmental challenges, meet our international obligations on climate change and respond to new technological developments'. The government's intention was to shore up public support for the 15% option.

However, the plan quickly fell apart. On July 15, at the second public hearing held in Sendai, 175 people applied to attend. Sixty-six supported the 0% option, 14 supported the 15% option and 13 supported the 20% to 25% option. Of the 130 people selected to attend the hearing, 105 showed up on the day. Nine people spoke during the meeting, with three speaking in favor of each option. The third was an executive from Tōhoku Electric Power's planning division. He gave his company's perspective on the scenarios and spoke of the necessity of maintaining the nuclear role at 20% to 25%. The sixth speaker was also a former Tōhoku Electric Power employee and he also spoke in favor of the higher option. A further two participants were Tokyo residents. They spoke in favor of the 15% option, while another speaker from Kanagawa Prefecture supported the higher option. The hearing descended into chaos as participants heckled the organizers, claiming that 'the selection process is rigged' and complaining that 'this hearing was meant to give people from the disaster zone a chance to speak'.

Of the 352 people who applied to attend the hearing in Nagoya on the following day, 106 had indicated that they supported the 0% option, while only 18 supported the 15% option and 37 the higher option. Of the 120 selected to attend, 86 turned up on the day of the hearing. As had happened in Sendai, three people were asked to speak in favor of each of the three scenarios. The third speaker to address the hearing was a section manager from Chūden. He claimed that 'not a single person has died as a direct result of the Fukushima accident', earning him the ire of people who had lost family members during the evacuation. He went further, saying that 'if there was a 35% scenario I would go for that one, or a 45% I would go for that one. That would be the safest option.' The crowd heckled the Chūden manager, with one private news broadcaster who covered the event characterizing his intervention as a form of 'sabotage'.

Even staff from the Agency for Natural Resources and Energy were baffled, with one official exclaiming, 'are the electric utility people stupid or something?' The NPU increased the number of people permitted to speak in favor of the 0% option at the next public hearing and placed a ban on speeches by electric utility employees. Of the 11 hearings held, 1,542 people applied to speak, of whom 68% supported the 0% option, 11% supported the 15% option, and 16% supported the 20% to 25% option. In the responses to the questionnaires participants were asked to complete after the hearing, 35% supported the 0% option, 2% supported the 15% option and 6% supported the higher option (Ōshika 2013: 584–92; Yamaoka 2015: 135–6; Cabinet Secretariat website a; National Policy Unit website a; *Asahi Shimbun*, July 16, 2012, morning edn, and July 17, morning edn; *Nihon Keizai Shimbun*, July 16, 2012, morning edn).

Let us now turn to the Deliberative Poll on Energy and Environmental Policy Options. Deliberative polling is a method of gauging public opinion on an issue. It has been hotly debated in recent years as a new method of democratic expression. Deliberative polling begins with a standard public opinion survey. This is supplemented by deliberation sessions. Before the sessions begin, participants are provided with a set of information materials prepared by experts that contain a balance of perspectives. After studying these materials, participants attend deliberation sessions, where they break up into smaller groups and discuss the issue. They are also free to ask questions of an expert panel during joint sessions. After deliberating the issue, participants are once again surveyed to see how their opinions have changed. This allows pollsters to gauge how people's opinions change as a result of careful consideration of the issue. The deliberative poll began with a national telephone poll of men and women over the age of 20, conducted from July 7 to 22 (T1). There were 6,849 respondents to the telephone poll, and 286 respondents agreed to take place in a deliberative forum held in Tokyo on August 4–5. Participants in the forum were asked to complete a second questionnaire on arrival (T2) and then a third one when the forum was over (T3) (one participant returned home on the 5th, so the final number of respondents was 285).

Rather than being presented with simple 'yes' or 'no' questions, participants were asked to rate each of the three scenarios on a scale of 0 to 10, with 0 being 'complete opposition' and 10 being 'complete support', and 5 indicating neutrality. The majority of participants supported the 0% scenario on arrival (T2) and their number only increased when participants were surveyed again after the intensive discussions (T3). However, support for the 15% option decreased between T1 and T3. Support for the 20% to 25% option was the lowest of all the three scenarios and this scenario also received the greatest amount of opposition (Cabinet Secretariat website b).

The final component of the public consultation process was the public comment phase. This took place between July 2 and August 11 and generated 89,124 submissions. Support for the 0% option was as high as 87%, while support for the 15% option was just 1%. Only 8% of submissions were in support of the 20% to 25% option (National Policy Unit website b). From the middle of June, when the government decided to restart the Ōi Nuclear Power Station, weekly demonstrations outside the Prime Minister's official residence grew in size and began to be reported in the media.[3] It seemed that popular opinion was firmly in favor of a complete phase out of nuclear power.

On June 26 the House of Representatives passed a bill to raise the consumption tax. On July 2 Ozawa's group quit the DPJ in protest and formed its own party. On August 8 Noda promised to call elections in the short-term in order to gain the support of the LDP and Komeito, which were calling for the dissolution of the House of Representatives, and force the bill through the upper house. Support for Noda's government and for the DPJ in general continued to plummet. Panic was spreading within the party about its prospects at the next election. Edano and Furukawa responded by trying to steer the party towards an antinuclear position.

The Innovative Energy and Environmental Strategy

On August 6 Noda asked the relevant government agencies to examine the implications of a zero nuclear policy and to investigate solutions for any possible challenges. Edano ordered Director General Gotō Osamu of the Agency for Natural Resources and Energy to prepare a draft report on 'zero nuclear: issues and solutions'. However, Gotō replied that a complete phase out of nuclear power was impossible and failed to produce the report. Furukawa then assembled an expert committee of public servants and academic experts who favored the abolition of nuclear power and asked them to prepare the draft for a new Innovative Energy and Environmental Strategy. Furukawa asked the team to address a number of issues in the draft strategy, including phasing out nuclear power, renewable energy, reducing energy consumption and increasing the efficiency of thermal power, limiting the operating life of nuclear reactors to 40 years, no new construction of nuclear reactors, limiting restarts to reactors whose safety could be confirmed, abandoning the nuclear fuel cycle project, decommissioning the Monju fast breeder reactor, establishing a new system for subsidies provided to nuclear host communities under the three laws for electric power development, and placing nuclear power under complete government control (Ōshika 2013: 592–7, 606–9).

In April former Prime Minister Kan established a caucus for Diet members who supported getting rid of nuclear power and proposed a complete phase out by 2025. Kan telephoned Noda almost daily, arguing that the party 'cannot win unless we adopt a zero nuclear position'. On August 22 he organized a face-to-face meeting between Noda and a group of antinuclear activists who were responsible for organizing the weekly protests outside the Prime Minister's Office.

In October 2011 an Energy Project Team was established inside the DPJ. At the conclusion of its 59th meeting, it adopted the following three principles. First, to implement a strict 40-year limit on the working life of all nuclear reactors. Second, to only allow the restart of those reactors whose safety had been confirmed by the Nuclear Regulation Authority (NRA). Third, that no new nuclear power stations be constructed and that no new reactors be constructed at existing power stations. The Energy Project Team accepted the complete phase out of nuclear power in principle. However, it did not provide a specific timeline for achieving this goal but suggested instead that this would be determined in the process of implementing this policy (Sengoku 2013: 18–20, 34). This was the position adopted by Sengoku Yoshito, executive acting chairperson of the DPJ's Policy Research Council.

Kan then tried to set up an Energy and Environment Research Commission within the party. However, Sengoku saw that Kan was planning to have himself appointed as chairperson of the commission and use it to advocate for the party to adopt an antinuclear position. He met with Maehara Seiji, the chair of the party's Policy Research Council. Together, they placed the Energy and Environment Research Commission under the umbrella of the Policy Research Council and appointed Maehara as chair and Sengoku as secretary. The research commission began meeting on August 24. However, the commission was divided between Kan and the party's nuclear opponents on the one hand, who advocated phasing out nuclear power by 2025, and Sengoku and the party members who were receiving support from the Federation of Electric Power Related Industry Workers' Union of Japan (Denryoku Sōren) on the other, who were opposed to a hasty abolition of nuclear power. This led to a stalemate. Eventually, on September 6, Maehara forced the commission to accept an extended phase out for nuclear power up until 2040. The research commission then published its proposal to 'strictly apply the three principles [developed by the Energy Project Team] and use every available policy resource to realize zero nuclear power', bringing the discussions to a close (Ōshika 2013: 598–606; Sengoku 2013: 25–9).

The six members of the commission appointed by Furukawa were preparing the draft Innovative Energy and Environmental Strategy. On September 4 the NPU opened negotiations over the draft strategy with METI, the Ministry of the Environment, the Ministry of Land, Infrastructure, Transport and Tourism, the Ministry of Agriculture, Forestry and Fisheries, and the Ministry of Education, Culture, Sports, Science and Technology (MEXT). Each ministry requested amendments to the draft, which changed on a daily basis. The Agency for Natural Resources and Energy forced the addition of the phrase, ‘those nuclear power stations that have passed the appropriate safety checks will be utilized as important energy sources’. MEXT forced the plan to decommission the Monju fast breeder reactor project to be dropped from the draft document (Ōshika 2013: 611–2).[4] Opposition from Aomori Prefecture also thwarted the plan to suspend the nuclear fuel cycle project. On September 6 the municipal council of Rokkasho (Kamikita District, Aomori Prefecture), where a number of nuclear fuel cycle facilities are located, made a submission to the government, threatening retaliation if the government dropped the nuclear fuel cycle project. The submission outlined specific retaliatory measures such as refusing to accept ‘any further shipments of reprocessed waste on their way back from Britain and France’ and returning ‘spent nuclear fuel that is currently being stored in a temporary facility in the village’ (Komori 2016: 154–5). If all the spent nuclear fuel stored in the reprocessing facility at Rokkasho was to be returned to nuclear power stations around the country, there is a risk that the spent fuel pools for each reactor would be filled and the reactors would therefore no longer be able to function. Japan’s electric utilities send their spent fuel to Britain and France for reprocessing. This creates an even more highly radioactive waste product that is shipped back to Rokkasho for storage in the Vitrified Waste Storage Center. If Aomori Prefecture refused to receive the radioactive waste returning from overseas, ships filled with vitrified waste would be unable to dock in any port. Rokkasho is the site of Japan’s only storage facility suitable for the interim storage of high-level radioactive waste. Within the government, there was a growing sense that renegotiating the relationship with Aomori as part of a broader rethinking of reprocessing was impossible. The DPJ therefore deferred making any change to the nuclear fuel cycle project.[5]

The United States government also placed pressure on the DPJ over its planned changes to nuclear energy policy. From the beginning of September, Japan had numerous consultations with United States government representatives on the Innovative Energy and Environmental Strategy. The Americans questioned the plan to continue reprocessing spent nuclear fuel while maintaining that nuclear power would be phased out by 2040. If Japan was to reduce the size of its nuclear reactor fleet and to abandon plutonium-thermal generation while continuing to reprocess spent fuel, its stockpile of plutonium would increase, posing an international security risk. From a nuclear non-proliferation perspective, the United States has had longstanding concerns over Japan's accumulation of plutonium due to its potential for use in the manufacture of nuclear weapons. The United States also expressed concern that if resource-poor Japan was to abolish nuclear power, then global fossil fuel prices might increase. Without an operating nuclear reactor fleet, Japan's technical knowledge in this field would also deteriorate. This would also have a negative impact on the United States nuclear industry. American reactor manufacturers General Electric and Westinghouse have already exited the manufacturing business because the poor prospects for future orders, a result of the liberalization of the domestic electricity sector in the United States, has made the business unprofitable. These companies now only provide planning and design services for nuclear reactors. The government's plan to export nuclear technology from Japan involves Japan's manufacturers building and maintaining the reactors, while paying the American manufacturers patent fees. If Japan eliminates nuclear from its domestic market, Japanese manufacturers will lose the expertise they have developed in this area, casting doubt on the joint Japanese–American venture to sell nuclear reactors on the international market. With these issues in mind, high-level United States officials called on Japanese policymakers to 'consider the consequences of legislative changes or Cabinet decisions that limit their future options', arguing for a strategy that would leave the door open to future governments reversing the phase out of nuclear power. Japanese officials responded to the American concerns by removing a line from the strategy document that referred to 'submitting a bill to implement the innovative energy and environmental strategy…to the Diet as quickly as possible' (Ōta 2015: 106–9; Ōshika 2013: 613; Suzuki 2014: 72, 82–6; *Tokyo Shimbun* (TOKYO Web) September 22, 2012, and October 20, 2012).

When the Innovative Energy and Environmental Strategy document was completed on September 14, it was full of contradictions. It retained the three principles that had been developed by the Energy Project Team and the call by the DPJ's Energy and Environment Research Commission to 'develop policy resources such as to enable the phase out of nuclear power by 2040'. However, the target for renewables in the 2030 energy mix was set as per the NPU's 15% nuclear scenario, making a complete phase out impossible. Despite the strategy's goal of a complete phase out, it retained support for the nuclear fuel cycle project. Of Japan's existing reactor fleet, only five reactors will still be within their 40-year operating life by 2040. Therefore, even if the plan to decommission all nuclear reactors after 40 years is strictly adhere to, some nuclear reactors will still be operating in the 2030s.

Furthermore, on September 15 Edano suggested the government would allow J-Power, whose Ōma Nuclear Power Station (Ōma Town, Shimokita District, Aomori Prefecture) is 38% built, and Chūden, which has almost finished building reactor No. 3 at its Shimane Nuclear Power Station (Matsue City, Shimane Prefecture), to complete construction. In his statement he noted that 'there is currently no legal mechanism for halting construction of nuclear facilities that have already obtained development consent'. The decision directly contravenes one of the three principles, 'that no new nuclear power stations be constructed and that no new reactors be constructed at existing power stations'. If these new facilities were to be operated for 40 years, it would violate the goal of phasing out nuclear power by 2040 (Ōshika 2013: 613–4; Sengoku 2013: 39).

The real intention of the Innovative Energy and Environmental Strategy was to garner support for the DPJ in the forthcoming election, even though the government knew that it would be unable to phase out nuclear power. The purpose of devising the strategy was also to prevent a split in the party, in which the antinuclear faction would leave. The strategy was little more than a series of political slogans. In reality, the DPJ was powerless to solve the many difficulties involved in formulating nuclear energy policy, which is wrapped up in a web of special interests.

Nuclear energy policy under the Abe government

In the House of Representatives elections on December 16, 2012, the LDP–Komeito coalition achieved a major victory over the DPJ and returned to power. With numerous parties promising to abolish nuclear power, the LDP went into the election promising 'to build an economy and a society that does not depend on nuclear power'. At the same time, however, it also promised to 'make a decision about whether to restart each nuclear power station in turn with the goal of deciding the fate of the entire nuclear fleet within three years'. The party promised 'to leave decisions on safety to the expert judgment of the NRA'. The wording of this promise suggested the LDP was planning to restart those nuclear reactors whose safety had been confirmed. Despite the DPJ's broader goal of phasing out nuclear power, in practice its policy was no different from that of the LDP and both parties planned to proceed with restarts of those reactors whose safety had been confirmed by the NRA. By promising 'to implement a "best mix of energy sources" within ten years that could be sustained into the future' (Liberal Democratic Party of Japan website), the LDP was effectively deferring medium- to long-term decisions on the future of nuclear power. The party did not address the issue of whether it would authorize the construction of new nuclear reactors.

Revising the plan to phase out nuclear power

As president of the LDP, during the election campaign Abe had criticized the DPJ's policy of phasing out nuclear power by 2040 as 'irresponsible'. On January 30, 2013, Abe placed the Innovative Energy and Environmental Strategy under review and announced that his government would develop a new energy policy. On February 28 he announced that his government would 'restart those reactors whose safety has been confirmed'.

The nuclear village's hopes for the future of the technology were buoyed by the LDP's return to government. METI moved to reorganize the Fundamental Policy Committee. METI established a Coordination Subcommittee within the Advisory Committee for Natural Resources and Energy and reduced its size from 25 to 15 members. The ministry got rid of five of the eight members of the committee who were seen as holding an antinuclear position (*Asahi Shimbun*, January 14, 2014, morning edn). The Coordination Subcommittee was later renamed the Basic Policy Subcommittee and began formulating a new Basic Energy Plan. On February 25, 29 members of the Society for a Nuclear Power Renaissance presented the Prime Minister with a document recommending the government restart the nuclear reactor fleet.[6] Members of the society included managerial staff from the reactor manufacturers, trading companies, television stations and newspapers. The document urged the government 'to reaffirm the importance of nuclear power as a means of tackling climate change' and 'to recognize that renewables are not currently capable of providing an inexpensive and stable supply of electricity'. It also recommended that the government 'carry out thorough safety checks and then restart those reactors that are currently offline' and 'not to hesitate over exporting nuclear technology in order to maintain the high regard with which Japan's nuclear technology is held internationally, despite the Fukushima accident'. A number of METI officials were also involved in writing this document (Komori 2016: 190; Asahi Shimbun Keizaibu 2013: 115–22, 127–30; *Asahi Shimbun*, May 19, 2013, morning edn; *Nihon Keizai Shimbun*, February 26, 2013, digital edn).

Prime Minister Abe, like the DPJ government before him, made exporting nuclear technology into a key plank of his policy platform and strove to promote the new industry. In January 2013 the Abe government concluded an agreement with Vietnam to construct a nuclear power station there. The plan had been under negotiation with the previous DPJ government since before the Fukushima disaster. In May Abe embarked on a tour of the Middle East, where he concluded a cooperation agreement on nuclear energy with the United Arab Emirates and Turkey. Abe also made an agreement with Saudi Arabia to proceed with working-level consultations on nuclear power cooperation. While Abe claimed that 'Japan has a responsibility to share the lessons of the Fukushima accident with the world and contribute to improving nuclear safety worldwide', a source close to the Prime Minister said that the real reason behind the planned exports was that 'if we are not going to build nuclear reactors, we have to export them in order to maintain our technical capabilities' (*Nihon Keizai Shimbun*, February 10, 2013, morning edn; *Asahi Shimbun*, May 19, 2013, morning edn). The government also backed restarting Japan's domestic nuclear reactor fleet as part of its strategy for economic growth. On June 14 the Abe Cabinet signed off on a new growth strategy known as 'Japan is Back'. One aim of the strategy was to make Japan a country in which it is easy to do business. In order to achieve this, the government planned to 'reduce energy costs and ensure a stable electricity supply by restarting those nuclear power stations that the NRA decides have met its safety standards' (*Asahi Shimbun*, May 31, 2013, morning edn).

Prime Minister Abe, the LDP, METI and business leaders were all eager to restart the reactor fleet. However, tough new safety standards implemented by a more independent NRA led to long delays. Ironically, it was the LDP that had advocated giving the NRA more independence from the government during its time in opposition.

The establishment of the NRA

In 2011 Kan's government decided to separate NISA from METI and merge it with the functions of the Cabinet Office's Nuclear Safety Commission to create a new external nuclear regulatory agency attached to the Ministry of the Environment. Shiozaki Yasuhisa (LDP, House of Representatives) criticized the proposal, arguing that the nuclear regulator had to be completely independent from government. The IAEA has long maintained that nuclear regulators need to be effectively independent from government as specified in its own safety standards. In December 2007 the IAEA had told the Japanese government that NISA was not sufficiently independent from the Agency for Natural Resources and Energy. At that time, the LDP–Komeito government had ignored the warning.

The DPJ leadership transition following Kan's resignation led to a delay in the government submitting a bill for the new regulator to the Diet. In the meantime, Shiozaki developed a counter proposal that would establish the NRA as a highly independent body under Article 3 of the *National Government Organization Act* (an Article 3 committee). Shiozaki was certainly not opposed to nuclear power. His devotion to the cause of 'establishing a nuclear safety regulator that is completely independent as required by the IAEA guidelines' was aimed squarely at enabling the restart of Japan's domestic reactor fleet (Shiozaki 2012: 110). His aim was to limit the Prime Minister's prerogative on nuclear matters. After the Fukushima accident, Kan had intervened in disaster management onsite by ordering TEPCO to vent the reactor containment vessels and inject water into the damaged vessels in an attempt to maintain cooling. Kan later ordered the shutdown of the Hamaoka Nuclear Power Station and prevented the restart of the Genkai Nuclear Power Station. For Shiozaki, what he called the 'Kan Naoto risk' had to be contained. He insisted that 'the intervention of an amateur prime minister with a head full of half-baked knowledge did nothing but create confusion and chaos during the disaster response' (*Asahi Shimbun*, May 30, 2012, morning edn). In order to forestall any future 'Kan risk', Shiozaki wanted to restrict the Prime Minister's prerogative on matters of nuclear regulation, during both normal and emergency conditions but particularly during an emergency.

However, others inside the LDP were wary of granting too much independence to the nuclear regulator. Nevertheless, the bill the LDP eventually submitted to the Diet contained the following points. The NRA would be established as an Article 3 committee attached to the Ministry of the Environment as an external agency. It would be completely independent of the ministry in terms of human resources and all other matters. The Secretariat of the Nuclear Regulation Authority would also be established to act as the NRA's secretariat. The NRA would have full control over its personnel and budget. The chairperson and four NRA commissioners would be appointed by the Prime Minister, with the consent of the Diet. Once appointed, their position would be guaranteed. Employees of the secretariat of the NRA would be ineligible for return if they were seconded from another agency (except during the first five years of the new agency's life) and would also be prohibited from parachuting into a related organization upon retirement (*amakudari*). Technical decisions related to the clean-up of nuclear accidents inside nuclear power stations (onsite) would be the responsibility of the NRA. The Prime Minister would have no power to direct operations. Responsibility for evacuating nearby residents and other matters outside the power station (offsite) would remain with the government. All work related to nuclear safety, safeguards and nuclear security that had previously been divided between METI and MEXT would be brought under the single umbrella of the NRA. The NRA would have the authority to give independent advice to the heads of related agencies on matters under its jurisdiction. The objectives of the *Act on the Regulation of Nuclear Source Material, Nuclear Fuel Material and Reactors* would be amended to include 'protecting people's lives, health, and property, preserving the environment, and assuring national security'. Authority to grant permits and approvals under the Act was also unified under the NRA.[7]

On June 20 the *Act for Establishment of the Nuclear Regulation Authority* was passed by the Diet. The DPJ made almost no changes to the LDP–Komeito bill. In the context of a divided Diet, where the opposition had a majority in the upper house, the DPJ had little choice but to act in accordance with the wishes of the opposition and prioritize the early establishment of a new nuclear regulator.[8]

The selection of NRA commissioners proved to be a difficult process due to the strict new conditions. The regulations excluded anyone who had been a manager or an employee at an electric utility or a nuclear energy-related body in the past three years and anyone who had received more than 500,000 yen in remuneration from any of these bodies. Most nuclear energy specialists could not meet these criteria. Eventually, the government appointed Tanaka Shun'ichi, a specialist in radiation physics and a special adviser to the Research Organization for Information Science and Technology, to chair the NRA. Tanaka, who was former vice-chairperson of the board of the Japan Atomic Energy Research Institute, chairperson of the Atomic Energy Society of Japan and acting chairperson of the Japan Atomic Energy Commission, was certainly a member of the 'nuclear village'. However, after the Fukushima accident he had been working on decontamination in Iitate Village (Sōma District, Fukushima Prefecture) and elsewhere as a 'decontamination adviser' (*Asahi Shimbun*, July 20, 2012, evening edn; July 21, morning edn; July 26, morning edn). Antinuclear Diet members remained critical of Tanaka, whom they referred to as a 'central member of the nuclear village' (*Asahi Shimbun*, August 2, 2012, morning edn). In order to get around the rebels inside his party, Noda used the 'exception powers' provided by Article 2, paragraph 5, of the appendix to the Act for Establishment of the NRA. This gives the Prime Minister the authority to appoint the chairperson and commissioners to the NRA without Diet approval when a state of emergency has been declared under Article 15 of the *Act on Special Measures Concerning Nuclear Emergency Preparedness*. With a state of emergency from the Fukushima disaster still in effect, Noda was able to bypass the Diet and appoint Tanaka, launching the NRA on September 19.

The NRA's new safety standards

In a move seemingly calculated to address the fears of nuclear opponents, the NRA adopted a tough stance on the establishment of new safety standards and the investigation of possible seismic faults beneath existing nuclear power station. On January 31, 2013, the NRA completed the draft outline of the new safety standards. Under the new standards, nuclear power station operators would be legally obligated for the first time to have a plan in place to deal with a severe accident such as might result from a terrorist attack or a natural disaster that exceeded the plant's design basis. The draft outline would require nuclear operators to create an Emergency Response Center able to withstand a design-basis earthquake and be used as a command center in emergencies, as well as a second control room, where reactor cooling operations could be conducted from a distance. Other compulsory measures under the new guidelines included systems to ensure reactor cooling in exceptional circumstances, the installation of filtered containment venting systems, water storage tanks to ensure make-up water would be available for reactor cooling, the provision of multiple external power sources, and vehicle-mounted electricity generators and fire trucks, as well as spray equipment capable of injecting coolant from a distance. Fireproof cables were not required in nuclear reactors approved for construction before 1975. These would now have to be replaced with fireproof alternatives. Countermeasures for earthquakes and tsunamis included formulating a design-basis tsunami at all nuclear power stations. Reactors that are prone to tsunami risk would need tsunami protection facilities covering the entire power station site, including a breakwater and watertight doors to prevent water intrusion into openings in the reactor housing.

In terms of earthquake preparedness, the NRA adopted an expanded definition of an active fault and decided that if the status of a fault is in doubt then comprehensive studies must be carried out. The new safety standards forbade the construction of a nuclear power plant above an active fault.[9] Operators were also required to anticipate and prepare countermeasures in case of volcanic eruption, tornado, bushfire or intentional aircraft strike (*Asahi Shimbun*, July 18, 2015, morning edn). However, the standards do permit the regulator to grant an extension of time for operators to implement some of these countermeasures, meaning that restarts could potentially go ahead without them. The result of the NRA's adoption of these strict standards, 'the highest safety standards in the world', was that the restart of the existing reactor fleet would be delayed and operators would have to make significant investments in order to comply.

There was some dissent inside the LDP over the NRA's tough stance. On May 14 a number of LDP members formed the Parliamentary Network for a Stable Energy Supply. The new group's supporters included numerous Diet members who represented nuclear host communities. These representatives aired their concerns during the network's meetings in statements such as, 'we need a stable electricity supply as part of our growth strategy', 'this is...hard on the electric utilities' and 'perhaps it was a mistake to create the NRA' (*Asahi Shimbun*, May 15, 2013, morning edn; May 19, morning edn).

On June 19 the NRA endorsed the new standards, which were to come into effect on July 8. On the day the new standards came into effect, Hokkaidō Electric Power, KEPCO, Shikoku Electric Power and Kyūden all submitted applications to the NRA, asking it to conduct safety inspections that would permit the restart of ten nuclear reactors at five nuclear power stations. On July 5 Abe stated that 'we have a responsibility to ensure a cheap and stable energy supply. If the regulator determines that the reactors are safe, then we will proceed with restarts.' Some LDP politicians called for an increase in the number of inspectors to enable a speedier restart of the fleet. However, the NRA responded that the number of people who had the requisite level of specialist knowledge to carry out inspections was limited. Tanaka stated publicly that 'the utilities' financial situation will not be taken into account when we are making safety assessments' (*Asahi Shimbun*, July 9, 2013, morning edn).

The regulator also adopted a tough stance on seismic inspections. Before the establishment of the NRA, NISA had ordered additional studies at six nuclear power stations where doubts had been raised as to whether certain fracture zones (faults) were active. The NRA determined that there was a high probability that there were active faults beneath Japan Atomic Power Company's Tsuruga Nuclear Power Station (Tsuruga City, Fukui Prefecture), Tōhoku Electric Power's Higashidōri Nuclear Power Station (Higashidōri Village, Shimokita District, Aomori Prefecture) and Hokuriku Electric Power Company's Shika Nuclear Power Station (Shika Town, Hakui District, Ishikawa Prefecture). Under the new guidelines, if an active fault was discovered underneath an existing nuclear reactor, it would have to be decommissioned. The electric utilities asked the NRA not to recognize the faults as active and to authorize the restart of the reactors.

The LDP, the electric utilities and local governments in nuclear power station host communities all criticized the NRA for the strictness of the safety standards and the thoroughness of its investigations into seismic faults. In November the Parliamentary

Network for Stable Energy Supply called senior officials from the secretariat of the NRA to a meeting. There, the network's secretary Takagi Tsuyoshi (LDP, House of Representatives), whose electorate is in Fukui Prefecture, pressed the officials to speed up the process, stating, 'once we've confirmed that it is safe, we must make effective use of nuclear energy. We hope that you will complete your safety inspections as quickly as possible.' At the network's first meeting on May 14, it had 43 members. By December its numbers had swelled to more than 140, one-third of the LDP's total of 410 Diet members. In December the network put forward a proposal to allow the construction of new nuclear power stations and to enable reactors to continue operating for more than 40 years (*Asahi Shimbun*, December 30, 2013, morning edn). On September 18 the LDP established a Project Team on Nuclear Regulation with Shiozaki Yasuhisa as chairperson. On December 3 the Project Team released a statement strongly criticizing the NRA. On December 25 Shiozaki met with NRA Chairperson Tanaka and submitted the Project Team's statement. For the chair of the NRA to meet one on one with a Diet member was unprecedented.

The creation of the Basic Energy Plan

On November 12, 2013, former Prime Minister Koizumi Jun'ichiro urged the government to abandon nuclear power at his first press conference since his resignation in 2006. At the press conference, Koizumi explained that a visit to Finland's Onkalo spent nuclear fuel repository with executives of reactor manufacturers had made him realize the scale of the challenge posed by nuclear waste and brought him round to an antinuclear position. Prime Minister Abe responded with a statement that 'it would be irresponsible at this point to promise to abandon nuclear power'. In an apparent rebuke to Abe, Koizumi noted that 'even if we do restart them, we have nowhere to put the waste' and argued that 'it is overly optimistic and actually far more irresponsible to think that we can find a final disposal site for radioactive waste anywhere in Japan'. He called on politicians to rethink their commitment to nuclear energy, stating, 'I think that if our politicians could agree to phase out nuclear power, then smart people will certainly come forward to find ways of moving forward. We have to look not only to our Diet representatives, but utilize expert knowledge to come up with hitherto unknown alternative energy sources' (*Asahi Shimbun*, November 13, 2013, morning edn; November 18, morning edn).

During his time in office, Koizumi commanded a high level of popular support and was still seen as an influential figure. His comments sent a shockwave through the LDP. At the Tokyo gubernatorial elections scheduled for February 9, 2014, former Prime Minister Hosokawa Morihiro ran on an antinuclear platform. Koizumi lent his support to the challenger. The government had intended to formally endorse the new Basic Energy Plan in January. However, it decided to defer a Cabinet decision on the plan for fear of its influence on the election. Hosokawa lost the gubernatorial election by a large majority, and the LDP–Komeito-supported candidate Masuzoe Yōichi was elected Tokyo Governor. On April 11, fresh from its victory in Tokyo, Cabinet formally endorsed the new Basic Energy Plan.

The new Basic Energy Plan noted that the Fukushima nuclear accident had caused a number of problems for the country's energy policy. First was the growing trade deficit (METI calculated that fuel imports had increased by 3.6 trillion yen). Second was energy security (energy self-sufficiency had decreased from 19.9% to 6%). Third were increasing electricity prices (average household electricity costs were up by approximately 20%). Fourth was an increase in carbon emissions (by 83 million tons). The plan designated nuclear energy as 'an important source of baseload power' and stated that 'we must continue to maintain our nuclear energy capacity at a level sufficient to ensure stability of supply, reduce energy costs, counter climate change and retain our technical and personnel resources in order to ensure nuclear safety' (Agency for Natural Resources and Energy website a).

However, even if Japan was to restart some of its existing reactors, if the policy of decommissioning after 40 years of continuous operation as specified by the NRA was to be maintained, then by 2027 only half the current fleet would still be viable. Furthermore, by 2037 the number of operable reactors would drop to five and by 2049 there would be no nuclear reactors operating in Japan. The use of the phrase 'continue to maintain' in the Basic Energy Plan implied that the government would proceed with new construction or rebuild existing power stations. However, on January 6, 2014, Abe stated that 'we are not currently contemplating any new construction' (*Asahi Shimbun*, January 7, 2014, morning edn).

According to the Basic Energy Plan:

> the safety of our nuclear power stations will be left to the expert judgment of the NRA. The government will proceed with restarts based on the decisions of the NRA, which are made in accordance with the strictest safety standards in the world. Where the NRA finds that restarts are possible while maintaining these standards, the government will take the lead in gaining the support and understanding of local governments in the affected areas. (Agency for Natural Resources and Energy website a)

With regard to the issue of what to do with high-level radioactive waste, the plan stated the government would 'take the lead in finding a final disposal site' (Agency for Natural Resources and Energy website a), and with regard to the nuclear fuel cycle project, it stated that the government would:

> continue to work with local governments and the international community to obtain their understanding while also giving due consideration to their contributions thus far. We will proceed with both the reprocessing and the plutonium-thermal projects while remaining flexible as to our options in the medium-to-long-term. (Agency for Natural Resources and Energy website a)

In the section on Japan–United States cooperation on energy, the plan noted:

> in terms of the peaceful use of nuclear energy, we have already created a system that will support the development of joint business ventures between Japanese and American reactor manufacturers in the commercial sphere. As partners, Japan and the United States have an important role to play in strengthening the system that underpins the use of nuclear energy while guaranteeing its peaceful use, ensuring non-proliferation, and guaranteeing international nuclear security. (Agency for Natural Resources and Energy website a)[10]

As part of the process of formulating the Basic Energy Plan, the government sought to gauge public opinion by accepting submissions for public comment. In the one-month public comment period that opened on December 6, 2013, the government received

18,663 submissions. METI publicly released the main opinions and its response. However, it did not publish a breakdown of opinions for and against nuclear power. *Asahi Shimbun* journalist Komori Atsushi submitted a freedom of information request for all the submissions and conducted an analysis. He found that 94.4% of the submissions could be classified as antinuclear, including those calling for decommissioning and those opposing reactor restarts. Only 1.1% of the submissions were pronuclear, such as those calling for restarts. Another 4.5% were difficult to classify as for or against (Komori 2016: 82–9).

The Basic Energy Plan described the NRA's new safety standards as 'the strictest in the world'. Kan submitted a request for clarification in the Diet, asking the government to clarify the basis for this claim. On April 25, in a written response that had received Cabinet endorsement, the government replied that 'while comparing the standards of other countries, the NRA's standards have been formulated as the world's highest'. Kan complained that this response 'does not provide any evidence for the claim. The reasoning that the standards are the world's highest because they are the world's highest is simply a tautology.' A senior LDP official also noted that 'the idea that these are the world's highest standards is simply a lie. The phrase is just paying lip service to the notion of safety while trying to support reactor restarts.' NRA chair Tanaka acknowledged that, in terms of terrorist attacks, 'our standards are not necessarily the world's highest' and stated, 'The claim that they are the world's strictest standards is a political or a linguistic problem, not a concrete fact.' One NRA official stated, 'we have not surpassed the European and American safety standards. Rather, we have simply caught up to them.' The new standards were adapted for the existing nuclear reactor fleet. It was not possible to take into account the latest reactor designs (*Asahi Shimbun*, April 26, 2014, morning edn; July 22, morning edn).

Personnel changes at the NRA

By 2014 the restart of the nuclear reactor fleet was still stalled and the LDP continued to criticize the NRA for the length of time it was taking to carry out safety inspections. The electric power industry, too, indicated its growing dissatisfaction with the process, accusing Shimazaki Kunihiko, the NRA acting chairperson with responsibility for seismic issues, of failing to listen. However, the NRA continued its investigations into seismic risks at the nuclear power stations, recognizing a number of faults as active in spite of opposition from the industry. The NRA continued to point out problems with the utilities' seismic studies and earthquake predictions. The NRA's policy was to prioritize applications to restart idle reactors that had already passed its major tests, such as earthquake risk. Those that failed these initial checks were then pushed back in the queue for many months. Nevertheless, the industry continued to use old estimates for a standard seismic motion in its applications. However, previous earthquakes at TEPCO's Kashiwazaki-Kariwa Nuclear Power Station (Kashiwazaki City, Niigata Prefecture, and Kariwa Village, Kariwa District, Niigata Prefecture) and at Fukushima Daiichi (Ōkuma Town and Futaba Town, Futaba District, Fukushima Prefecture) had already exceeded the standard seismic motion specified in the design, causing major damage. This led to concerns that unrealistically small estimates of standard seismic motions had been used in the design of other power stations. However, if the operators increased the size of a standard seismic motion in their designs, they would have to reinforce structures, including reactor housings and pipes. These additional seismic protection measures would take time and cost additional money (*Asahi Shimbun*, January 14, 2015, morning edn; July 18, morning edn).

In principle, commissioners were appointed to the NRA for five years. However, immediately after the NRA was established, Shimazaki Kunihiko and Ōshima Kenzō (a former diplomat) were appointed for shorter terms of just two years, meaning that their tenure would expire in September 2014. Their replacements were to be announced at the end of May. In the lead up to the announcement, the LDP intensified its criticism of Shimazaki. At a meeting of the Project Team on Nuclear Regulation on May 9, the secretariat of the NRA executives was called for questioning. During the hearing, one representative from Fukui Prefecture expressed his dissatisfaction with Shimazaki's determination that a fault under the Tsuruga Nuclear Power Station was active. Another representative questioned the NRA's uncompromising implementation of the safety standards. Project Team chair Shiozaki claimed that Shimazaki lacked experience in nuclear regulation, openly calling for his replacement. Senior government leaders also criticized Shimazaki (*Asahi Shimbun*, May 28, 2014, morning edn; July 19, morning edn).

On May 27 the Abe Cabinet decided not to reappoint Shimazaki Kunihiko and Ōshima Kenzō. Instead, it nominated University of Tokyo Professor Tanaka Satoru, an expert in nuclear engineering with a particular interest in radioactive waste, and Tōhoku University Professor Ishiwatari Akira, a geologist who had few ties to the nuclear energy industry. Ishiwatari had served as a chairperson on committees investigating faults underneath nuclear power stations and enjoyed the confidence of Shimazaki. Tanaka, however, was another story. Tanaka was the former chair of METI's nuclear energy subcommittee and a major figure in the nuclear village. Even senior officials at the Agency for Natural Resources and Energy said that he was 'far too conspicuous to be appointed as a government official'. Tanaka was also an honorary director of the Japan Atomic Industrial Forum from 2010 to 2012. He had received more than 1.6 million yen in research funding from sources such as reactor manufacturer Hitachi-GE Nuclear Energy and the TEPCO Memorial Foundation in 2011. In the previous three years, he had helped ten of his students find employment in the electric utilities and reactor manufacturers. Critics pointed out that these industry ties represented a conflict of interest that contravened the NRA guidelines. However, Environment Minister Ishihara Nobuteru refused to adhere to the previous DPJ government's guidelines. He rescinded the guidelines and appointed Tanaka to the NRA (*Asahi Shimbun*, May 28, 2014, morning edn; May 30, morning edn; June 7, morning edn; July 18, morning edn; July 19, morning edn; Nuclear Regulation Authority website).

The energy mix

On November 21, 2014, Prime Minister Abe dissolved the House of Representatives. At a press conference held the same day, he restated his previous intention to 'restart those reactors whose safety has been determined by the regulator'. However, NRA chair Tanaka continued to maintain the position that 'the standards are fit for purpose but we cannot say that the reactors are safe'. The struggle over who would guarantee the safety of the nuclear reactor fleet continued (*Asahi Shimbun*, November 29, 2014, morning edn).[11] The election resulted in a major victory for the LDP–Komeito coalition.

On January 30, 2015, a subcommittee of the Advisory Committee for Natural Resources and Energy's Strategic Policy Committee, which is charged with determining long-term estimates for energy supply and demand, began debating the 2030 energy mix. The 2010 energy mix targets had included 28.6% for nuclear. The government planned to 'reduce nuclear's role in the mix as much as possible'. The question on everybody's mind was how far it would be reduced. The subcommittee did not contain a single member who advocated reducing the role of nuclear in the mix to zero. It was dominated by industry figures and nuclear energy researchers who wanted to maintain some level of nuclear power (*Asahi Shimbun*, January 31, 2015, morning edn). METI acted as the secretariat for the subcommittee. On March 30, emphasizing that energy mix targets in America and Europe allocated from 60% to 90% of the total to nuclear, hydroelectricity and coal-burning, the ministry suggested that Japan should aim to allocate 60% or more to sources of 'uninterruptable baseload power'. In the 2013 energy mix, the 40% of the mix allocated to baseload power sources was divided between nuclear (1%), coal-burning (30%), and hydroelectricity and geothermal (9%). However, coal-burning power produces excessive carbon emissions, constructing new hydroelectric dams would be difficult and geothermal power would require environmental assessments. It would therefore be difficult to increase the proportion of any of these sources in the energy mix. What this meant in practice was that METI wanted to maintain the proportion of nuclear energy in the mix at approximately 20% (*Asahi Shimbun*, March 31, 2015, morning edn).

The LDP, following Abe's lead, was also putting pressure on METI to maintain nuclear's role in the mix. In the middle of March, Abe asked Nukaga Fukushirō (LDP, House of Representatives), chair of the LDP's Research Commission for Nuclear Energy Policy and Supply and Demand, to 'find a way to lower electricity prices'. As part of his economic growth strategy, Abe wanted to avoid the negative influence that rising electricity prices might have on households and small- to medium-size businesses. At a closed meeting of the Research Commission on April 2, Nukaga prepared a draft proposal for the government, asking it to guarantee that baseload power from nuclear energy, coal-burning, hydroelectricity and geothermal would make up 60% of the energy mix (*Asahi Shimbun*, March 31, 2015, morning edn; April 3, morning edn; July 9, morning edn).

On July 16 the subcommittee formally endorsed the document, 'Outlook for long-term energy supply and demand', which set a target of reducing annual energy demand by 17% in order to conserve energy. On top of this, it set renewables at 22%–24% (8.8%–9.2% for hydro, 7% for solar, 1.7% for wind, 3.7%–4.6% for biomass and 1.0%–1.1% for geothermal). Nuclear's role in the mix was set at 20%–22%, while liquefied natural gas was set at 27%, coal at 26% and oil at 3%. Unless new nuclear reactors were to be constructed, approximately ten reactors would have to be maintained beyond their 40-year operating life in order to maintain nuclear's proportion of the energy mix at 20% in 2030. Extending the operating life of existing reactors by up to 20 years was supposed to be possible only as an 'exceptional measure'. However, Economy, Trade and Industry Minister Miyazawa Yōichi urged the NRA to endorse such exceptional measures, stating, 'if the operators want to use a system for prolonging the life of the reactor fleet, then we should let them' (*Asahi Shimbun*, April 29, 2015, morning edn; May 27, morning edn; July 3, morning edn; Agency for Natural Resources and Energy website b).

Decommissioning and reactor restarts

While the electric utilities may have been chafing at the length of time taken by the NRA to complete its inspections, they were in fact proceeding steadily. On July 16, 2014, the NRA completed a draft inspection report for reactors No. 1 and No. 2 at Kyūden's Sendai Nuclear Power Station in which it found that they did meet the new safety standards. In October Satsumasendai's municipal government gave its approval to restart the reactors, as did Kagoshima Prefecture in November. The remaining procedures to restart the reactors, while they took longer than initially expected, proceeded steadily. On August 11, 2015, reactor No. 1 at Sendai was restarted, the first restart to take place under the new safety standards. Since the shutdown of reactors No. 3 and No. 4 at the Ōi Nuclear Power Station for routine maintenance inspections in September 2013, Japan had been effectively nuclear free. The restart of the Sendai reactor brought this to an end. By August 2016 the NRA had signed off on reactors No. 3 and No. 4 at KEPCO's Takahama Nuclear Power Station (Takahama Town, Ōi District, Fukui Prefecture) and reactor No. 3 at Shikoku Electric Power's Ikata Nuclear Power Station (Ikata Town, Nishiuwa District, Ehime Prefecture), as well as reactors No. 1 and No. 2 at Takahama, in that order.

The Abe government had been pushing for restarts, but chose not to position itself on the front lines. On August 4, Economy, Trade and Industry Minister Miyazawa explained that the government's position on individual reactor restarts was 'to approach restarts using the legal framework developed under the DPJ government' and to 'see whether they meet the NRA's strict standards' before 'making a final decision once the operators have received permission for change of usage'. He characterized the process as one 'in which the technical decisions are made independently by a third party and there is no space for a political decision to intervene'. He also stated that the government had 'no immediate plans' to modify this system (*Asahi Shimbun*, August 4, 2015, evening edn).

While the restarts were proceeding, METI was also working towards decommissioning those reactors that had reached the end of their operating lives. The NRA set a deadline of July 2015 for applications to extend the operational life of reactors that would reach 40 years of operation in July 2016. The deadline forced the utilities to decide whether to decommission the reactors or attempt to prolong their lives. The utilities want to continue operating their reactors for as long as possible. Compared with purchasing fossil fuels, they can obtain uranium relatively cheaply. Furthermore, when the legal depreciation period ends, their liability for property taxes will also decrease. However, under the NRA's new safety standards the utilities would have to implement additional safety measures in order to extend the operating lives of the reactors. These measures will cost hundreds of billions of yen. The utilities therefore had to consider whether it might not be cheaper to decommission those older reactors that have a relatively low output. However, if the utilities decide to decommission their reactors, their unused plant and nuclear fuels will be worthless, creating a loss in their ledgers of approximately 21 billion yen per reactor. This would represent a considerable burden on the utilities' finances.

On January 14, 2015, METI completed a proposal for a new accounting rule that would lessen the burden on the utilities of decommissioning. The new rule would allow utilities to spread the loss incurred by the useless plant and fuel over a period of ten years. METI also decided that even after the liberalization of the retail electricity market was complete, the utilities could still add a levy to the electricity price to cover the cost of decommissioning. A senior METI official explained that 'restarts are gaining a lot of attention but we need to show that we are also moving forward with decommissioning'. METI anticipated that by decommissioning aging reactors it could demonstrate a willingness to reduce the number of nuclear reactors and thereby lessen some of the public opposition to restarts (*Asahi Shimbun*, January 11, 2015, morning edn; January 15, morning edn).

The new accounting rule came into effect on March 13. On March 17 KEPCO and Japan Atomic Power Company decided to decommission some of their reactors. Chūden and Kyūden followed on March 18. At the same time, KEPCO decided to seek a 20-year extension for the relatively high output reactors No. 1 and No. 2 at Takahama, which commenced operation in November 1974 and November 1975 respectively, and reactor No. 3 at Mihama Nuclear Power Station (Mihama Town, Mikata District, Fukui Prefecture), which commenced operation in December 1976. Generally, in order to obtain permission to extend the operating lives of these reactors, the company would have needed to apply within the initial 40-year operating period. However, as a transitional measure, the deadline for applying to extend reactors No. 1 and No. 2 at Takahama Nuclear Power Station was pushed back to July 7, 2016. However, if the operator could not obtain permission by this deadline, then the reactors would have to be decommissioned. The LDP and local government leaders urged the NRA to decide quickly whether to extend the operating lives of the reactors.

The NRA had previously stated that permission to extend the operating life of a reactor would only be granted once and for up to 20 years. Such permission would only be granted in exceptional circumstances and would be extremely difficult to obtain. In the case of Takahama, when KEPCO submitted its application for an extension late, the regulator initially stated that 'we may decide to discontinue the inspection process'. However, the NRA gave in to outside pressure. Inspections at other nuclear power plants were put on hold while the inspection team focused on reactors No. 1 and No. 2 at Takahama. The NRA agreed to conduct the inspection first and defer seismic testing until after construction works had been completed. It seems that the NRA has compromised on the issue because of the Abe government's target of 20%–22% for nuclear in the 2030 energy mix targets. If no new reactors were built, then at least ten older reactors would have to be granted extensions. It also seems that if the regulators refused the extension based on the company's failure to meet the deadline alone, they might be open to litigation by the utility.

Finally, on June 20, 2016, the NRA decided that, providing some strengthening works were undertaken and a number of aging pipes and cables were replaced, Takahama reactors No. 1 and No. 2 could be safely operated for up to 60 years and the operator was granted permission for the extension. However, KEPCO decided that it would take more than three years to complete the additional safety measures, such as replacing ageing cables and complete seismic testing. The utility plans to restart Takahama reactors No. 1 and No. 2 sometime after the fall of 2019 (*Asahi Shimbun*, June 21, 2016, morning edn).

Local government evacuation plans

Another issue that has arisen in the debate over restarting the reactor fleet is that of local government evacuation plans. In 2012 the NRA finalized its Nuclear Emergency Response Guidelines, which accept the possibility that nuclear accidents might occur. Drawing on the IAEA's 2002 safety requirements publication, *Preparedness and Response for a Nuclear or Radiological Emergency*, the guidelines establish a system for a disaster planning zone around each nuclear facility. Before Fukushima, the area within an eight- to ten-kilometer radius of a nuclear power station was designated as an Emergency Planning Zone. Under the new guidelines, the area within a five-kilometer radius of a nuclear power station is designated as a Precautionary Action Zone. The area between this zone and a 30-kilometer radius of the plant is designated as an Urgent Protective Action Planning Zone. All local governments within a 30-kilometer radius of a nuclear power station are legally obligated to prepare an evacuation plan in case of a nuclear accident. As early as 2006, a working group at the Nuclear Safety Commission had looked at implementing the IAEA standards but NISA opposed the move, fearing that 'it would give the impression that our existing measures are inadequate and create anxiety among the populace'. Therefore, the implementation of the international standards was postponed. However, during the Fukushima nuclear accident, evacuations and instructions to remain indoors had taken in some regions that were more than 30 kilometer from the power station, prompting the NRA to expand the Emergency Planning Zone in its new guidelines. This meant that the number of local governments that would now be required to formulate an evacuation plan increased from 15 prefectures and 45 municipalities to 21 prefectures and 135 municipalities. Seven times as many people, or 4.8 million, would now be potentially affected by evacuations.

Preparing evacuation plans involves working out the best method for evacuating residents based on population and geographical factors, making provision for evacuation shelters and transportation, and determining several evacuation routes. In February 2016 only 96 local governments had completed these plans. None of the municipalities located near Chūden's Hamaoka Nuclear Power Station or Japan Atomic Power Company's Tōkai Daini Nuclear Power Station (Tōkai Village, Naka District, Ibaraki Prefecture) have completed their evacuation plans due to the length of time it has taken to find somewhere for the large population of potential evacuees to go (*Nihon Keizai Shimbun*, March 25, 2016, morning edn). The United States Nuclear Regulatory Commission reviews evacuation plans for nuclear reactors in the United States to determine their effectiveness. If evacuation plans do not contain an adequate method for safely evacuating local residents, then permission to operate a nuclear reactor is withheld. In Japan evacuation plans are not required in order to gain permission to operate a nuclear reactor. It is left to local governments to formulate their own evacuation plans and as a result the plans are inadequate.

The evacuation plans for communities located near the Sendai Nuclear Power Station are one such example. Kagoshima Governor Itō Yūichirō, who actively pushed for the restart of the Sendai Nuclear Power Station, declared that 'ten kilometers is enough for evacuation plans for people in care. A 30-kilometer evacuation plan is unrealistic.' This was despite the fact that there are medical and social welfare facilities located within 30 kilometers of the power station that are included in the prefecture's general disaster preparedness. The resulting plan contains no provision for evacuating medical and welfare facilities beyond a ten-kilometer radius of the nuclear power station.

Another example is the Ikata Nuclear Power Station reactor No. 3, which was restarted in August 2016. It is doubtful whether residents could be evacuated quickly from the surrounding area in an emergency. Ikata is located at the root of the Sadamisaki Peninsula. This long, thin peninsula stretches 40 kilometers east to west and is only six kilometer across at its widest point. Approximately 4,700 people live on the peninsula west of Ikata. According to the emergency response plan, residents are to be evacuated along National Route 197, the highway that runs along the length of the peninsula. However, this would mean passing within one kilometer of the nuclear power station and would potentially also lead to traffic congestion. If the highway is cut off, evacuations are supposed to proceed by boat from the peninsula's harbors to Ōita Prefecture. However, in case of rough weather or a tsunami, it would not be possible to evacuate by sea. If radiation dose levels at the time of an accident were to exceed one millisievert, private ferry operators would be unable to assist with the evacuation. Furthermore, most of the peninsula's settlements are not located on National Route 197 but on sloping land by the sea. There is a risk that they might be cut off due to landslides. It is possible to join National Route 197 from these communities via narrow, steep roads that are a few kilometers long but it might not be possible to use these roads in the event of an earthquake. During torrential rains in June 2016, some of them collapsed. If evacuation becomes impossible, residents would be asked to remain indoors. According to the Cabinet Office, there are 44 earthquake-proof indoor evacuation shelters located to the west of the power station, with capacity for 14,476 people. The presumption is that the 7,600 local residents would be able to take shelter here in case of a tsunami. However, in April 2016, when the Kumamoto earthquake struck, concerns were raised as to whether it would really be possible to remain indoors if an earthquake was the cause of a nuclear disaster (*Asahi Shimbun*, August 13, 2016, morning edn; *Nihon Keizai Shimbun*, August 13, 2016, morning edn).

A third example is that of Shiga Prefecture, which has two cities within a 30-kilometer radius of nuclear power stations in neighboring Fukui. Shiga Prefecture, anticipating the spread of radiation, calculated that 1,000 buses would be needed to evacuate the 57,000 people who live within 43 kilometer of the power plants, the area the prefectural government calculated might be irradiated in the event of a nuclear accident. However, the prefecture was only able to secure 500 buses and so it asked the national government to secure the remaining buses. The national government formally wrote to the Nihon Bus Association asking for assistance, but it could not guarantee that the additional buses would be available. During the Fukushima nuclear accident, a number of bus companies refused to assist with evacuations due to fears of radiation exposure. Prefectural governors lack the legal authority to compel bus drivers to work in a radioactive environment. The national government plans to ask them to respond if radiation dose levels are less than one millisievert; if dose rates rise above this, however, then there is nothing they can do. Furthermore, most residents evacuated from Fukushima by car, so the idea of using buses is itself doubtful.

A fourth example is Tōhoku Electric Power's Onagawa Nuclear Power Station (Onagawa Town, Oshika District, Miyagi Prefecture). There is only one road that residents could use to evacuate; if they were to evacuate north, they would have to cross a bridge spanning the Kitakami River. However, following the Great East Japan Earthquake, roads were cut off in many directions and a number of bridges became impassable due to earthquake damage (*Asahi Shimbun*, June 18, 2014, morning edn; July 21, morning edn; July 23, morning edn).

The Sendai Nuclear Power Station is the first to have been restarted since Ōi went offline. There were calls to make it a requirement of restarting Sendai that municipalities within a 30-kilometer radius that are required to complete evacuation plans agree to the plan, in addition to the host communities of Satsumasendai City and Kagoshima Prefecture. The municipal councils of Ichikikushikino and Hioki submitted formal proposals demanding that they be consulted prior to the restart but their demands were ignored. Governor Itō said that 'it would be unwise to allow people with little knowledge of nuclear power to make such a decision. It would only lead to complications.' Chief Cabinet Secretary Suga Yoshihide also stated that the government was planning to model future reactor restarts 'on the process that took place at Sendai' (*Asahi Shimbun*, November 8, 2014, morning edn).

The courts wake up

As the government and utilities moved forward with reactor restarts, opponents moved to stop them. The courts, which had rarely handed down a decision to stop a nuclear reactor before the Fukushima accident, 'woke up'. On May 21, 2014, the Fukui District Court issued a judgment that suspended reactors No. 3 and No. 4 at KEPCO's Ōi Nuclear Power Station. When a problem occurred with the spent fuel pool above Fukushima reactor No. 4, the government had to consider evacuating everyone who lived within a 250-kilometer radius of the plant. Presiding Judge Higuchi Hideaki took this into account in issuing the ban sought by 166 plaintiffs who lived within the same distance of the Ōi power station. He also considered the fact that weak safety measures and a lack of equipment meant the plant was not adequately prepared to deal with a potential earthquake.

Judge Higuchi stated that he had based his decision 'not on whether the plant meets the NRA's new safety standards, but on whether there is a real danger that an incident similar to Fukushima might take place' or not. He looked at whether it would be possible to maintain reactor cooling in the event of a large earthquake. In his decision, he noted that since 2005, four reactors had been subject to five earthquakes that exceeded the standard seismic motion they were designed to withstand. He found that there was 'no reliable basis' for KEPCO's estimate of a standard seismic motion of 700 Gal at Ōi. KEPCO claimed that there was no risk of a meltdown, unless an earthquake at the plant was 1.8 times that size or 1,260 Gal. However, Judge Higuchi found that 'there is a danger that an inland crustal earthquake of that scale might occur at Ōi'. The judge also pointed to flaws in KEPCO's handling of the spent fuel storage pool. He wrote that 'it can only be said that all of the necessary precautions are in place if both the spent fuel pool and the reactor containment vessel are contained within a robust enclosure'. The judgment also noted:

> while KEPCO argues that it is necessary to restart their nuclear power station in order to ensure a stable supply of electricity, from a legal standpoint, we cannot make a decision based on comparing the right to life of the huge number of people who live near the plant with the issue of whether electricity prices will go up or down…even if the suspension of this nuclear power station leads to a large trade deficit, the true measure of a nation's wealth is whether the people can put down roots and live out their lives on their own land. If we lose this, then we have truly lost our national wealth. (*Asahi Shimbun*, May 22, 2014, morning edn)

A KEPCO executive, commenting on the decision, stated that 'based on this judgment, there is no way to determine what level of additional safety measures would be required. It is in effect, a rejection of reactor restarts', showing that the company was shaken by the decision. Chief Cabinet Secretary Suga stated that there was 'absolutely no change' in the government's position to proceed with reactor restarts where they had been found to meet the NRA's new safety standards. The NRA, too, noted that it was not responsible for deciding whether or not to restart the reactors, only whether or not they met its safety standards, and insisted that it would continue with its inspections as normal (*Asahi Shimbun*, May 22, 2014, morning edn).

KEPCO appealed the decision, opening up the possibility that it might be overturned. Depending on the results of the appeal, the restart of Ōi might go ahead, if it can pass the NRA's inspections and secure the consent of local governments. In December 2014 nearby residents applied to the Fukui District Court for a provisional disposition, which would have greater legal force, in order to stop the restart of the Ōi and Takahama nuclear power stations. Presiding Judge Higuchi dealt first with Takahama, which the NRA had prioritized for safety inspections. On February 12, 2015, the NRA formally signed off on Takahama reactors No. 3 and No. 4 as having met the new safety standards.

On April 14 Judge Higuchi found in favor of the nine plaintiffs who lived 50–100 kilometers from the Takahama Nuclear Power Station and issued a provisional disposition preventing the restart of reactors No. 3 and No. 4. This was the first time a court had issued a decision requiring the operation of a nuclear reactor to be suspended immediately. The decision to issue a provisional disposition was legally binding, meaning that unless it was overruled in a future legal action, it would be impossible to restart the reactors. Judge Higuchi gave three reasons for his determination that the operation of the nuclear power station posed 'a real and imminent risk'. First, he noted that since 2005 there have been five earthquakes that exceeded the standard seismic motions that nuclear power stations were designed to withstand and, therefore, these design estimates were not credible. Second, he recognized a possibility that, even if an earthquake not exceeding the Takahama Nuclear Power Station's design basis was to occur, there was still a possibility that external power supply might be lost and that the injection of cooling water into the reactor pressure vessel might be interrupted, leading to a loss of core cooling. Third, he stated that the spent fuel pool was not contained in a secure facility of equal durability to that of the reactor. He also stated that he would not rescind the provisional disposition unless the weak points in the Takahama Nuclear Power Station were remedied, which would involve significantly increasing the company's current determination of the standard seismic motion to which the power station might be subjected and completing additional construction work to increase the plant's seismic resistance to meet that level. The operators would also have to ensure that reactor cooling systems and the spent fuel pool were strengthened to the highest possible level. Judge Higuchi found that the new safety standards, which did not require these radical measures, 'were far too lax and cannot guarantee safety'. He therefore ruled that a danger existed that the personal rights of nearby residents might be infringed.

Chief Cabinet Secretary Suga commented on the ruling, stating that the restart of nuclear reactors 'was only decided after an independent NRA had judged after due consideration that they meet the new safety standards which are the strictest standards in the world' and that 'we want to proceed without fuss with the restarts' (*Asahi Shimbun*, April 15, 2015, morning edn). In a routine interview, NRA Chairperson Tanaka opposed the decision, stating that the Fukui District Court's ruling was mistaken in matters of fact and citing a number of specific examples (*Asahi Shimbun*, April 16, 2015, morning edn).

On April 22 the Kagoshima District Court rejected a request by local residents for a provisional disposition to prevent the restart of the Sendai Nuclear Power Station. In its decision, the court noted that the new safety standards and the NRA inspections based on them were 'rational, and could not be faulted even when compared with the latest scientific knowledge'. The Kagoshima District Court's decision privileged the knowledge of nuclear energy specialists, following in the footsteps of earlier Supreme Court decisions that found the courts could not rule on technical matters and could only rule on whether a step in the inspection process was missing (*Asahi Shimbun*, April 22, 2015, evening edn; April 23, morning edn).

In April 2015 Judge Higuchi was transferred to the Nagoya Family Court. He was replaced on the Fukui District Court bench by three judges who had a background working in the General Secretariat of the Supreme Court (*Asahi Shimbun*, March 3, 2016, morning edn; interview with Kawai Hiroyuki, co-convener of '*Datsugenpatsu bengodan zenkoku renrakukai*'). One of the judges, Presiding Judge

Hayashi Jun, rescinded the provisional disposition that required the immediate suspension of the restart of Takahama reactors No. 3 and No. 4 on December 24. He also dismissed the application for a provisional disposition to stop the restart of Ōi reactors No. 3 and No. 4, stating that 'the restart of these reactors is not imminent'. Judge Hayashi ruled that, after reviewing the validity of the new safety standards, he found that they did incorporate appropriate measures for dealing with earthquakes based on the latest scientific knowledge and that they required particularly high standards to guarantee the earthquake resistance of important facilities inside the plant. He noted that the content of the standards was rational. He also ruled that the standard seismic motion used in the power station's design had been determined based on detailed data from geotechnical studies provided by KEPCO and decided that there was 'adequate redundancy' in the earthquake resistance of the facilities at the power station (*Asahi Shimbun*, December 25, 2015, morning edn). After receiving this judgment, KEPCO restarted Takahama reactor No. 3 on January 29, 2016. This was the first plutonium-thermal plant to be restarted under the new safety standards. Reactor No. 4 was restarted on February 26, but three days later a problem occurred and led to its emergency shutdown.

On March 9 the Ōtsu District Court in Shiga Prefecture, neighboring Fukui Prefecture, where the Takahama Nuclear Power Station is located, recognized the objections of 29 residents living within 30–70 kilometers of the Takahama Nuclear Power Station and issued a provisional disposition that suspended the operation of reactors No. 3 and No. 4 at Takahama. This was the first time a court order had led to the shutdown of an operational nuclear reactor. Presiding Judge Yamamoto Yoshihiko determined that it is the responsibility of the utilities, which have the necessary documentation, to prove that nuclear reactors are safe. If a utility cannot justify why a particular reactor is safe, then it can be inferred that its claim has an irrational basis. He also found that the investigation into the cause of the Fukushima nuclear accident was 'still underway' and that the NRA's decision to set new safety standards in that context was 'concerning'. The judge therefore concluded that 'we have to hesitate when asked whether [the new standards and the inspections based on them] can provide a basis for maintaining the public peace'. After examining the countermeasures in place for a severe accident at Takahama, he also found that the operator's determination of a standard seismic motion was based on an inaccurate assessment of the length of the fault line and that it could not be said that adequate redundancy was in place. He also found that even based on the new safety standards, the earthquake resistance of the cooling system for the spent fuel pool was not up to the same standard as that for the reactor and that KEPCO could not adequately explain its plan for responding if damaged spent fuel pools were to leak. The judge also found that evacuation plans ought not to be left to local governments and that 'there is an urgent necessity for the state to take the lead in preparing more detailed evacuation plans'. He decided that 'we need safety standards that also take evacuation plans into account and the state must show good faith and take responsibility for formulating them'. Yamamoto also said that KEPCO 'must give due consideration to evacuation plans as part of its safety measures', calling on the utilities to get involved rather than leave planning for evacuation to national and local governments.

The judgment caught the NRA by surprise. When a memo about the decision was circulated during an NRA meeting, many commissioners responded with a surprised, 'what?' One executive said, 'does this mean KEPCO will have to shut down reactor No. 3?' They could not contain their surprise. Chairperson Tanaka said, 'I do not think there are any flaws in the new standards. There is no need to revise the view that these are nearly the highest safety standards in the world.' Chief Cabinet Secretary Suga stated, 'we maintain our original position. We will proceed with restarts where reactors have met our worldclass new safety standards.' A senior METI official also said, 'As happened in the Fukui case, it is likely that a future judgment will reverse this one' (*Asahi Shimbun*, March 10, 2016, morning edn). However, the decision had a major impact and on March 10 KEPCO shut down the No. 3 reactor. On July 12 Ōtsu District Court Presiding Judge Yamamoto dismissed KEPCO's application to lift the provisional disposition. The suspension of operations at Takahama is likely to remain in place for some time to come.[12]

Analysis

Why did nuclear policy remain largely unchanged?

Why was the DPJ government unable to bring about significant policy change and phase out nuclear power? Over many years, the LDP government's promotion of nuclear power created powerful interests that profited from the technology. These actors strongly opposed nuclear abolition. Among the strongest actors in the pronuclear camp were local governments in reactor host communities. Economic and public financial structures that depend on nuclear power have become entrenched within the regions that host nuclear power stations. That is why Fukui Prefecture lobbied for the national government to designate nuclear energy as a key source of electric power and to take responsibility for ensuring the continuation of nuclear power. For similar reasons, Aomori Prefecture and Rokkasho municipality called for the maintenance of the nuclear fuel cycle project and threatened to return their stored spent fuel to nuclear power stations throughout the country and to refuse to accept high-level nuclear waste from overseas. The DPJ government was unable to gain the consent of these local governments for a change of policy.

The utilities also insisted on maintaining nuclear energy. Nuclear power is an extremely expensive method of generating electricity, in terms of both maintenance and management costs. In 2012 only Hokkaidō Electric Power's Tomari Nuclear Power Station (Tomari Village, Furu'u District, Hokkaidō), which operated until May 5, and KEPCO's Ōi reactor No. 3, which was restarted on July 5, and reactor No. 4, which was restarted on July 21, were operating, leaving the country almost entirely without nuclear power. Nevertheless, according to the *Asahi Shimbun* economics bureau, the total cost associated with nuclear energy (including maintenance, management, repairs and human resources) for the nine electric utilities was still 1.2657 trillion yen. This equals 74% of their total nuclear costs for 2010. Even without operating the nuclear reactor fleet, it cost an extraordinary amount simply to maintain it. The utilities will do whatever they have to do to get the reactors working again (Asahi Shimbun Keizaibu 2013: 10–3). In addition, Japan Nuclear Fuel Limited, which manages the nuclear fuel cycle project, and the Japan Atomic Energy Agency (the former Power Reactor and Nuclear Fuel Development Corporation), which manages the Monju fast breeder reactor, were absolutely opposed to any revision of the nuclear fuel cycle project because it would make them unviable or lead to staff cuts. The electric power union, Denryoku Sōren, one of the DPJ's biggest donors, worked with the industry to advocate for nuclear power and put pressure on individual DPJ representatives. DPJ Diet representatives who had been elected with the support of Denryoku

Sōren and those who represent electorates that host nuclear power stations opposed an antinuclear policy, complicating the debate on the issue within the party.

METI steadfastly opposed the attempts by Prime Minister Kan, and later by Economy, Trade and Industry Minister Edano and National Strategy Minister Furukawa in the Noda Cabinet, to change the government's nuclear energy policy and phase out nuclear power. The economic bureaucracy had developed a cooperative relationship with the electric power industry that was symbolized by the practice of *amakudari* (descent from heaven). It was also sensitive to calls from business leaders concerned about electricity shortages and price increases who wanted to be able to rely on a secure energy supply. The bureaucracy was also concerned that electricity shortages and price increases might lead to deindustrialization and that purchasing fossil fuels for thermal power generation would create a trade deficit. The potential impact of these factors on the economy was a major concern for the bureaucracy, and it therefore opposed a phase out of nuclear power on energy security grounds.

Within the government, there were also those who wanted to maintain Japan's right to extract plutonium by reprocessing spent nuclear fuel. Under the Japan – United States Nuclear Cooperation Agreement, Japan is the only nonnuclear weapons state that is permitted to extract plutonium that could potentially be used for weapons. Japan originally adopted nuclear power in order to increase its energy self-sufficiency via the nuclear fuel cycle project. Some industry figures are still very attached to this idea. At the same time, in terms of national security, some proponents of nuclear energy also valued the fact that by maintaining the technical ability to reprocess uranium, Japan was also maintaining the ability to develop nuclear weapons. Ishiba Shigeru (House of Representatives), a powerful LDP politician, spoke in favor of maintaining the nuclear fuel cycle project after Fukushima on the basis that Japan's latent ability to manufacture nuclear weapons at short notice acts as a deterrent to potential enemies. This idea of 'technological deterrence' was also popular in the public service. In 2014 a senior Japanese government official told the head of the United States Department of Energy that Japan's ability to produce nuclear weapons was extremely important in terms of its relationship with China and North Korea. This was his unofficial way of communicating to the Americans that the nuclear fuel cycle project was a matter of life and death for Japan (Ōta 2015:

180–2). Japan has a vested interest in the nuclear fuel cycle project at the international level.

The United States also opposed Japan abandoning nuclear power, both because it wanted to support nuclear technology exports by joint Japanese–American ventures and because of nuclear nonproliferation concerns over plutonium stockpiling. The DPJ government was unable to persuade these powerful interest groups of its antinuclear agenda and it lacked the power to push it through without them.

Why has the Abe government failed to restart the reactor fleet swiftly?

As I have discussed in this chapter, Prime Minister Abe and many other LDP politicians (particularly those who represent nuclear host communities), local governments in nuclear host communities, METI, the electric power industry and business leaders have all pressed for the restart of Japan's existing reactor fleet. While TEPCO may have lost some of its former political clout, the nuclear village is still in good health. Prime Minister Abe, METI and Japan's business leaders all wanted to restart the nuclear reactor fleet in order to lower electricity prices (at least temporarily) and to prevent a blowout in the trade deficit caused by expensive fossil fuel imports. Both the local economy and public finances in nuclear host communities depend on nuclear power. These local governments and the Diet members who represent these areas are eager to restart the reactors. The electric power industry is spending an enormous amount of money on maintaining its reactor fleet, even without generating electricity. Uranium fuel is also much cheaper than fossil fuels. In order to increase their profits, the utilities are keen to restart the reactors. Nevertheless, despite their hopes for a full-scale restart, the only nuclear reactors in operation as at August 2016 are reactors No. 1 and No. 2 at the Sendai Nuclear Power Station and reactor No. 3 at the Ikata Nuclear Power Station. The Fukushima accident has had a significant impact and there have been major changes to nuclear energy policy.

So why has the restart of the reactor fleet not proceeded as smoothly as its advocates would like? First, the establishment of a highly independent NRA has meant that reactors must meet strict safety standards before they can be restarted. For opponents of nuclear power, the new standards are still inadequate. The fact that the NRA has granted permission to extend the operating life of at least one nuclear reactor beyond 40 years – permission the regulator had previously stated would only be granted in exceptional circumstances – does cast doubt on how independent the new regulator really is. Nevertheless, compared with the regulatory regime that existed before Fukushima, the inspection regime is much tougher. This is the main reason that reactor restarts have not proceeded in line with the nuclear village's expectations. It has also led to ongoing criticism of the NRA by LDP politicians, as well as by utilities and business leaders who want to restart the fleet as soon as possible.

Ironically, it was the LDP in opposition that opposed the DPJ's original proposal that the government establish the NRA as an external agency of the Ministry of the Environment. The LDP insisted on making the regulator an Article 3 committee, the National Regulation Authority, giving it the highest possible level of independence, and forced the then governing DPJ to adopt its proposal. The reason the LDP insisted on giving the NRA so much independence can be understood by examining the term 'Kan risk' coined by the main agitator for the plan, the LDP's Shiozaki Yasuhisa. The LDP was critical of Kan for intervening politically in the accident response and accused the DPJ government of making the situation worse. By insisting on increasing the regulator's independence as a means of attacking the DPJ politically, the LDP ended up being caught in its own trap when it returned to government. Had the LDP been in government at the time, it is unlikely that the NRA would have been given such a high degree of independence. This example demonstrates the impact of the change of government on nuclear policy. Furthermore, if the DPJ government had not faced a divided Diet in which it lacked a majority in the upper house, its original proposal would likely have passed. This demonstrates the effect of a divided Diet.

Second, the issuing of a provisional disposition by a district court judge to prevent the Takahama Nuclear Power Station from being restarted was extremely significant. Before the Fukushima accident, judges tended to respect the technical expertise of nuclear experts and follow the precedent established by the Supreme Court that courts could only rule on gaps in the inspection process. No court had ever stopped a nuclear reactor from operating. After the Fukushima accident, Judge Higuchi Hideaki of the Fukui District Court and Judge Yamamoto Yoshihiko of the Ōtsu District Court ruled that the NRA's new safety standards were insufficient to prevent damage from a nuclear accident and issued judgments that shut down a nuclear power station. The Supreme Court, however, has not changed its pre-Fukushima position on nuclear issues. The Supreme Court appointed three judges from its General Secretariat to replace Judge Higuchi after he was transferred out of the Fukui District Court. One of these judges rescinded the provisional disposition that had prevented the nuclear reactors from operating. However, until the case makes it to the Supreme Court and the court makes a ruling that establishes a precedent on the issue, any district court can potentially issue a provisional disposition that shuts down an operating nuclear reactor. This means that operators now face the risk that their reactors might be shut down by a court decision.

Third, while it has yet to become clear, it is possible that opposition from prefectural governors opposed to nuclear power may prevent the utilities from restarting reactors in their jurisdiction. Up until now, prefectural governors have assented to restarts where the NRA has certified that they have passed inspection. However, the Kashiwazaki-Kariwa Nuclear Power Station is currently undergoing safety inspections by the NRA. Niigata Governor Izumida Hirohiko, who harbors a deep distrust of TEPCO, has not relaxed his antinuclear posture. However, Izumida announced on August 30, 2016, that he would not contest the Niigata gubernatorial election scheduled for October 16.[13]

At the July 10, 2016, gubernatorial elections in Kagoshima Prefecture, Mitazono Satoshi, who promised to suspend the operation of the Sendai Nuclear Power Station, beat the incumbent Itō Yūichirō and was elected governor. Once he became governor, Mitazono tried to force Kyūden to shut down the Sendai Nuclear Power Station on August 26 and undertake further safety tests. Kyūden chose not to comply with the order. In reality, the governor lacks the legal authority to shut down a nuclear power station. However, Kyūden has a safety agreement with Kagoshima Prefecture and Satsumasendai City. The agreement recognizes the prefecture's right to enter the facility to conduct inspections and to request specific measures based on the results of those inspections. Reactor No. 3 at the Sendai Nuclear Power Station will have to be shut down for routine maintenance in October, as will reactor No. 4 in December. If the governor continues to oppose restarts after that, it might be difficult to proceed (*Asahi Shimbun*, July 12, 2016, morning edn; August 27, morning edn).[14]

Commonalities between the DPJ government and the Abe Cabinet

Both the DPJ and the Abe government support restarting nuclear reactors that have passed the NRA's safety inspections. The major differences between them relate to the latter's designation of nuclear energy as an 'important source of baseload power' in the new Basic Energy Plan and changing the 2030 energy mix target so that nuclear makes up 20% to 25%. The new plan also sets a target for renewables at 22% to 24%. Because this is the upper limit, the government might actively prevent further expansion of the renewable sector. However, Prime Minister Abe has also stated more than once that he does not intend to build new nuclear power stations. In order to meet the 2030 targets, it will therefore be necessary to extend the operating life of some older reactors beyond 40 years. These older reactors will pose a greater accident risk and may operate less efficiently due to technical problems. Based on the current rate of reactor restarts and factoring in the risk that a court or a prefectural governor may suspend operations at one or another of the nuclear power stations, it seems unlikely that nuclear will be able to fulfill 20% to 22% of Japan's energy mix in 2030.

So why has the Abe government failed to promote new construction? Opposition to nuclear power among the public has become firmly entrenched. Nevertheless, as the 2014 Tokyo gubernatorial election made clear, nuclear issues have very little influence over most voters' choice of candidate. The LDP–Komeito coalition has continued to win national elections, including the 2012 lower house elections, the 2013 upper house elections, the 2014 lower house elections and the 2016 upper house elections. Nevertheless, the Abe government is keenly aware of the mass opposition to nuclear power and so is unable to support the construction of any new nuclear capacity. Of course, in an environment in which most existing reactors are not being restarted due to close inspection by the NRA, attempting to build new reactors seems unrealistic. Furthermore, with the liberalization of the electric power industry, building new nuclear reactors does not make economic sense for the utilities. In an attempt to address the effects of liberalization and enable operators to continue to operate their existing reactors, METI is considering allowing the cost of nuclear power to be recovered by adding it to transmission charges or paying for it via taxation (*Asahi Shimbun*, March 12, 2016, morning edn). However, the provision of special government subsidies contradicts the guiding logic of METI's liberalization agenda. Unless something changes, there will be a de facto phase out of nuclear power.

Nevertheless, the Abe government is continuing with the nuclear fuel cycle project despite continuing failures and little chance of its ever being realized. The government is leading the quest to find a viable final disposal facility for spent nuclear fuel. However, even if the state nominates a place for such a facility, it seems unlikely that any region will accept the nomination. While the Abe government adopts a pronuclear posture, in effect it is continually deferring the important political decisions that would enable it to implement such a policy.

Outwardly, the DPJ and the Abe governments have very different positions on nuclear policy but in practice they have both been making policy on the run in response to the prevailing antinuclear mood among the public. The DPJ government promoted its antinuclear credentials but lacked a concrete plan for phasing out nuclear power and was forced to approve the continuation of the nuclear fuel cycle project. The Abe government has avoided taking responsibility for nuclear safety by leaving the issue to the NRA's judgment. It has put off building the new nuclear power stations that would be necessary to realize its energy policy and has deferred making a decision about the nuclear fuel cycle project and finding a final disposal facility for spent nuclear fuel. As Paul Pierson (1994) has observed for welfare policy, nuclear policy is caught up in a web of complex interests. Past policies have produced interest groups that restrict current options, making it difficult to effect policy change. Both the DPJ and the Abe governments have failed to exercise political leadership and have deferred difficult long-term decisions due to the difficulty of dealing with the complex interests that are bound up in nuclear policy domestically and internationally, between the government and opposition, and between central and local governments.

4 3.11 and the 2012 General Election: Political Competition and Agenda Setting

Kubo Yoshiaki

Disasters provide opportunities for policy actors to set new agendas (Birkland 1997). However, the kinds of agenda set and the policy changes they may lead to depend on the learning processes after disaster (Birkland 2006). In particular, the agendas set in political competition depend on the stance and behavior of political parties and politicians. This is because parties and politicians have incentives to set issues that are favorable in political competition.

This chapter analyzes the party leadership elections held after the Great East Japan Earthquake and the Fukushima Daiichi nuclear accident (collectively referred to as 3.11) and the 2012 general election in order to show policy agendas, political competition and electoral consequences after 3.11, which was a large-scale compound disaster involving earthquake, tsunami and nuclear accident crises. How did Japanese political parties and politicians view 3.11, and what kind of political competition occurred?[1]

In Japanese politics the centralization (or nationalization) of politics and elections has been pointed out since the 1990s (Muramatsu 2006; Hirano 2007; Maeda 2009; Imai 2011; McElwain 2012). For example, regarding the voting behavior of the electorate, the weight of the party vote has increased compared to the personal vote (Hamamoto 2007; Hamamoto and Nemoto 2011; Imai 2011), the influence of the party leaders' image has increased (Kabashima and Imai 2002; McElwain and Umeda 2011) and it has become easier for nation-wide swings to occur (Rosenbluth and Thies 2010; McElwain 2012). The party system, with the decrease in the effective number of candidates in the single-member districts, shifted towards a two-party system in which the LDP[2] and the DPJ compete (Maeda 2008; Kohno 2009; Scheiner 2012).

As a result, in 2009 the DPJ formed a coalition government with the SDP (until 2010) and the PNP (until 2012). However, the DPJ's administration did not progress smoothly.[3] Hatoyama Yukio, Kan Naoto and Noda Yoshihiko served as prime minister until the DPJ lost office in 2012. Kan, who was prime minister at the time of 3.11, resigned in August 2011, and Noda took office as the next prime minister. Kan and Noda cooperated with the LDP and Komeito under the divided Diet (parliament), where the majority in the House of Representatives (lower house) cannot control the House of Councillors (upper house) (see Chapter 1 of this book). Some DPJ members who opposed Noda's policy left the party one after another.

During the Noda administration, some new parties were formed, such as the JRP and the TPJ. In the 2012 general election, the JRP and TPJ, as well as YP, attracted attention as 'third force parties' (see Reed 2013). However, this did not mean that third force parties had set candidates in every district. Voters who had no candidates of their preferred political party or its cooperating parties in their districts could not vote based on political parties.[4]

The 2012 general election resulted in the LDP and Komeito returning to power. After the 2012 victory, the LDP–Komeito coalition under the second Abe administration won the upper house election in 2013, the general election in 2014 and the upper house election in 2016.[5] Compared to after the 2012 general election, Japanese party politics from 3.11 to the 2012 election was more unstable. Despite being based on national political competition centering on political parties, the DPJ, which had acted as one of the two major political parties, had split and new parties were formed.

In the political competition from 3.11 to the 2012 general election, what kinds of agenda were set and what were the consequences? According to previous research that focused on agenda in the party system, the strategies of each party and their mutual relationships determine agenda. First, each party strategically ignores or selectively emphasizes specific issues (Meguid 2005; Riker 1996). Second, the interaction between parties further influences agenda, such as between the ruling and opposition parties (Green-Pedersen and Mortensen 2010) and between the party blocs (groups formed by cooperating parties) (Green-Pedersen and Mortensen 2015).

In the case of Japan, the formation of a two-party system since the 1990s was actually the formation of two main party blocs. With the rise of third force parties, cooperative relationships were built in the 2012 general election between (1) the DPJ and PNP, (2) the

LDP and Komeito, (3) the JRP, YP and NRP, and (4) the TPJ, SDP, JCP and NPD.[6] Therefore, it is necessary to focus on the degree of *intraparty*[7] and *intrabloc* consensus-building process, and also on the *interbloc* consensus-building process.

From the above, the relationship between post-disaster political competition and agenda setting can be organized into the following three paths.

1. Issues without consensus within each party or bloc, regardless of any differences between blocs, will not cause competition.
2. Issues with consensus within each party or bloc, and without differences between blocs, will become valence issues, and competition will not occur.
3. Issues with consensus within each party or bloc, and with differences between blocs, will become position issues, and competition will occur.

The issues of particular interest in this chapter are disaster construction and prevention, and nuclear power and energy policy. The conclusion is anticipated to be as follows. On the one hand, disaster reconstruction and prevention followed path 2. In the two main party blocs that cooperatively worked on disaster recovery and reconstruction through the three-party agreement of the DPJ, LDP and Komeito, disaster recovery and reconstruction became a valence issue. As a result, competition over disaster recovery and reconstruction did not occur. On the other hand, nuclear power and energy policy followed path 1. Although the bloc of the TPJ, SDP, JCP and NPD built a consensus, the other three blocs had no consensus. As a result, competition over nuclear power and energy policy did not occur.

However, interpreting the vote data analysis of the 2012 general election (Mori 2016), the disaster recovery and reconstruction efforts by the DPJ, LDP and Komeito were not necessarily seen in a positive light. The vote for JRP or TPJ, which challenged the two main party blocs as new political parties, shows disconnect and interlocking of urban areas and afflicted areas.

The structure of this chapter is as follows. The first section shows the degree of consensus within parties and blocs by examining the issues and results of party leadership elections held after 3.11 and the opinion distribution of candidates in the 2012 general election. The next section examines differences among parties or blocs in party manifestos and party leader debates. The final section, after reviewing the results of the 2012 general election, discusses what kinds of political competition occurred in Japan after 3.11 as a conclusion.

The structure of intraparty and intrabloc competition

This section describes the party leadership election process that each party implemented after 3.11 and examines the distribution of opinions on disaster reconstruction and transition away from nuclear power, based on a questionnaire survey of the 2012 general election candidates. What were the issues debated in each party's leadership election, and what were the results? In particular, what were the attitudes of each candidate towards the Great East Japan Earthquake and the Fukushima Daiichi nuclear power plant accident?

The party leadership election process after 3.11

There were five leaders newly appointed between 3.11 and the 2012 election: Noda Yoshihiko (DPJ), Abe Shinzō (LDP), Jimi Shōzaburō (PNP), Ishihara Shintarō (JRP), and Kada Yukiko (TPJ). Of these party leaders, Noda and Abe were chosen as a result of their party leadership elections. The DPJ held two leadership elections, about half a year after 3.11 in August 2011 and the next year in September 2012. The LDP held a leadership election in September 2012. I will describe those processes in order.[8]

2011 DPJ leadership election

There were five candidates in the 2011 DPJ leadership election: Maehara Seiji, Mabuchi Sumio, Kaieda Banri, Noda Yoshihiko and Kano Michihiko. The number of votes obtained in the first round was Kaieda, 143; Noda, 102; Maehara, 74; Kano, 52; and Mabuchi, 24. The runoff was Noda, 215, and Kaieda, 177. According to news reports, the majority of the DPJ Members of Parliament (MPs) who voted for Maehara and Kano in the first round voted for Noda in the final round. Kaieda was only able to gain support from MPs who had voted for Mabuchi and a small portion of MPs who had voted for Kano (*Asahi Shimbun* and *Mainichi Shimbun*, August 30, 2011, morning editions). This means that the 'Anti-Ozawa' group that distanced itself from Ozawa Ichiro voted for Noda, and the 'Pro-Ozawa' group surrounding Ozawa voted for Kaieda (*Yomiuri*

Shimbun, *Asahi Shimbun*, *Nihon Keizai Shimbun* and *Mainichi Shimbun*, August 30, 2011, morning editions).

What did each candidate appeal about? In the political views submitted when filing for candidacy,[9] disaster recovery and reconstruction was a high priority for all candidates. Specific issues of debate were tax increase for reconstruction and restarting of nuclear power plants. Only Noda was positive about tax increase for reconstruction, while Maehara, Kaieda and Noda were positive about restarting nuclear plants. In contrast, future energy policy did not emerge as a position issue (*Yomiuri Shimbun*, *Asahi Shimbun* and *Nihon Keizai Shimbun*, August 28, 2012, morning editions).

However, the biggest issue was the management of government and party administration. A significant difference was in the relationship with the opposition party under the divided Diet. Maehara and Noda showed positive attitudes towards ruling and opposition party negotiations, while Kaieda was reluctant.

This conflict shows that the DPJ was faced with the task of coordinating Diet management and party management. The DPJ was struggling to form an agreement with the opposition party under the divided Diet. While the predecessor, Kan, was able to negotiate a three-party agreement with the LDP and Komeito, Ozawa's group instead strengthened its opposition. In other words, the 'Anti-Ozawa' versus 'Pro-Ozawa' situation overlapped about whether to continue cooperation between the DPJ, LDP and Komeito in order to overcome the divided Diet. The new leader was forced to choose between the two. Ultimately, Noda, who emphasized cooperation between the DPJ, LDP and Komeito agreement, was elected. After this election, Noda proceeded with disaster reconstruction, consumption tax increase and restarting nuclear power plants.

2012 DPJ leadership election

There were four candidates in the 2012 DPJ leadership election: Noda Yoshihiko, Akamatsu Hirotaka, Haraguchi Kazuhiro and Kano Michihiko. Unlike the previous election, this time an election by party members and supporters was also held, as well as by MPs. Noda won 818 points (of which the party members and supporter vote was 296 points), Haraguchi, 154 (party members and supporter vote, 72), Akamatsu, 123 (party members and supporter vote, 24) and Kano, 113 (party members and supporter vote, 17). Noda won the majority and was re-elected.

Noda stressed during the election campaign that he would implement policies through agreements between the DPJ, LDP and Komeito. In response, other candidates criticized Noda's management of government affairs, which had caused Ozawa and others to leave the party. However, Noda's stance to focus on cooperation between the DPJ, LDP and Komeito was highly regarded among members and supporters of the DPJ.[10] Noda's election victory indicates that party and government administration policy did not become a point of contention, unlike in the 2011 DPJ leadership election.[11]

I would like to point out two points concerning the handling of the disaster reconstruction and nuclear accident. First, although Akamatsu, Haraguchi and Kano mentioned these in specific policy arguments, Noda referred to them only before each policy debate. This was a clear change from 2011. In other words, disaster reconstruction, which was at the top of policy attention in 2011, had become a prerequisite for advancing other policies.

Second, there was the possibility that Noda would be subject to criticism about restarting the Ōi Nuclear Power Plant. However, on September 19, 2012, during the leadership election campaign, a Cabinet decision was made on 'future energy and environmental policy'. This made it difficult for candidates to turn energy policy into a point of contention.

2012 LDP presidential election

There were five candidates in the 2012 LDP presidential election: Abe Shinzō, Ishiba Shigeru, Machimura Nobutaka, Ishihara Nobuteru and Hayashi Yoshimasa. The voting results by the LDP party members and MPs were 199 votes for Ishiba (of which 165 were from local branches), Abe, 141 (local branches, 87), Ishihara, 96 (local branches, 38), Machimura, 34 (local branches, 7) and Hayashi, 27 (local branches, 3). The runoff election by MPs resulted in Abe winning 108 votes and Ishiba 89 votes. Although it was Ishihara who gathered stronger support from elder MPs of each faction within the LDP, he was unable to proceed to the final ballot.

What attracted attention before the announcement of the 2012 LDP presidential election was that Tanigaki Sadakazu, the former LDP president, was not standing for election. He had promoted cooperation with the DPJ government under the divided Diet. There was also the possibility that the new president would abandon the three-party agreement, but all five candidates announced that they would uphold the agreement. This was seen as a choice by the LDP to battle the general election on issues other than those concerning the three-party agreement, since highlighting the agreement would make it difficult to differentiate the LDP from the DPJ.[12]

What did each candidate appeal for? Although there was almost no difference in terms of the main issues, Abe and Ishiba, who proceeded to the final ballot, focused on issues such as diplomacy, security and constitutional amendment.[13] There was a difference of 20 votes from MPs between Abe and Ishiba in the first round, and 19 in the runoff. These numbers mean that the votes that went to the other candidates in the first round were split almost equally between Abe and Ishiba in the final ballot. One of the factors that can be pointed out as having gathered support for Abe and Ishiba was their stance of focusing on issues such as diplomacy, security and constitutional amendment.

Candidates addressing the issue of disaster recovery and reconstruction were Abe, Machimura (also mentioning nuclear accident) and Ishihara. Therefore it is unlikely that addressing the issue of disaster recovery and reconstruction would have had a direct effect on the voting result. However, it was noted that the winner, Abe, had mentioned the Great East Japan Earthquake not as a part of his policy but as a reason for running, due to the similarities with the 2012 DPJ leadership election winner, Noda, who had positioned disaster recovery and reconstruction as a prerequisite for various policies. The common stance of Abe and Noda is thought to be the background leading to the two parties raising the issue of disaster recovery reconstruction ahead of their priority policy during the 2012 general election.

Opinion distribution of the 2012 general election candidates

As described above, there was no debate over disaster recovery and reconstruction, and energy policies, in the DPJ leadership election and the LDP presidential election in 2012. Therefore, did the candidates of the 2012 general election share a common stance with Noda or Abe? I examine this based on the University of Tokyo Taniguchi Laboratory and *Asahi Shimbun* joint survey of 2012 House of Representatives election candidates.[14]

What policies were highlighted by candidates in the 2012 general election? Candidates were asked to number issues from 1 to 3 in order (N = 1,388). Top choices were fiscal and monetary, 14.3%;

nuclear power and energy, 14.0%; and jobs and employment, 12.5%. The total percentage of choices 1 to 3 were nuclear power and energy, 47.3%; jobs and employment, 36.1%; fiscal and monetary, 29.6%.

Categorizing by political party, the DPJ and PNP candidates placed emphasis on jobs and employment. The DPJ candidates also emphasized pensions and healthcare. The LDP and Komeito candidates placed emphasis on different policies between the two parties: the LDP emphasized fiscal and monetary, and diplomacy and security, and Komeito emphasized disaster reconstruction and prevention. In particular, the fact that 70% of Komeito candidates prioritized disaster reconstruction and prevention distinguished them from other parties. The JRP, YP and NRP had many candidates who emphasized political and administrative reform. Most candidates in the TPJ, SDP, JCP and NPD placed emphasis on nuclear power and energy. The proportion of candidates who chose it in the top three exceeded 80%.

However, there were differing opinions about the disaster and nuclear power generation between regions that were directly affected by the triple disaster and those that were not. Table 4.1 shows the percentages of emphasized policies (to third place) for each party nationally and for Iwate, Miyagi and Fukushima. Regional factors play a stronger role in emphasis on disaster reconstruction and prevention, whereas political parties play a stronger role in emphasis on nuclear power and energy.

Even though the influence of political parties was greater than regional factors for nuclear power and energy, the political parties did not necessarily control candidates. Figure 4.1 shows the proportion of candidates, by political party and region, divided into four categories based on their opinions on the ratio of nuclear power to total energy and restarting nuclear power plants. The four categories are (1) aiming for zero nuclear power in the future and opposed to resuming nuclear plant operations (Zero nuclear + opposed to restart), (2) aiming for zero nuclear power in the future but allowing nuclear plant to resume (Zero nuclear + allow restart), (3) accepting both nuclear power generation and restarting of plants (Allow nuclear power generation + restart), and (4) none of the above (Other).[15]

Considering Figure 4.1 together with Table 4.1, parties are divided into three groups. First are the TPJ, SDP and JCP. Opinions for 'Zero nuclear + opposed to restart' occupied a majority, and in the

Table 4.1: Policies emphasized during the 2012 general election (to third place) by party and region

	Disaster reconstruction & prevention		Nuclear power & energy		Number of observations	
	National	Iwate Miyagi Fukushima	National	Iwate Miyagi Fukushima	National	Iwate Miyagi Fukushima
DPJ	23.7	100.0	36.6	41.7	257	12
PNP	50.0		0.0		2	
LDP	24.5	93.3	6.4	0.0	298	15
Komeito	71.1		22.2		45	
JRP	11.5	100.0	16.0	20.0	156	5
YP	6.0	33.3	77.6	100.0	67	3
NRP	100.0		0.0		1	
TPJ	23.4	90.0	80.4	90.0	107	10
SDP	12.5	66.7	93.8	100.0	32	3
JCP	17.8	86.7	87.5	80.0	320	15
NPD	0.0		0.0		7	
NPN	0.0		0.0		1	
Total	20.5	88.9	47.3	52.4	1,388	63

Source: The UTokyo-Asahi Survey (UTAS) conducted by Masaki Taniguchi of the Graduate Schools for Law and Politics, the University of Tokyo and the *Asahi Shimbun*, the 2012 House of Representatives election candidate survey.

election there were many candidates who focused on nuclear power and energy across the nation. Although the proportion was somewhat lower, this was the same for YP.

Second are the DPJ, Komeito and JRP. The opinions ‘Zero nuclear + opposed to restart’ and ‘Zero nuclear + allow restart’ occupied a certain percentage within each party. The DPJ and JRP fielded candidates for single-member districts in Iwate, Miyagi and Fukushima, and the proportion of focus on nuclear power and energy policy was higher than the national percentage.

Third is the LDP. ‘Allow nuclear power generation + restart’ was common, and candidates in Iwate, Miyagi and Fukushima did not highlight nuclear power and energy. However, it is shown that in Iwate, Miyagi and Fukushima, 20% of candidates were for ‘Zero nuclear + opposed to restart’.

Thus attitudes towards future nuclear power generation and the dispersion of attitudes towards restarting nuclear power plants were

Figure 4.1: Opinions of 2012 general election candidates on zero nuclear power and restarting nuclear power plants

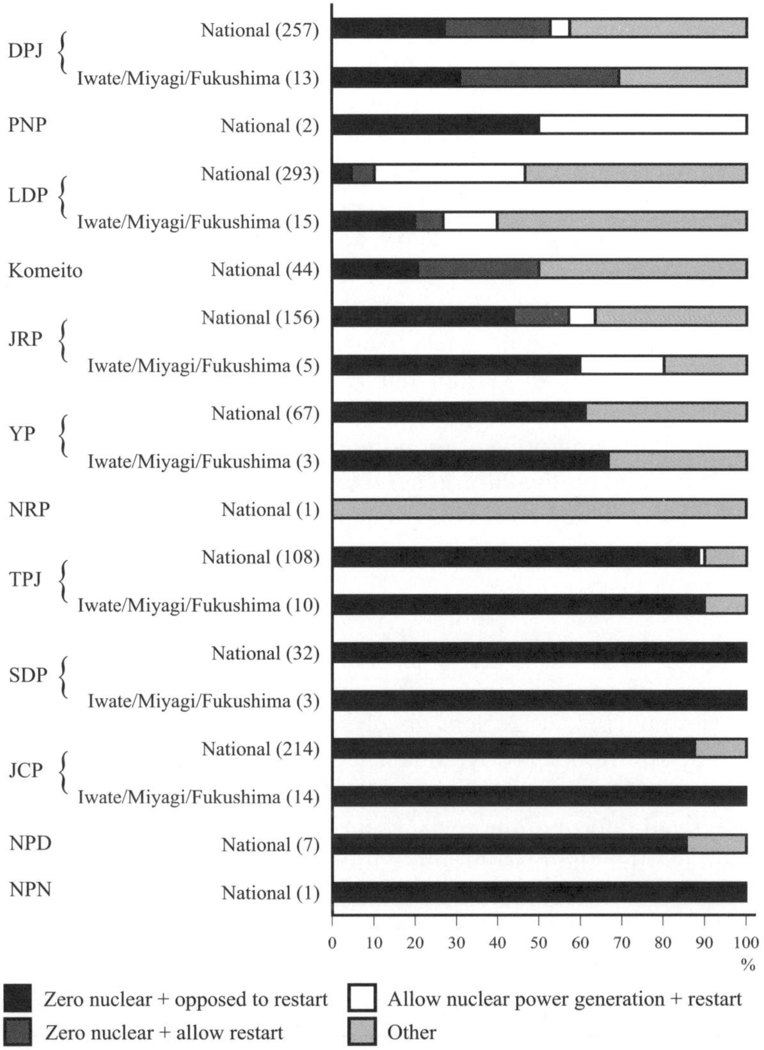

Note: The bracketed numbers after 'National' or 'Iwate/Miyagi/Fukushima' show the number of observations.
Source: As per Table 4.1.

especially significant in the DPJ, Komeito and JRP. While some DPJ members, such as Kan Naoto, were against nuclear power generation, there were also members, such as Noda, who were open towards restarting nuclear plants. The trade union of the electric power industry would have influenced this. Although the Komeito power base, the Soka Gakkai (which means value creation society), was opposed to nuclear power, the party's cooperation with the LDP, which had a positive stance towards nuclear power, resulted in Komeito being caught in the middle. The JRP merged with the SP in November 2012. While the former JRP was against nuclear power, the former SP had a positive attitude towards nuclear power.[16]

To summarize this section, in the party leadership elections held after 3.11, disaster recovery and construction was mentioned as a prerequisite for various policy debates, whereas debates on nuclear power generation were not widespread. As a result, disaster construction and prevention policies did not attract nationwide attention among the 2012 general election candidates. While the TPJ, SDP, JCP and NPD bloc tried to raise it as an issue for debate, others were reluctant to do so because of non-consensus within parties or blocs.

The structure of interparty and interbloc competition

On November 14, 2012, Noda dissolved the House of Representatives. The dissolution directly concerned the three-party agreement (August 2012) between the DPJ, LDP and Komeito surrounding the integrated reform of the social security and tax systems. However, the day after the dissolution, editorials in national newspapers argued that the reputation of the DPJ itself should be questioned in this election.[17] Also, the three leaders had avoided disputing any particular issue during the party leader debate in the Diet when Noda announced the dissolution.[18] What kinds of debates were held in the 2012 general election? In this section, I look at the content of manifestos and party leader debates.[19]

Comparison of manifestos

Comparing the manifestos[20] of each party showed a large difference in style and quantity. There were differences in content even among cooperating parties. For the purpose of understanding what each party treated as key points, I would like to pay attention to the use of headlines and their order in the manifestos.[21]

The DPJ, LDP, Komeito and PNP all listed disaster recovery and construction first. It can be said that disaster recovery and construction had become a valence issue in the blocs surrounding the so-called 'two major parties'. This is consistent with Noda's and Abe's respective appeals during the DPJ leadership election and the LDP presidential election in 2012.

How about the other two blocs? The JRP, YP and NRP focused on the economy. The JRP and NRP listed it first, and YP listed it second. This stance shows similarities with the LDP, which emphasized the phrase 'regain the economy'. On the other hand, the TPJ, JCP, SDP and NPD focused on issues such as withdrawal from nuclear power, opposing the increase of consumption tax and opposing the Trans-Pacific Strategic Economic Partnership Agreement (TPP).

To summarize in relation to disaster reconstruction and nuclear power plants, seven parties (the DPJ, PNP, LDP, Komeito, YP, SDP and JCP) headlined earthquake recovery and reconstruction. The JRP and TPJ, which had nominated candidates for the first time, did not have headlines in relation to disaster recovery and reconstruction. All parties except the LDP and NPD had a headline for energy policy. It is particularly clear that the TPJ and SDP had listed 'graduate from nuclear power' and 'goodbye nuclear power' first, and were trying to make these debates topics in the election.

Next, I look at the attitudes towards disaster reconstruction and prevention, and nuclear power and energy policies in more detail from the contents of each party's manifesto. The candidate survey shows that the emphasis on disaster reconstruction and prevention was due to region, whereas nuclear power and energy policy differed largely between political parties. How was it addressed in the manifesto?

In their manifestos, every party mentioned recovery and reconstruction from the Great East Japan Earthquake. Attitudes were divided over two points: future measures for disaster prevention and reference to specific regions other than Tōhoku. Future disaster prevention measures were mentioned by seven parties: the PNP, LDP, Komeito, NRP, SDP, JCP and NPD.

Areas other than the Tōhoku region (the Nankai Trough, directly under the capital, etc.) were mentioned by three parties: the PNP, LDP and Komeito. These parties seem to have assumed a position to actively advance disaster prevention measures in other regions and across the nation, in response to the great earthquake that occurred in East Japan. What is particularly noteworthy is that the PNP has shown a proactive stance towards disaster prevention measures. The PNP's common stance on disaster prevention with the opposition LDP and Komeito, rather than the coalition partner DPJ, had suppressed making a position issue in the general election.

Finally, positions in nuclear power were split into three: (1) the position seeking the best mix of power supply with the premise of using nuclear power, such as LDP and PNP; (2) the position aiming for zero-nuclear, such as the TPJ, JCP, SDP and NPD; (3) the position to allow nuclear power plants to decide on a target timing to achieve zero-nuclear but allow resumption of plant operations for the time being, such as the DPJ and YP.

In contrast to these parties, Komeito and JRP used subtle expressions. Komeito avoided declaring a time period, most likely considering its relationship with the LDP. Although indicating a time, the JRP also showed a passive attitude. According to reports at the time, the reason for this passive expression was due to its merger with the SP.[22] As a result, this made it fall out of pace with the YP, which aimed for a withdrawal from nuclear power.

Positions were also divided on the pros and cons of electricity market reform as a specific measure to achieve a withdrawal from nuclear power. This issue was mentioned by the DPJ, YP, JRP, SDP and Komeito. It was, however, not mentioned by the LDP, PNP, TPJ, JCP etc.

To summarize the above, attitudes towards disaster prevention measures differed within the DPJ and PNP bloc, and attitudes towards withdrawal from nuclear energy and electricity market reform differed in three blocs: the DPJ and PNP; the LDP and Komeito; and the JRP, YP and NRP. These differences can be incentives to avoid disputing issues in the election campaign. The next section considers the party leader debates.

Disputes in party leader debates

During the 2012 general election, two party leader debate sessions were hosted, an online debate by Nico Nico Douga on November 29 and a press club debate by the Japan National Press Club on November 30.[23] Before considering the content, it is important to be aware that the themes were set by the moderator and audience. Three themes were set for the online debate (the TPP, consumption tax increase and nuclear power) and eight for the press club debate (disaster reconstruction, nuclear power, energy, diplomacy, security, economy, finance and the TPP). Common themes were the TPP, consumption tax increase and nuclear power. These received a high degree of focus from the media and viewers.[24]

However, the themes prepared by the moderator and audience may not necessarily have matched the topics that the party leaders wished to appeal. This section focuses on (1) opening remarks of party leaders in the debate and (2) questions asked between party leaders in cross-debate, because in these party leaders could speak more freely.

Opening remarks

Comparing the opening remarks, including two debates, shows that no theme was mentioned by all the leaders. However, when considered by bloc, the leaders of all four blocs mentioned economic policies, growth strategies, management of the DPJ government, political or administrative reform, and disaster response and prevention.

Conversely, many themes were mentioned by some blocs but not others. Of particular interest in this chapter are the following three points that suggested conflict between blocs.[25] First, the blocs that referred to employment and finance were completely reversed. The DPJ, TPJ, SDP and NPD referred to employment, while the LDP, YP and NRP talked about finance. This would have been a reflection of the difference in approaches towards economic measures and economic policy.

Second, the DPJ and PNP and the LDP and Komeito did not mention tax and public finance, such as the consumption tax increase. Parties that did mention it were the YP, NRP, SDP, JCP and NPD, all taking a critical stance. This would have been a reflection of the framework of the three-party agreement on integrated reform of social security and tax systems.

Third, out of all the blocs, the parties that did not mention nuclear power at all were the LDP and Komeito. This must have been a reflection of different views on nuclear power plants between the LDP and Komeito, as is clear from the manifesto comparisons in the previous section.

Figure 4.2: Composition of cross-debates at the Japan National Press Club

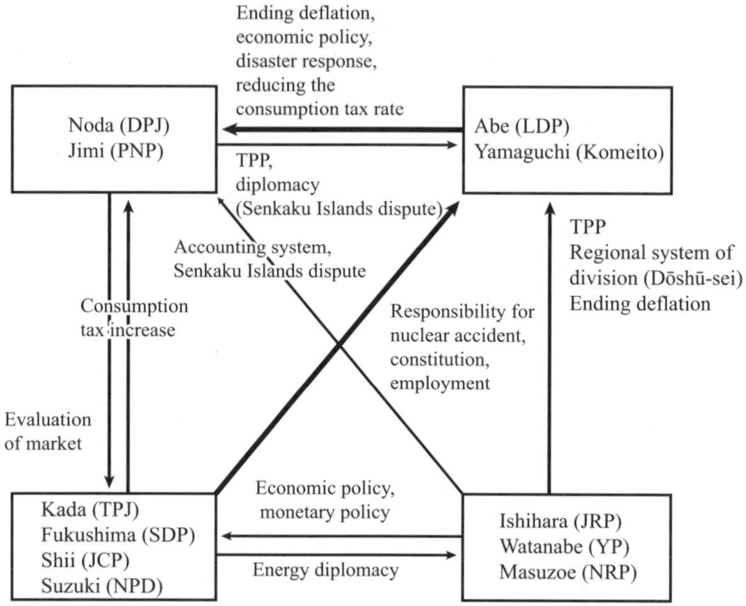

Note: A thin line means one questioner and a bold line means two questioners.
Source: Created by the author based on Japan National Press Club, 'The 46th general election of the members of the House of Representatives debate and press conference with 11 leaders of political parties'.

Cross-debate

In the press club debate, time was provided to ask questions (a so-called cross-debate). Organizing the leaders who were questioned shows that there were no internal nominations within the four party blocs. Therefore, in Figure 4.2, each bloc is set as a unit, and arrows indicate which bloc questioned which. The thin line means there was one questioner and the bold line means there were two questioners. The question topic is set next to the arrow.

Though there were many issues, this section focuses on the following three points because they imply three rifts when I consider the manifesto comparisons and the debate opening remarks. First, the three issues that had multiple questions in relation to them were economic policy, ending deflation and the TPP. Economic policy was a major issue, including finance and trade.

Second, the only question about consumption tax increase was from the JCP directed to the DPJ. Komeito questioned the DPJ about its attitude towards reducing the consumption tax rate, but it was not a question about consumption tax increase itself. Attitudes were divided on whether to criticize the three-party agreement, including consumption tax increase, or whether to not mention it at all.

Third, the only question about nuclear power plants was from the TPJ directed at the LDP. The DPJ, JRP and YP did not ask questions relating to withdrawal from nuclear power. Among them, the DPJ and YP made pledges to withdraw from nuclear energy, and also mentioned this issue in their debate opening remarks. However, the DPJ questioned the LDP about the TPP and diplomacy (the Senkaku Islands dispute), while the YP raised questions about the TPP, the regional system of division (Dōshū-sei) and ending deflation. Although it was a limited case of cross-debating within a debate, the DPJ and YP were reluctant to raise disputes with the LDP about nuclear power plants.

The findings of this section can be summarized as follows in terms of disaster reconstruction and withdrawal of nuclear power generation. In the manifestos, disaster reconstruction had become a valence issue among the two main party blocs, and only the LDP and PNP showed passive attitudes about withdrawing from nuclear power plants. However, at the party leader debates, many political parties, including the DPJ, were reluctant to argue about nuclear power. It was only the TPJ that was actively making this a point of contention.[26]

We should pay attention to each misalignment in the manifestos and the party leader debate statements of the DPJ, Komeito and JRP. In other words, despite appealing for a withdrawal from nuclear power plants in the manifesto, they were reluctant to argue for this at the debate. Considering the reasons from the standpoint of political competition gives two possibilities: (1) non-consensus within the party and (2) non-consensus between cooperating parties within the blocs.

Regarding (1), within political parties both the DPJ and JRP had candidates who had different stances on the issue of nuclear power. Regarding (2), within intrabloc cooperating parties there were differences in attitudes towards nuclear power between the LDP and Komeito, the DPJ and PNP, and the JRP and YP. The absence of consensus seems to have promoted non-confrontation on nuclear power issues at the party leaders debates.

The 2012 general election results and its implications

Up to this point, I have examined the agendas that were set after 3.11, describing the party leadership elections and the 2012 general election. In this last section, after reviewing the 2012 general election results based on the vote data analysis by Masuyama (2013) and Mori (2016), I examine the relationship of findings from the two previous sections to the 2012 general election results. In conclusion, I consider what kinds of political competition occurred in Japan after 3.11.

The setback of the two-party system and the rise of new parties

The 2012 general election voter turnout was 59.3%. The voter turnouts in the 2005 and 2009 general elections were recorded as exceeding 60% because the postal privatization in 2005 and the change of administration in 2009 became points at issue. In contrast, the turnout in 2012 remained at the same low level as in 2003.

The election resulted in the LDP and Komeito winning, and Abe Shinzō taking office. Table 4.2 shows a summary of the election results. Results showed a setback for the formation of a two-party system centered on the LDP and DPJ that had been built up in previous elections (Masuyama 2013; Mori 2016). The efficient number of candidates, which is the reciprocal of the sum of squares of the relative vote share of each candidate, had gradually decreased following the 1996 general election, which was the start of the mixed-member electoral system. However, the efficient number of candidates in 2012 returned to the 1996 level. Other indicators of competitiveness in the election also returned to 1996 levels (Masuyama 2013).

Comparing the absolute vote share of each party with the result of past elections shows that the DPJ, which lost power, collapsed in the basic vote itself. The LDP and Komeito, which won power, also failed to maintain basic votes. Compared to the poor performance of these established political parties, the YP and JRP showed great progress. In particular, the JRP exceeded the DPJ's vote share and became the second party (Mori 2016: Section 2).

Looking at each vote structure of 300 districts, although there was variation due to the personal attributes of some candidates, there was a nationwide slump in the established parties such as the DPJ and LDP (Mori 2016: Section 3). This can be understood as following

Table 4.2: Results of 2012 general election (seats won and vote share)

	Number of seats won & percentage of seats				Relative percentage of votes (%)		Absolute percentage of votes (%)	
	Single-member districts	**Proportional representation**	**Total**	**Percentage of seats (%)**	**Single-member districts**	**Proportional representation**	**Single-member districts**	**Proportional representation**
LDP	237	57	294	61.25	43.01	27.62	24.67	15.99
DPJ	27	30	57	11.88	22.81	16.00	13.08	9.26
JRP	14	40	54	11.25	11.64	20.38	6.68	11.80
Komeito	9	22	31	6.46	1.49	11.83	0.85	6.85
YP	4	14	18	3.75	4.71	8.72	2.70	5.05
TPJ	2	7	9	1.88	5.02	5.69	2.88	3.29
JCP	0	8	8	1.67	7.88	6.13	4.52	3.55
SDP	1	1	2	0.42	0.76	2.36	0.43	1.37
NPD	0	1	1	0.21	0.53	0.58	0.30	0.33
PNP	1	0	1	0.21	0.20	0.12	0.11	0.07
Minor parties	0	0	0	0.00	0.28	0.58	0.16	0.34
Independents	5	–	5	1.04	1.69	–	0.97	–
Total	300	180	480	100.00				

Source: Mori 2016: 297, Table 13-1.

the trend of centralization (or nationalization), which has been noted in Japanese political research since the 1990s.

On the other hand, a regional trend was also observed for obtaining votes, according to absolute vote share ranking by prefecture of the proportional representation election (Mori 2016: Section 4). First, the 3.11 afflicted areas are ranked high for the DPJ, JCP, SDP and the TPJ, in contrast to being ranked lower in the LDP, Komeito and JRP. Second, among the so-called third force parties, votes were garnered for the JRP mainly in the Kansai region, including Osaka; the YP gained votes in the metropolitan areas, including Tokyo; and the TPJ garnered votes in the afflicted area and the Tokyo metropolitan area.

Furthermore, according to absolute vote share in the proportional representative election by the population size of single-member districts, parties can be divided into three groups (Mori 2016: Section 4). First, parties with a high share in rural areas and a low share in urban areas were the LDP, DPJ and Komeito. Second, parties with a high share in urban areas and a low share in rural areas were the JRP and YP. Third, parties that collected votes evenly despite electorate size were the SDP, JCP and TPJ. When narrowed down to political parties that were formed before 2009, only the YP increased its vote share, while established political parties lost votes mainly in urban areas.

As mentioned above, the 2012 general election was an election in which the LDP and Komeito won, the DPJ collapsed in the basic vote across the nation, and the JRP and YP made great gains. On the other hand, when considering the structure of votes by regions, the JRP and YP showed significant gains in urban areas, and it was the TPJ that gained votes in both urban areas and afflicted areas. Each can be understood as a disconnection and a connection between urban areas and afflicted areas (Mori 2016: 314–15).

Political competition after 3.11: local and national

How do the results of the 2012 general election relate to what has been discussed in this chapter? I examine this after summarizing the findings from each section.

As pointed out in the first section, schematically organizing the disputes between political parties and politicians leading to the 2012 general election, it was possible to largely divide them into four blocs: (1) the DPJ and PNP, (2) the LDP and Komeito, (3) the JRP,

YP and NRP, and (4) the TPJ, SDP, JCP and NPD. In addition to this, the next two sections revealed four additional cracks.

The section on the structure of intraparty and intrabloc competition revealed the linkage between (1) internal division of the DPJ and (2) evaluation of the three-party agreement between the DPJ, LDP and Komeito. In the 2011 DPJ leadership election there was a dispute over whether to focus on the three-party agreement or the 2009 general election manifesto. Due to the victory of Noda, who was for the three-party agreement, Ozawa and others departed the party, which fostered a flow towards dissolution. As a consequence of the three-party agreement, disaster recovery and reconstruction became a valence issue between the two major political parties. It can be understood that the entry of the JRP and TPJ challenged the coordination between the two main party blocs.

The section on the structure of interparty and interbloc competition revealed the linkage between (3) the interrelationships between the third force parties and (4) disagreement over the pros or cons of the nuclear power plant. In terms of the interrelationships between third force parties, the JRP played a key role. Although it had originally appealed to withdrawal from nuclear energy, its merger with the SP weakened this claim. The DPJ and Komeito also had disagreements regarding nuclear power among their party MPs, and there was also a fissure between the cooperating parties of the PNP and LDP. These parties limited their reactions to popular will, demanding abolition of nuclear power, and there was no confrontation to resolve the differences in opinions within the parties or between cooperating parties. The bloc that was formed by parties such as the TPJ had attempted to instigate debate, but its influence was limited.

In the 2012 general election the DPJ lost due to its collapse of basic votes, and at the same time other established parties such as the LDP failed to maintain their basic votes. The fact that not only the DPJ but the LDP and Komeito struggled to gain votes suggests that the three-party agreement, which was upheld under the DPJ administration, had not necessarily been judged as positive. In more general terms, it may be seen as an assessment of the change towards cooperation between the two main party blocs, which were fierce competitors in past general elections.

Instead, there was a rise of the so-called third force parties. While the TPJ won votes in the Tokyo metropolitan area and afflicted area by positioning itself to transition away from nuclear power, the JRP and YP, which leaped in urban area votes, assumed a passive stance towards the removal of nuclear power plants. It also did not mention disaster recovery and reconstruction. In other words, agendas that were newly set after 3.11 did not link directly with changes in the structure of political competition caused by the rise of the JRP and YP in urban areas. This means that the issues of disaster reconstruction and prevention, and nuclear power and energy policy were suppressed as local points of contention. As a result, nationwide competition on these issues did not occur.

Thus, in the political competition after 3.11, disaster reconstruction became a valence issue, and nuclear power became an issue of non-competition despite being a position issue. Of the three paths shown in the opening section, disaster recovery and reconstruction followed path 2, and nuclear power and energy policy followed path 1. In sum, while the agendas set after 3.11 had positive effects on the challenge of third force parties against the two main party blocs, they also had negative effects on nationalization of competition caused by triple disaster.

5 Nuclear Power and the Will of the People

Yamamoto Hidehiro

Following the unprecedented Great East Japan Earthquake and the accident at the Fukushima Daiichi Nuclear Power Station, the direction our society ought to take has been thrown into question. The political decision-making process that will determine this is the issue of the moment. In a democratic society, political decisions must reflect the will of the people. In this context, how has the popular will expressed itself in the formation of disaster recovery and energy policy? Do the political decisions that have been made reflect the popular will?

Following the Great East Japan Earthquake, popular opposition to nuclear power became abundantly clear. The public's distrust of nuclear power was expressed as thousands of people took action to call for an end to nuclear power. Frequent demonstrations took place throughout Japan, leading to the establishment of weekly protests in front of the prime minister's official residence. The largest of these demonstrations were attended by hundreds of thousands of people. The growing movement attracted widespread attention and aroused public opinion on the restarting of idle nuclear reactors and the future of energy policy. No recent social movement has had such a major social impact as the antinuclear movement.

Japan's civil society is weak in terms of advocacy (Pekkanen 2006; Aldrich 2008). Compared with other countries, relatively few people have ever taken part in a social movement and few have a positive regard for such movements (Yamamoto 2014; Yamada 2016). In contrast to such a feature of Japan's civil society, antinuclear movements provided a counter example.

Despite the outburst of popular opposition to nuclear power, however, there has not been a fundamental shift away from its use at the national level. Not long after the earthquake, Japan recommenced its export of nuclear power technology and there were moves to restart the domestic nuclear reactor fleet. Following the victory of the Liberal Democratic Party (LDP) over the Democratic Party of Japan (DPJ) in the December 2012 elections, these efforts have only increased. A number of commentators have tried to explain the seeming intransigence of Japan's nuclear policy in the face of the Fukushima Daiichi nuclear accident (Samuels 2013; Aldrich 2014; Kingston 2014).

In terms of energy policy, there is little doubt that we are witnessing a reversion to the pre-Fukushima norms. Can we therefore conclude that the will of the people, as it appeared in response to the nuclear accident, has left no impression on society? In this chapter I address this question by examining how the debate on nuclear power developed in the nexus between social movements, the government and political parties, the mass media and public opinion. I demonstrate that, while popular interest in the issue of nuclear power peaked in the summer of 2012, antinuclear sentiments have become more pervasive than ever. I show that demands for a political decision-making process that reflects the will of the people have become deep-rooted and that the weakness of Japan's civil society may be beginning to change.

Models to understand social issues

Researchers have developed two principal models for understanding how social issues come to capture popular attention: the issue-attention cycle and punctuated equilibrium. The issue-attention cycle describes a series of processes whereby an issue that is taken up by the mass media or by a social movement gains broader recognition in society and becomes a point of contention before ultimately fading from public view (Downs 1972; Peters and Hogwood 1985). During the first stage, experts or interest groups take up the issue in question but it is not yet widely known. Then, due to an accident or some other factor, the issue becomes the subject of public attention. Finally, it fades from view, either due to the growing realization that it will be difficult to resolve or competition with other issues (Downs 1972). This series of processes is shaped by a number of factors, including changes in people's knowledge and values concerning the issue, reports in the mass media and political conflict (Jasper 1988). While social movements can initiate the issue-attention cycle, the influence of the mass media has also been highlighted (McCarthy et al.

1996). The debates over nuclear weapons and nuclear power also demonstrate how the influence of various actors shapes the issue-attention cycle (Jasper 1988; Joppke 1991).

The second model is the punctuated equilibrium model (Baumgartner and Jones 1991, 1993; Howlett 1997), which derives conceptually from evolutionary biology. This model proposes that when people come to see a given issue differently, the existing dominant policy framework change leads to the establishment of a new equilibrium. Change does not occur gradually, but suddenly as a result of external shocks or the interventions of various actors. For example, when more people become aware of the ethical, social and political dimensions of a policy issue that has hitherto been monopolized by experts on the basis of its technical complexity, the number of people who are concerned with the issue swells. This opens up space for policy change. For this reason, actors who want to see a change in existing policy try to change the way people perceive the issues at hand and expand the space for policy debate. Baumgartner and Jones (1991, 1993) explain the punctuated equilibrium model using nuclear weapons and nuclear power as examples.

The two models differ over the question of whether growing interest in a particular social issue is a temporary phenomenon that will ultimately dissipate or whether it will lead to the attainment of a new state of equilibrium (Howlett 1997; Holt and Barkemeyer 2012). The fact that both models have been developed using the issues of nuclear weapons and nuclear power as examples suggests that at different points in time popular concern with nuclear issues has led to both these outcomes. It has sometimes been a transient phenomenon and at other times has led to the formation of a new social consciousness. How did the issue-attention process develop after the Fukushima nuclear accident?

In this chapter I refer to both these models in order to examine how various actors contributed to making the demand for the abolition of nuclear power a focus of public attention. I also consider the implications of this manifestation of the will of the people. In describing the development of the antinuclear power movement I focus on three main areas: the actions of the government and of the political parties that constitute the political opportunity structure for the movements; the mass media and its role in the formation of the discursive opportunity structure; and shifts in public opinion.

The concept of the political opportunity structure refers to the political environment that influences a social movement's expectations of success or failure (McAdam 1999; Kriesi et al. 1995; Tarrow 1998). This theory suggests that people will only consider starting a movement when the political environment presents a considerable prospect of success (McAdam 1999; Tarrow 1998). The political opportunity structure comprises various elements, including the ability to access the political system, the internal relationships between political elites (alliances, antagonisms etc.) and their degree of stability, and the existence of political elites who are open to the demands of the movement (McAdam 1996).

With regard to the antinuclear movement, members of the Diet who were sympathetic to the abolition of nuclear power emerged after the Fukushima nuclear accident. At the same time, lines of antagonism between political elites regarding nuclear power also appeared. This political environment probably had a significant influence on the formation and development of the movement.

The notion of the discursive opportunity structure incorporates the political culture and political trends that provide opportunities for social movements (Feree 2003; Koopmans and Statham 1999; McCommon et al. 2007). This theory suggests that a movement will grow if its demands are adapted to the main values and ideas of society as a whole. This also makes it difficult for the political system to ignore it. The need to maintain a supportive discursive opportunity structure means that movements have to advocate for their claims in a manner that is likely to be well received by society.

For the antinuclear power movement, the widespread suspicion towards nuclear power following the Fukushima nuclear accident can be interpreted as presenting a discursive opportunity structure that was open to the movement.

I now explore how the movement developed in responsc to thcsc opportunity structures and how the people expressed their will concerning nuclear power.

The debate on nuclear abolition

Let us first review the situation prior to the accident at the Fukushima Daiichi Nuclear Power Station. Numerous authors have already described the powerful network of interests surrounding nuclear power that is known as the 'nuclear village' (*genshiryoku mura*)

(Honda 2005; Samuels 2013; Kingston 2014). The structure of this network evolved during the long period of uninterrupted LDP rule and involved balancing the interests of the Ministry of International Trade and Industry with those of the electric power utilities in the development of nuclear power. Regions where nuclear power stations were constructed received subsidies under the three laws for electric power development (among other inducements to compensate them for hosting the reactors). This made regional communities increasingly dependent on nuclear power (Nakazawa 2005; Kainuma 2011; Aldrich 2008, 2014; Hasegawa 2015).

Movements protesting nuclear power emerged alongside the antinuclear weapons movements. The formation of these movements depended mostly on the support of the Japan Socialist Party and the General Council of Trade Unions of Japan (Sōhyō) (Honda 2005). An independent citizens' antinuclear power movement only formed when a radiation leak from the nuclear-powered ship *Mutsu* in 1974 made people pay attention to the issue. More people began to speak out against nuclear power following the 1979 Three Mile Island accident, the 1986 Chernobyl disaster and several nuclear accidents in Japan. However, the main actors in these movements were outside the structures of political power. As a result, they had little effect on the formation of policy.

While fisherfolk and residents in host communities did organize movements opposing the construction of nuclear power stations, the provision of compensation and other inducements prevented these movements from spreading (Aldrich 2008, 2014). Nevertheless, there were movements that sought to reflect residents' voices in politics on this important issue for the local community. For example, during the 1990s in the town of Maki in Niigata Prefecture, a powerful movement emerged calling for a citizens' referendum on nuclear power station construction (Nakazawa 2005; Itō et al. 2005).

The antinuclear power movement

The situation changed dramatically after the March 2011 Fukushima nuclear accident. This external shock brought nuclear power to the attention of the public and triggered the issue-attention cycle. Protests took place immediately after the nuclear accident, but they only began to increase in number and geographical reach in April. In the Tokyo metropolitan region, the largest action was the No Nukes

Demonstration that took place in the youth cultural hub of Kōenji. Organizers put the number of participants at 15,000 (Oguma 2013; Ogawa 2014). A variety of protest actions then spread across Japan. On June 11, 121 demonstrations and other protest events took place across the country. Organizers estimated that, all together, these actions involved some 60,000 people (*Asahi Shimbun*, June 20, 2011, morning edn). In September another No Nukes Demonstration organized by young Kōenji-based activists attracted about 10,000 people. An estimated 60,000 people attended a protest rally called by the Japan Congress against Atomic and Hydrogen Bombs, one of Japan's leading antinuclear organizations. In October, 13 antinuclear power groups established the Metropolitan Coalition against Nukes (MCAN) in order to coordinate their activities more effectively. On March 11, 2012, one year after the nuclear accident, MCAN organized the 3.11 Tokyo Big March. Following the march, 30,000 people formed a human chain encircling the National Diet.

Following on from these successful actions, MCAN began to hold regular protests outside the prime minister's official residence every Friday. The first took place on March 29 in opposition to deliberations that were taking place over restarting the Kansai Electric Power Company's Ōi Nuclear Power Station in Fukui Prefecture. At first they were attended by no more than a few thousand people but they grew rapidly following the government's decision to restart the Ōi Nuclear Power Station in June. Organizers estimated that 200,000 people attended the action on June 29 (Oguma 2013).

As the protests grew, MCAN representatives held discussions with sympathetic Diet members on July 31. Then, on August 22, they met face to face with then Prime Minister Noda Yoshihiko. During the meeting, they called on the Prime Minister to revoke his decision to restart the Ōi Nuclear Power Station and challenged his government's choice of commissioners for the newly formed Nuclear Regulation Authority. However, they received little sympathy for their demands (*Asahi Shimbun*, August 23, 2012, morning edn). Thereafter, the weekly demonstrations have continued, though they have shrunk in scale.

While these antinuclear movements were formed in response to the urgency of the situation after Fukushima, they clearly displayed characteristics that were common to many other social movements that have emerged since 2000. Their protests were focused not on political arguments but on freedom of expression and possessed a unique style (Itō 2012; Gonoi 2012; Ogawa 2014). The organizers held sound demonstrations that featured music and participants banged drums, wore costumes and waved colorful placards. This trend has been observed in Japan since around the time of the 2003 anti-Iraq war demonstrations (Mōri 2009).[1]

The movements also made use of social media (Itō 2012; Gonoi 2012; Hirabayashi 2013). Information about the demonstrations was distributed via Twitter and Facebook. Many people were therefore able to share information about the movement even without its being reported by the mass media. Social media also enabled participants to become the media, and to share information about the demonstrations from the field while they were taking part. As a result, rather than a mass action over a particular demand, the demonstrations tended to become a space where participants could express themselves freely as part of a collective and participation became an end in itself (Itō 2012). These characteristics were common to other prominent movements around the world at that time, such as the Arab Spring and the Occupy Wall Street Movement in New York.

Another point of commonality with these movements was the participation of people who had no previous involvement in social movements. Hirabayashi (2013) found that 48% of people who took part in the antinuclear demonstrations held on June 11, 2011, were attending a political demonstration for the first time. By September 11 the proportion of first-time participants remained relatively high at 35%. The free style and the use of social media described above seem to be connected to the participation of a new social layer that had not previously shown much interest in demonstrating.

These characteristics, along with the increased awareness about nuclear power that was generated by the nuclear accident, contributed to participation on a scale that has not been witnessed in recent years.

Government and political parties

Next I examine the actions of the government and political parties that constitute the political opportunity structure for the antinuclear power movement. In 1993 the LDP's rule, uninterrupted during most of the post-war period, collapsed. In 2009, following various political reconfigurations, a genuine change of government took place when the DPJ took power from the LDP. Many DPJ Diet members have relationships with social movements and trade unions. Their overall policy stance is more favorable to social movements than that of the LDP. Furthermore, the DPJ administration advocated the development of a New Public Commons, whereby it aimed to revitalize civic and political participation. As a result, the DPJ administration produced a more open political opportunity structure for social movements.

Nevertheless, the DPJ made no significant changes to the nuclear energy policies established during the previous LDP administration. Immediately prior to the accident at Fukushima Daiichi, nuclear power was being promoted as a form of clean energy that could contribute to containing carbon dioxide emissions from fossil fuels and tackle global warming.

After the Fukushima Daiichi nuclear accident, then Prime Minister Kan Naoto adopted a clear antinuclear position. In May 2011 Kan asked Chūbu Electric Power to shut down two reactors (No. 4 and No. 5) at its Hamaoka Nuclear Power Station in Shizuoka Prefecture that were thought to be located on a fault line. Eventually, the company agreed. Kan also announced his goal of increasing the role of renewable energy in Japan's overall energy mix to 20% by the early 2020s. At a July press conference, he voiced his personal desire for a society that would, in the future, no longer depend upon nuclear power.

However, Kan was soon replaced by his DPJ colleague Noda Yoshihiko. The new Cabinet, while still aiming for a reduction in the degree of reliance on nuclear power in the long term, supported the restart of nuclear reactors that had passed safety inspections. Attempts to restart nuclear reactors began in earnest in April 2012. The decision to restart reactors No. 3 and No. 4 at the Ōi Nuclear Power Station, which had cleared the newly created stress test regime introduced after the Fukushima accident, was controversial. As Noda's Cabinet, concerned over the potential for electricity shortages during the summertime, worked towards restarting the reactors, host communities and the public expressed fierce opposition to the plan. As a result, the decision to restart the reactors had to be continuously postponed.

Within the DPJ, opinions were divided. While the Energy Project Team approved of the restarts, the Nuclear Incident Countermeasures Project Team declared them to be premature. In April 2012 forces within the DPJ that continued to oppose nuclear power came together to create a road map for ending the country's reliance on nuclear power. While trying to block the restart of the Ōi Nuclear Power Station, they also proposed that Japan eliminate nuclear power by 2025. DPJ members were not alone in their opposition to nuclear power. In March a cross-party alliance of Diet members opposed to nuclear power established the Group for Zero Nuclear Power. In June the group called for 24 reactors to be decommissioned immediately and released a scale that ranked nuclear power stations in terms of risk.

These debates demonstrate the opposition to nuclear power that existed within the Diet. There were clear lines of antagonism within the political process and political elites were in conflict with one another over the issue. The antinuclear movement found allies among some of them. This political opportunity structure created a growing sense that the abolition of nuclear power was a real possibility. This helps explain why the crowds outside the prime minister's residence during the weekly Friday protests reached such a peak. Nevertheless, Noda authorized the restart of the Ōi Nuclear Power Station on June 16. In July reactors No. 3 and No. 4 were restarted (although they were shut down once again in September 2013 for routine inspections).

During this period, the government carried out a deliberative poll on the future of nuclear power. Deliberative polling is a method that aims to create a reference for policymakers based on the considered opinions of the people. Participants in a deliberative poll engage in debate with one another and have the opportunity to question experts. This method thereby captures the entire deliberative process, including the way people's opinions change in the course of the discussions (Fishkin 2009).

Participants in the poll were asked to consider one of three proposals put forward by the government's Energy and Environment Council on nuclear power's future role in Japan's energy mix: 0%, 15% and 20%–25%. The deliberative poll, carried out in August, found the highest level of support (46.7%) for the 0% option (Enerugii, Kankyō no Sentakushi ni Kansuru Tōrongata Yoron Chōsa jikkō Iinkai 2012).

In response to these poll results, the Noda government released the Innovative Energy and Environmental Strategy. The main aim of this strategy was that nuclear power would be phased out completely by the 2030s. Thus, while MCAN representatives met with Prime Minister Noda and the restart of the Ōi Nuclear Power Station was underway, there were some mechanisms where popular support for the abolition of nuclear power could find expression within the political process.

However, the three big business groups – the Japan Business Federation (Keidanren), the Japan Association of Corporate Executives, and the Japan Chamber of Commerce and Industry – expressed strong criticism of the strategy at a joint press conference. This opposition from business (combined with concerns for the relationship with the United States expressed in the Japan – United States Nuclear Cooperation Agreement and for Aomori Prefecture, where the Rokkasho Reprocessing Plant is located) meant that Cabinet did not formally adopt the Innovative Energy and Environmental Strategy (*Asahi Shimbun*, September 19, 2012, morning edn).

Support by DPJ Diet members for the antinuclear position also began to disappear. While the nuclear issue was being debated, internal conflict over a proposed consumption tax increase and Japan's participation in negotiations for the Trans-Pacific Partnership Agreement (TPP) deepened. A group around DPJ politician Ozawa Ichirō, who supported the antinuclear position, broke away from the party. Other DPJ lawmakers, including Kawauchi Hiroshi, Hatoyama Yukio and Yamada Masahiko, crossed the floor to vote against the consumption tax increase, leading to their suspension from the party. The Nuclear Incident Countermeasures Project Team, which had been a base of antinuclear support, struggled to keep going (*Asahi Shimbun*, July 25, 2012, morning edn).

Having left the DPJ, Ozawa Ichirō's group formed the People's Life First party and advocated for the abolition of nuclear power and against participation in the TPP negotiations and consumption tax hikes. Yamada Masahiko's group formed the Tax Cuts Japan – Anti-TPP – Zero Nuclear Party. These parties, alongside then Shiga prefectural governor Kada Yukiko's Tomorrow Party of Japan, contested the December lower house elections as a bloc on an antinuclear platform. However, they failed to become an electoral vehicle for popular opposition to nuclear power, winning only nine seats. In comparison to consumption tax hikes and social security, nuclear and energy policy issues did not seem a major electoral issue for voters.

The LDP's decisive victory in the December 2012 lower house elections ushered in a change of government. LDP leader Abe Shinzō's administration downgraded the DPJ's Innovative Energy and Environmental Strategy, with its goal of a nuclear-free Japan by the 2030s, to the status of a white paper and began to actively pursue nuclear reactor restarts (*Asahi Shimbun*, March 7, 2013, morning edn). The nuclear issue has not been a significant factor in subsequent elections and the LDP has continued to win them.

The Basic Energy Plan that was adopted by Cabinet in April 2014 designates nuclear power as one of the key sources of electricity and specifies that nuclear reactors that have passed the appropriate safety checks ought to be restarted. Under this policy, Kyūshū Electric Power's Sendai Nuclear Power Station, located in Kagoshima Prefecture, was restarted in August 2015 after having met the Nuclear Regulation Authority's new standards. The Abe administration has continued to maintain and support nuclear power and its policy stance on the issue has hardened.

The mass media

The discursive opportunity structure for the antinuclear power movement is constituted by the dominant mode of thought in the society. In the next two sections I examine the formation of this structure through the media environment and shifts in public opinion.

One characteristic of the post-Fukushima situation was the way information was distributed through channels that lay outside the mass media. Immediately after the accident, citizens used social media to disseminate information and reports from a perspective at odds with that of the mass media. As I discussed above, the antinuclear power movement's use of social media has also been widely noted. Nevertheless, many people obtained information about the nuclear accident, the antinuclear power movement, and the actions of the government and political parties primarily from mass media reporting in the newspapers and on television. These media can have an enormous influence on the formation of public opinion across a variety of generational and occupational categories.

However, there were major differences between newspaper companies in terms of their stance on nuclear power and they each had different ways of reporting the issue (Yamada 2013; Saitō et al. 2014). Let us examine the title of editorials concerning nuclear power that appeared in the major national newspapers (*Yomiuri Shimbun*, *Asahi Shimbun*, *Mainichi Shimbun, Nikkei Shimbun* and *Sankei Shimbun*).[2]

When Prime Minister Kan called for Japan to reduce its reliance on nuclear power in July 2011, the *Asahi* responded in an editorial with the headline, 'An agenda that ought to be embraced by the whole of government'. The *Mainichi* also adopted a positive attitude towards the speech in its editorial, 'A step in the right direction'. Taking the opposite tack, the *Yomiuri* reacted negatively, with 'Irresponsible, meaningless talk', as did the *Sankei*, with 'Ad hoc announcement fails to inspire confidence', and the *Nikkei*, with 'PM Kan's irresponsible "reduce reliance on nuclear power"' (all published July 14, 2011).

Responding to the September 2011 announcement of the Innovative Energy and Environmental Strategy, the *Asahi* called on the government to show its commitment to make the plan a reality, with 'Make zero nuclear power a reality', as did the *Mainichi*, with 'Let's make it happen'. On the other hand, the *Yomiuri*, the *Sankei* and the *Nikkei* criticized the move with their respective editorials, '"Zero nuclear power" is no strategy', 'We can't lead if we isolate ourselves from the world' and 'The national interest takes a hit' (all published September 15, 2011).

When the Abe government was elected in 2012 and announced its intention to revise the DPJ government's antinuclear power stance, the *Yomiuri* reacted positively, with 'Supporting a realistic nuclear energy policy' (January 8, 2013). The *Sankei* followed suit, with 'A welcome to realistic decision' (December 28, 2012), as did the *Nikkei*, with 'Guarantee nuclear safety to eliminate electricity shortages' (December 30, 2012). Taking the opposite stance, the *Asahi* was critical, with 'Have we learned nothing?' (December 29, 2012), as was the *Mainichi*, with 'Don't stifle the debate' (December 28, 2011). These editorials demonstrate that, of the major national newspapers, the *Asahi* and the *Mainichi* groups were in favor of reducing Japan's reliance on nuclear power, while the *Yomiuri*, *Sankei* and *Nikkei* groups all advocated for the necessity of nuclear power.

How did the newspapers report on the antinuclear power movement? In terms of the amount of coverage, there were also major differences between the papers. The *Asahi* and the *Mainichi* carried significant coverage. The papers in the *Yomiuri*, *Sankei* and *Nikkei* groups only covered it minimally, even during the 2012 peak (Yamamoto 2016). In terms of content, the *Yomiuri* and the *Sankei* presented the meeting that occurred in August 2012 between the prime minister and representatives from MCAN as an abnormal event that would have long-lasting negative consequences (*Yomiuri Shimbun*, August 23, 2012, morning edn; *Sankei Shimbun*, August 23, 2012, morning edn). The *Asahi* simply reported that the meeting took place. The *Tōkyō Shimbun*, a Tokyo regional paper that consistently reported favorably on the antinuclear demonstrations, praised the prime minister's decision to meet with the group because it would create an opportunity for him to listen directly to the concerns of a citizen group. The newspaper called on the prime minister to listen to the 'voice of the people' and suspend the restart of nuclear reactors (*Tōkyō Shimbun*, August 23, 2012, morning edn). These different approaches to reporting on the movement demonstrate the newspaper conglomerates' different positions on the direct participation of people in the political process by means of demonstrations.

Public opinion

Next I consider how ordinary people reacted to the accident at the Fukushima Daiichi Nuclear Power Station. Prior to the Fukushima accident, popular opposition to nuclear power had increased following the accidents at Three Mile Island and Chernobyl, as well as numerous smaller accidents in Japan. However, in the 1990s support for nuclear power as a form of green energy that does not emit carbon dioxide increased as part of the attempt to contain global warming (Shibata and Tomokiyo 1999). However, immediately after the Fukushima accident, a Gallup poll found that support for nuclear power had decreased, while opposition to the technology had increased around the world. In Japan, in particular, support for nuclear power decreased significantly (Endō 2012; Horie 2014).

The Japan Atomic Energy Relations Organization has been conducting regular public opinion polls on attitudes to the use of nuclear power since before the Fukushima accident. Its survey asks respondents to comment on the necessity of nuclear power from a variety of perspectives. Prior to the Fukushima accident, as many as 70% of respondents said that nuclear power was 'necessary' when they were asked, 'Is nuclear power generation necessary?' After 2011 this figure fell to less than 40%. This trend continues, even two years after the accident (Nihon Genshiryoku Bunka Shinkō Zaidan 2014: 62–3).

The *Asahi Shimbun*'s opinion polling captures changes in public opinion on nuclear power on a shorter timescale. The poll asks respondents, 'Are you for or against the use of nuclear power?' In April 2011 supporters outnumbered opponents 50% to 32%. However, by June there had been a dramatic shift, with only 37% in favor and 42% opposed. By December 2011 more than half the respondents were opposed to the technology (57%), while only 30% said they were in favor. While opposition has decreased slightly in subsequent polls, it remains steady at around 50% and opponents continue to outnumber supporters. These results suggest that negative perceptions of nuclear power have become more widespread and more firmly held as a result of the Fukushima accident.[3]

Despite these results, nuclear power is no longer the main focus of attention. In opinion polling conducted by the *Asahi Shimbun*, nuclear power and energy policy have never outranked the economy, employment or social insurance as the most important electoral issue. In the March 2013 survey, respondents were allowed to nominate up to three issues they had regarded as important in the previous year's lower house elections. Only 38% nominated nuclear power and energy policy, while 67% indicated the economy or employment, and 59% said pensions and social insurance (*Asahi Shimbun*, April 18, 2013, morning edn). In a survey conducted between May and June 2013, respondents were invited to nominate any and all of the policy issues they regarded as important during the upcoming upper house elections. Only 30% nominated nuclear power and energy policy (4% as their most important issue), while 76% nominated the economy and employment (38% as the most important), and 60% nominated social insurance and social welfare (17% as the most important) (*Asahi Shimbun*, June 26, 2013, morning edn). Considering that this survey lifted the restriction on the number of issues respondents could nominate, it seems that people were no longer as interested in the issue of nuclear power. Considered together, these opinion polls suggest that while many people do not regard nuclear power as necessary, their concern did not extend so far as to make nuclear energy into a major electoral issue.

Attitudes towards nuclear power

In this section I examine attitudes towards the antinuclear power movement and opinions on nuclear power more generally based

Figure 5.1: Attitudes to the antinuclear power movement

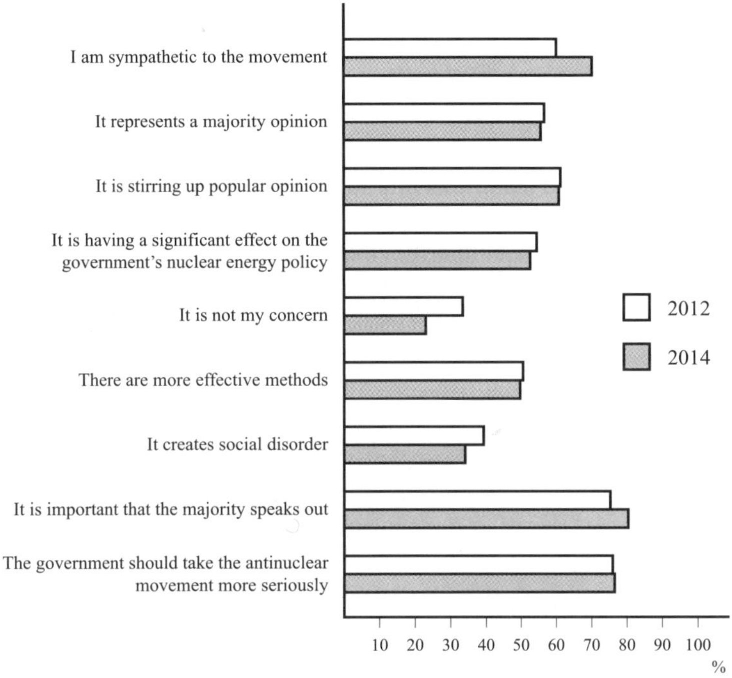

on the results of an online survey I conducted in October 2012 and December 2014. The survey targeted men and women aged 20–69 located all over Japan. I obtained 2,063 responses in 2012 and 521 in 2014. The October 2012 survey took place just after the protests outside the prime minister's residence and the Noda administration's announcement of the Innovative Energy and Environmental Strategy. It was a period when the nuclear issue was receiving considerable attention from ordinary people. By contrast, the December 2014 survey was carried out just after the LDP's overwhelming victory in the lower house elections when public support for abolishing nuclear power was already on the decline. By comparing these two surveys, we should be able to capture the way public opinion on the abolition of nuclear power has changed over time.

Figure 5.1 shows the proportion of respondents who had a favorable opinion of the antinuclear power movement. Despite the movement's loss of momentum between 2012 and 2014, the responses were mostly unchanged: about 50%–60% felt that it 'represents a majority opinion', 'is stirring up popular opinion' and 'is having a significant effect on the government's nuclear energy policy'. Favorable responses to the statement 'I am sympathetic to the movement' increased from 60.4% to 70.6%, and there was a decline in the number of people who felt that the movement 'is not my concern' (from 34.2% to 23.2%). This suggests that, despite having received less public attention in this period, the movement still had popular support.[4]

At the same time, 50% of respondents agreed with the statement, 'There are more effective methods'. This suggests they did not accept that demonstrations and other forms of protest were the most effective method of advocating against nuclear power. Furthermore, 30%–40% felt that the movement 'creates social disorder'.

In addition, as many as 80.6% agreed that 'It is important that the majority speaks out' and that 'The government should take the antinuclear movement more seriously'. The findings demonstrate that, as well as questioning the need for nuclear power, respondents supported people expressing their opinions on this issue.

Next I consider people's attitudes to nuclear power. The differences in stance between political parties and newspaper conglomerates discussed above also had an influence on people's opinions on nuclear power. Table 5.1 shows people's attitudes to nuclear power in relation to their support for particular political parties and their preferred newspapers.[5]

First, let us examine disagreement with the statement 'It is inevitable to rely on nuclear power'. There was little difference between the two surveys, with just less than 50% disagreement in both. However, opinions did diverge based on support for a political party and preferred newspaper. Approximately 70% of Japanese Communist Party supporters disagreed with the statement in both surveys. In 2014 a large percentage of DPJ supporters, 72.1%, also disagreed, while 48.3% disagreed in 2012. By contrast, fewer LDP supporters disagreed with this statement in 2012 (35.6%) or in 2014 (24.8%). The increase in disagreement among DPJ supporters and the decrease among LDP supporters over the survey period indicate that opinions on nuclear power have become more clearly linked to support for a particular political party.

With regard to preferred newspapers, a large number of *Asahi* and *Mainichi* readers disagreed, while few *Yomiuri*, *Sankei* and

Nikkei readers did so. The different editorial stances adopted by the newspaper conglomerates are also visible in their readers. Between the two surveys, negative responses among *Asahi* readers increased from 59.9% to 71.2% and among *Mainichi* readers from 50.0% to 66.7%. Conversely, negative reactions to the statement decreased for *Sankei* readers from 36.5% to 15.0% and for *Nikkei* readers from 40.7% to 22.2%. Attitudes to nuclear power became more clearly linked to preferred newspaper. *Yomiuri* readers were the exception, with a relatively low 40.1% of people disagreeing with the statement in 2012, increasing to 51.2% in 2014.

At the same time, 70.1% of all respondents agreed with the statement 'Restarting nuclear reactors needs a popular mandate' in 2012, a figure that increased slightly to 76.2% in 2014. More respondents considered that reactor restarts required a popular mandate than were opposed to reliance on nuclear power. In terms of support for a political party, fewer LDP supporters and those who did not support any party agreed with the need for a popular mandate for restarts. Even so, more than 60% of them agreed with this statement in both surveys. This percentage increased in 2014. Fewer *Sankei* readers agreed that a popular mandate was necessary (55%), but about 70% of readers of all the other newspapers did.

As discussed above, agreement with the statement 'I am sympathetic to the antinuclear power movement' increased in the 2014 survey. The number of Japanese Communist Party supporters who were sympathetic to the movement was particularly high (80%). Among LDP supporters and those who do not support any party, there was less sympathy (in the 50%–60% range). However, the percentage of DPJ and LDP supporters who were sympathetic to the movement increased in 2014. In terms of preferred newspaper, sympathy for the movement was high among *Asahi* and *Mainichi* readers (60%–80%) in contrast to a relative lack of sympathy among *Yomiuri*, *Sankei* and *Nikkei* readers (40%–70%). These results mirror the trend observed with regard to attitudes to reliance on nuclear power. There seems to be a definite connection between stance on nuclear power and both political party support and preferred newspaper.

These results suggest that, while there is definite support for the antinuclear power movement and skepticism regarding the necessity of nuclear power, there is a detectable difference in stance depending on political party support and preferred newspaper.

Table 5.1: Attitude to nuclear power by political party and preferred newspaper (%)

	It is inevitable to rely on nuclear power (Disagree)		**Restarting nuclear reactors needs a popular mandate** (Agree)		**I am sympathetic to the antinuclear power movement** (Agree)		***N***	
	2012	**2014**	**2012**	**2014**	**2012**	**2014**	**2012**	**2014**
Political party								
Liberal Democratic Party (LDP)	35.6	24.8	67.9	73.9	50.2	66.1	452	165
Komeito	57.6	52.4	83.1	76.2	71.2	76.2	59	21
Democratic Party of Japan (DPJ)	48.3	72.1	77.1	86.9	69.7	88.5	201	61
Japan Restoration Party	53.4	50.0	78.1	87.9	68.5	75.9	292	58
Social Democratic Party	83.3	100.0	77.8	83.3	88.9	100.0	18	6
Japanese Communist Party	67.6	76.5	77.5	86.3	78.9	86.3	71	51
Your Party	58.2	–	82.9	–	70.0	–	170	–
People's Life First	71.8	–	69.2	–	69.2	–	39	–
Other	33.3	18.2	61.9	72.7	38.1	45.5	21	11
No support for a political party	47.0	52.0	61.9	66.2	55.7	60.8	740	148
Preferred newspaper								
Yomiuri Shimbun	40.1	51.2	67.4	79.8	53.7	73.8	307	84
Asahi Shimbun	59.9	71.2	76.1	91.8	72.2	84.9	327	73
Mainichi Shimbun	50.0	66.7	69.7	73.3	64.5	80.0	76	15

Sankei Shimbun	36.5	15.0	55.4	55.0	37.8	45.0	74	20
Nikkei Shimbun	40.7	22.2	68.3	66.7	49.7	69.4	167	36
Regional or local newspaper	49.8	43.2	70.3	81.5	67.8	74.1	472	81
Other	50.0	44.7	80.0	76.3	75.0	71.1	20	38
Rarely/never read a newspaper	47.7	46.6	70.2	70.1	56.6	63.8	620	174
Total	48.1	47.8	70.1	76.2	60.4	70.6	2,063	521

However, there was little difference in terms of political party support or preferred newspaper on the question of the need to obtain a popular mandate for reactor restarts. The majority of respondents agreed with this statement. Support for a popular mandate only increased between 2012 and 2014, as did growing sympathy towards the antinuclear power movement. In the survey of attitudes to the antinuclear power movement, there was significant agreement that speaking out is important and that the government should take the movement more seriously.[6]

These results suggest that people want a system where decisions about nuclear power are made not by the sectional interests that are represented by the so-called nuclear village, but through a political system that reflects the will of the people. This continues to be the case even under the pronuclear LDP administration.

Conclusion

In this chapter I have considered the way the antinuclear power movement, the government and political parties, the mass media and public opinion responded to the Fukushima Daiichi nuclear accident. I have examined the shifting debate on nuclear power in the relationship between these spheres.

Prior to the Fukushima nuclear accident, nuclear power was gaining traction as a form of green energy and there was significant popular support for the technology. However, the nuclear accident changed the situation dramatically. A vigorous antinuclear power movement emerged and attracted widespread notice for its new style. At the same time, there was growing public support for the abolition of nuclear power. Clear antagonisms also appeared within the political process between pro- and antinuclear power factions. Among the latter were Diet members who actively allied themselves with the antinuclear power movement. Collectively, these responses produced political and discursive opportunity structures that were open to the antinuclear power movement.

The debate on nuclear power reached its peak between June and September 2012, when the Ōi Nuclear Power Station was restarted and the government released the Innovative Energy and Environmental Strategy. The nuclear reactors at Ōi were the first to be restarted after the Fukushima accident. As such, the restarts were of great concern to both pro- and antinuclear power factions.

The growth of the antinuclear power movement led to weekly protests outside the prime minister's residence, attracting up to 200,000 participants. Nevertheless, the Ōi Nuclear Power Station was ultimately restarted and Cabinet failed to approve the zero nuclear power plan. Despite a powerful display of the popular will, the movement was unable to have a concrete effect on the policymaking process.

During the same period, the DPJ experienced deepening internal conflict over the consumption tax hike and a number of antinuclear Diet members left the party. As a result, the antinuclear movement lost its allies within the governing political elite. Furthermore, the consumption tax hikes directly impacted people's everyday consumption habits and the economy. These issues have come to be more important than nuclear power to the majority of people. As a consequence, the antinuclear movement lost momentum. After the summer of 2012, the political and discursive opportunity structures were closed to the movement.

It would be possible to interpret this process in terms of the issue-attention cycle, whereby people lose interest in an issue as it comes into competition with other issues. However, as at July 2016 there have still been no major nuclear reactor restarts and opinion polls continue to show strong support for the abolition of nuclear power. After the Fukushima nuclear accident, opposition to nuclear power has become much more widespread. While it remains to be seen how the debate will unfold, we may have arrived at a new equilibrium, as suggested by the punctuated equilibrium model. If so, then the will of the people has become an essential component of future nuclear and energy policy.

As was demonstrated in the survey I conducted, even though opinion remains divided as to the necessity of nuclear power, the majority supports the idea that popular consent for reactor restarts is necessary. This suggests that people do not want decisions regarding nuclear power to be monopolized by sectional interests but to be decided through a transparent and democratic process. People recognize the importance of citizens speaking out as they have done through the antinuclear power movement. In the 2014 survey, when LDP rule had stabilized, support for the idea of popular involvement in the political process was clearer than it had been at the height of the antinuclear movement in 2012. These results indicate that the issue of nuclear power acted as a catalyst for the expression of long-held popular distrust and dissatisfaction with the monopolization of the political process by a minority of elite actors.

This was further demonstrated by the appearance and rapid growth of movements opposed to a package of national security-related legislation in the summer of 2015. Despite serious concerns that the national security-related legislation, which incorporates the right to collective self-defense, is unconstitutional, the legislation successfully passed through the Diet with the majority support of the ruling coalition. In response, approximately 120,000 people demonstrated outside the Diet on August 30 in defense of the principle of constitutional government (*Asahi Shimbun*, August 31, 2015).

The issue of nuclear abolition is no longer able to command the attention it once did, making it difficult to realize in practice. However, the ability of people to speak out and voice their will has increased. In this sense, the discursive opportunity structure is open. We must now watch with interest to see whether this will provide the basis for energizing Japan's social movements, increasing their role as advocates and transforming civil society in Japan.

6 Nuclear Damage Compensation: Mechanisms for Dispute Resolution

Ōkura Sae and Kubo Yoshiaki

Disasters destroy both lives and livelihoods, and appropriate support and aid are necessary to ease the burden on victims. However, when a corporation or individual is implicated in the causation or exacerbation of a disaster, a conflict of interest arises between perpetrators and victims with regard to compensation and reparations. The failure to resolve such conflicts can cause additional hardship for victims and prolong their suffering. In this chapter we discuss nuclear damage compensation after the Fukushima nuclear accident. We examine the mechanisms that were available to resolve disputes between the perpetrators (the Tokyo Electric Power Company (TEPCO) and the national government) to see how the conflicting interests have been reconciled and the effect different mechanisms had on the aid that victims received.

After the Fukushima accident on March 11, 2011, the government allowed TEPCO to continue operating. In order to compensate victims, it activated the provisions of the *Act on Compensation for Nuclear Damage* (1961 Compensation Act). Under the Compensation Act, TEPCO assumes unlimited and exclusive liability for nuclear damage without the need for proof of negligence or fault (Article 4, paragraph 1). The Act also compels nuclear operators to conclude an indemnity agreement for compensation of nuclear liability with the government (Articles 6 and 7). Under this agreement, TEPCO was required to lodge 120 billion yen with the government in anticipation of any future nuclear damage claims. The government is responsible for providing financial assistance to TEPCO to meet its obligations where they exceed this amount (Article 16, paragraph 1). As of January 2015, TEPCO's total liability for the 2.185 million claims arising from the disaster stood at 4.4604 trillion yen.[1] The government has contributed the 120 billion yen lodged with it under the terms of the government indemnity agreement and a further 4.5337 trillion yen of financial assistance under the terms of the Compensation Act.

Endō (2013) praised the rapid implementation of the compensation scheme under the Compensation Act. However, others have criticized the scheme for failing to clarify where liability for nuclear damage actually lies (Ōshima and Yokemoto 2012; Yokemoto 2013a; Honma 2014). Scholars of administrative law have sought to establish the extent of the state's responsibility for nuclear damage compensation (Isono 2011; Ōtsuka 2011; Kojima 2011; Hitomi 2011; Morishima 2011a, 2011b, 2011c; Harada 2013; Shimoyama 2014). In this chapter, we focus instead on the government's involvement in compensation mechanisms.[2]

The Tōkaimura nuclear accident and Justice System Reform

TEPCO's assumption of exclusive and unlimited liability under the Compensation Act means that nuclear damage compensation is dealt with under civil law. However, the government has intervened in the compensation procedures in two important ways. It has invested significant resources in developing a number of mechanisms whereby victims can access compensation. Where those resources are deployed is a matter of considerable interest. First, in order to mediate settlements between TEPCO and disaster victims, the government established the Dispute Reconciliation Committee for Nuclear Damage Compensation under the provisions of the Compensation Act. This committee established a set of guidelines defining the scope of TEPCO's liability for nuclear damage (Article 18, paragraph 2, point 2). The Dispute Reconciliation Committee was granted the ability to issue these guidelines by a 2009 amendment to the Compensation Act.

The only previous case where the government has established a Dispute Reconciliation Committee was after the 1999 Tōkaimura[3] nuclear accident. At that time, approximately 7,000 claims for compensation, worth a combined total of approximately 15.4 billion yen, were ultimately lodged. Only two applications were made for mediation through the Dispute Reconciliation Committee (0.03 % of the total number of claims), neither of which was settled. The guidelines for determining the scope of liability for nuclear damage were determined by a private organization, the Nuclear Damage Investigation Study Group.[4] The 2009 amendment to the Compensation Act based on the lessons of the Tōkaimura nuclear accident gave the Dispute Reconciliation Committee responsibility for setting general guidelines for nuclear damage compensation (Yaginuma 2009).[5]

The government also established the Nuclear Damage Compensation Dispute Resolution Center (ADR Center), under the Dispute Reconciliation Committee's umbrella. The purpose of the ADR Center was to provide a mechanism for alternative dispute resolution (ADR) outside the court system. The Justice System Reform[6] adopted a number of goals to facilitate greater and timelier access at all levels of the judicial system, recognized ADR as one possible mechanism for achieving this and designated it for expansion as part of the reform process. With the establishment of the ADR Center, disaster victims had access to three mechanisms to obtain compensation for nuclear damage after Fukushima. They could apply directly to TEPCO, apply to the ADR Center for assistance in obtaining a mediated settlement, or sue TEPCO and/or the government in court.

The operation of these mechanisms in the post-Fukushima context presents an opportunity to reflect on the amendments made to the Compensation Act after the Tōkaimura nuclear accident and on the Justice System Reform process. In this chapter we consider the public policy implications of nuclear damage compensation. We focus on the relationship between the mechanisms available for accessing compensation and the aid that victims ultimately received. As an organization, the ADR Center had limited resources for dealing with requests for mediation. The same could be said of TEPCO, as a for-profit corporation, and of the court system. Government agencies such as the ADR Center are emblematic of a more basic problem. If an organization lacks the resources to keep up with the quantity of administrative work it is required to perform, then the quality of the work will suffer (Lipsky 1980). In the case of nuclear damage compensation, a lack of organizational resources might lead to delays in access to compensation or make it more difficult for victims to apply for compensation in the first place. What level of resources was allocated to each of the three compensation mechanisms after

Fukushima? What influence did the level of resources have on the aid victims received? In addition to investigating the functioning of each of the compensation mechanisms, this chapter also considers the more general question of how resources are allocated in judicial and quasi-judicial systems in Japan.

The chapter is structured as follows. In the next section we present a quantitative analysis of the way victims utilized each of the three compensation mechanisms: TEPCO, the ADR Center and the courts. Then, in the following section, we discuss the results of a survey of legal professionals who represented disaster victims after Fukushima. We explain what they thought of the mechanisms available to their clients for accessing compensation. Finally, we discuss our conclusions.

Mechanisms for accessing nuclear damage compensation

In this section we use quantitative analysis to show how victims used each of the three mechanisms for obtaining compensation. We then examine how each of the mechanisms was established and how its staffing levels changed over time.

Estimating the total cost of compensation

Not all of the damage caused by a disaster can be repaired using money alone. It is, therefore, difficult to grasp the total cost of a disaster. However, it is possible to gain a sense of the scale of these costs by looking at the total amount of compensation paid out and the total number of claims.

Figure 6.1 shows both the actual amount of compensation TEPCO has paid out so far and an estimate of the final cost. The latter estimate was made by TEPCO, based on the Dispute Reconciliation Committee guidelines, as well as other sources. TEPCO has been paying out an ever greater sum since it first began accepting claims in 2011. In December 2014 the total stood at 4.5657 trillion yen. The company estimates that the final compensation bill will reach 5.4214 trillion yen.

Although lack of space prevents it from being shown in Figure 6.1, the total number of claims for compensation had reached 2.185 million by January 23, 2015. Approximately 1.919 million of these claims are from individuals (including from people who evacuated

Figure 6.1: Actual and estimated costs of nuclear damage compensation

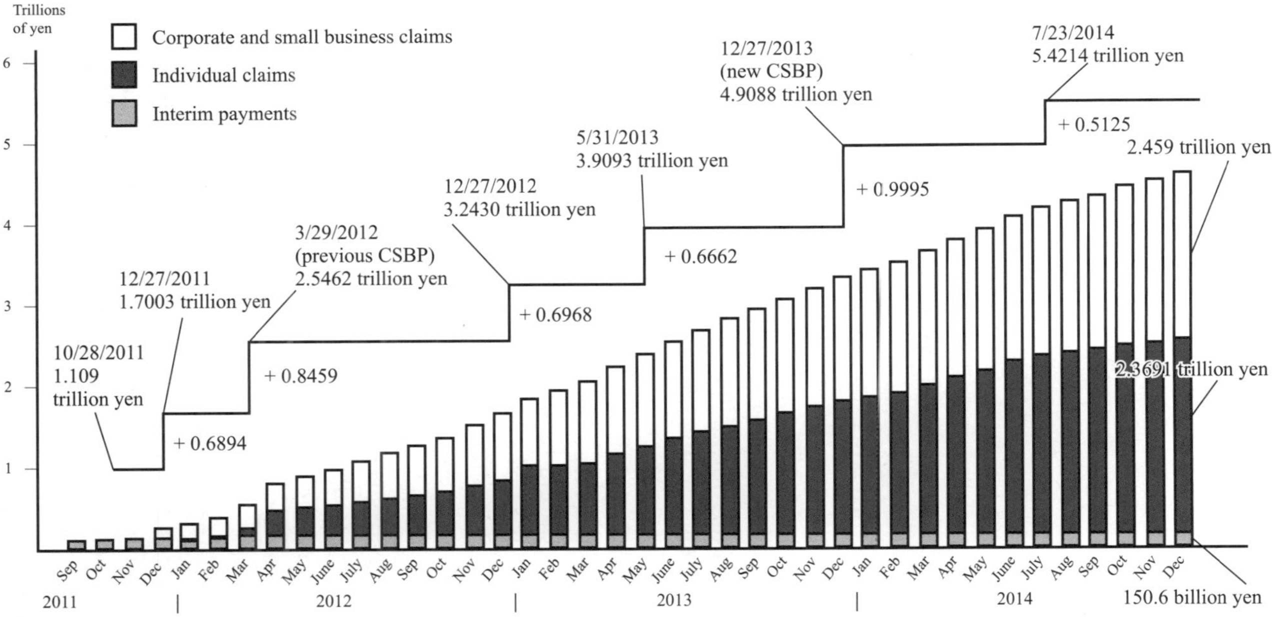

Note: CSBP = Comprehensive Special Business Plan. Line chart shows the estimated cost of nuclear damage compensation. Bar chart shows actual expenditure on nuclear damage compensation.
Source: TEPCO 2015: 3.

voluntarily without government order). A further 266,000 are from corporations and small businesses.

Utilization rates

How many claimants have utilized each of the three mechanisms in order to obtain compensation? The ADR Center is the only one of the three mechanisms to have publicly released data on the number of claims received. According to these figures, the ADR Center received 521 claims in 2011, 4,542 in 2012, 4,091 in 2013 and 5,217 in 2014 for a total of 14,371 claims. The parties have reached a settlement in about 80% of cases since 2013. The number of claimants involved in each case has also increased year on year with 1,206 in 2011, 12,055 in 2012, 25,914 in 2013 and 29,534 in 2014.[7] This is due to an increase in the number of group claims.

Neither TEPCO nor the courts have released data on the number of claims for nuclear damage compensation they have received. However, we can estimate the number of nuclear damage compensation cases before the courts by using the judicial system statistics published by the Supreme Court of Japan. If there has been a large number of lawsuits related to nuclear damages, we should be able to see an increase in the number of lawsuits in the three affected prefectures (Iwate, Miyagi and Fukushima, all located in Tōhoku) and in Tokyo, where many evacuees are now living. Figure 6.2 shows the number of civil and administrative suits lodged each month in the jurisdictions of the Sendai District Court (Miyagi Prefecture), the Fukushima District Court (Fukushima Prefecture), the Morioka District Court (Iwate Prefecture) and the Tokyo District Court (these figures include cases undergoing court-assisted mediation). In the three affected prefectures, the number of suits filed after the disaster actually decreased. In Tokyo, too, which hosts many evacuees, there was a slight reduction up until the end of 2012. Commenting before a Diet committee in August 2012, Tokura Saburō, then director-general of the General Secretariat of the Supreme Court, suggested that the reason relatively few nuclear damage-related lawsuits had been filed during this period was that plaintiffs' counsel were still preparing cases. He anticipated an increase in the number of cases coming before the courts. He also suggested that some claims

Figure 6.2: New civil and administrative court cases

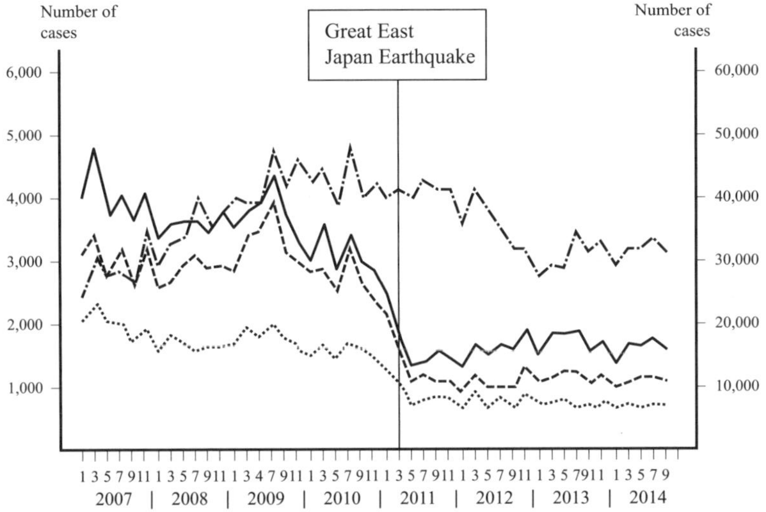

—— Sendai District Court jurisdiction (left-hand axis)
······· Morioka District Court jurisdiction (left-hand axis)
---- Fukushima District Court jurisdiction (left-hand axis)
—·— Tokyo District Court jurisdiction (right-hand axis)

Source: Based on Monthly Judicial Statistics.

currently undergoing mediation through the ADR Center would be likely to end up in the courts where the parties failed to agree to a settlement. Figure 6.2 shows that after 2012 the decline in the number of suits filed in Tokyo came to a halt in 2013, and there was a slight increase in the number in Sendai. Nevertheless, it is still difficult to identify a definite upward trend.

Now let us estimate the number of claims for compensation that were made directly to TEPCO. By January 23, 2015, TEPCO had received 2.185 million claims for compensation. The ADR Center, by contrast, had only received 14,371 up until the end of 2014. Based on these figures, the utilization rate for the ADR Center is just 0.66%. If we look at the court system, approximately eight million new civil suits were filed in the three affected prefectures and Tokyo in the 42 months prior to the disaster (April 2007 – September 2010). This is greater than the 5.75 million filed in the 42 months after the disaster (April 2011 – September 2014). It seems that the vast majority of claims for compensation have been made directly to TEPCO.[8]

Establishing the compensation system

Next we examine how TEPCO, the ADR Center and the court system prepared for the large number of compensation claims that were expected after the disaster.

TEPCO[9]

On April 28, 2011, TEPCO established the Fukushima Nuclear Compensation Consultation Room to oversee claims for nuclear damage compensation. In addition to its headquarters office staff, it incorporated a Compensation Operation Center to receive new claims, a Compensation Center (call center) to provide advice and receive new claims, and a series of Regional Compensation Centers to provide advice to local residents in person. By June 20 TEPCO had assigned 50 employees to head office, 300 to the compensation management center, 250 to the call center and 400 to the regional centers, for a total of 1,000 employees.

The centers began accepting claims from individuals in August and from businesses in September. However, the lengthy and complex application form for making a claim led to widespread confusion. TEPCO responded by increasing total staffing levels in the Fukushima Nuclear Compensation Consultation Room to 7,600. However, this was still insufficient to deal with claims promptly. On December 8, 2011, TEPCO admitted publicly to the delay in compensation payments. The company promised to address the issue by assigning more staff to deal with compensation claims and improving its processes for handling claims.[10] By April 2012 the company had assigned 13,100 staff to handle compensation claims.

On January 1, 2013, the company established the Fukushima Revitalization Headquarters, and all of its Fukushima operations were placed under this umbrella. The Fukushima Nuclear Compensation Consultation Room was included in the new headquarters and restructured into two subdivisions, the Compensation Promotion Unit and the Compensation Consultation Unit. The latter incorporated the call center and the regional centers. Approximately 50 staff

Table 6.1: The ADR Center workforce

	2011	2012	2013	2014
General Committee (*Sōkatsu Iinkai*)	3	3	3	3
Mediators	128	205	253	283
Secretariat	62	203	347	353
Research clerks (*Chōsakan*)	28	91	193	192
Administrative staff	34	112	154	161
(Located in the Fukushima office)	(8)	(25)	(26)	(28)

Note: These figures are based on the number of employees at the end of each year.
Source: ADR Center 2015.

members from the legal department who were responsible for handling TEPCO's involvement in the ADR Center process were also transferred to the Compensation Promotion Unit. The number of employees assigned to the Fukushima Nuclear Compensation Consultation Room peaked in 2012, and staffing levels were later reduced. Most of the cuts were made in the call center and in the Compensation Promotion Unit (the former Compensation Operation Center).

While staffing levels in the Compensation Consultation Unit (the former Regional Compensation Centers) have remained at around 2,000, in 2014 TEPCO once again increased staffing levels in the headquarters office and in the Compensation Promotion Unit. As of July 1, 2014, 250 staff members were assigned to the headquarters office, 500 to the call center, 2,000 to the Regional Compensation Centers and 7,000 to the Compensation Promotion Unit for a total of 9,750 staff.

Based on these changing staffing levels, we can surmise that the bulk of claims for nuclear damage compensation were submitted between the end of 2011 and the beginning of 2013. Then, in mid-2014, there was a second wave of compensation claims.

The ADR Center

Table 6.1 shows the number of staff employed by the ADR Center, including on the general committee (*Sōkatsu Iinkai*), as mediators, and in the secretariat (including research clerks (*Chōsakan*) and administrative staff). The number of mediators working for the ADR Center continued to grow until 2014, and the number of research clerks grew until 2013. The steady growth in the ADR Center proceeded as follows.

On April 11, 2011, the Ministry of Education, Culture, Sports, Science and Technology (MEXT) established the Dispute Reconciliation Committee for Nuclear Damage Compensation. On July 22 a number of additional 'special members' were appointed to the committee to assist with the large number of anticipated requests for mediation. Under the Compensation Act, designated Dispute Reconciliation Committee members or special members can be appointed as mediators to assist the parties to a nuclear damage compensation claim to reach a mediated settlement. The government revised the regulations governing the operation of the Dispute Reconciliation Committee prior to the establishment of the ADR Center in August to provide a mechanism to achieve mediated out-of-court settlements for nuclear damage compensation claims. The ADR Center consists of a general committee, a panel of lawyers who can act as mediators in the mediated settlement process, and a secretariat located in the Research and Development Bureau of MEXT. The secretariat provides administrative support to the mediators. Research clerks attached to the secretariat assist mediators by carrying out any investigations that are needed as part of the mediation process.

By December 2011 the ADR Center had employed 130 part-time mediators and 30 research clerks. However, despite the ADR Center's initial goal of resolving disputes within three months, by February 2013 settlements were taking an average of more than six months. During that time, the number of staff assigned to the ADR Center was increased, and efforts were made to simplify the process. The ADR Center also published a set of general standards and a number of settlements to serve as examples in future cases. In 2012 a second office was established. The number of mediators grew to approximately 200 and the number of research clerks to approximately 100. In 2013 there were 250 mediators working in the ADR Center, in addition to the 200 research clerks. In 2014 there were 280 and 200 respectively. As a result, the unresolved case load for each research clerk decreased from more than 30 in 2012 to fewer than 15 cases in 2013. In April 2014 the ADR Center announced that it had appointed all the additional research clerks that it needed for the time being.

The number of staff at the ADR Center was increased in response to the growing number of applications for mediation and the center's inability to reach settlements in a timely fashion. A comparison of these changes in staffing levels with those taking place at TEPCO shows that the system expanded over approximately the same period of time. However, TEPCO later reduced the number of staff it had handling compensation claims, while staffing levels at the ADR Center have remained stable since 2014.

The courts

Staffing levels in the court system are determined by the *Act Concerning the Limit of Number of Court Officials (1951).* In 2010, before the Fukushima accident, the court system employed eight presidents of the high courts, 1,782 judges, 1,000 assistant judges, 806 summary court judges and 22,089 administrative staff (not including court execution officers, casual staff, people on holiday or those who were employed for a period of less than two months). Since then, the number of judges and administrative staff has changed. In 2011 the number of judges was 1,827. This increased to 1,857 in 2012, 1,889 in 2013 and 1,921 in 2014. The number of administrative staff decreased from 22,089 in 2011, to 22,059 in 2012, 22,026 in 2013 and 21,990 in 2014.

However, the number of judges was already increasing before the accident. What impact, if any, did the earthquake and the nuclear disasters have on judicial appointments? Discussions concerning the nuclear accident took place in the Justice Committees of both houses of the Diet during the annual deliberations to vary staffing levels under the Court Officials Act in both 2011 and 2012.[11] In 2012 Tokura Saburō discussed the measures adopted by the Supreme Court to deal with the issue in some detail. He told the committees that the Supreme Court had begun preparing for an anticipated increase in litigation arising from the disaster immediately after it had occurred. This involved allocating 'a considerable number' of additional court clerks and secretarial staff to the affected prefectures and the creation of one-stop shops in some of the larger courts, such as Sendai, where people could access family court, district court and the summary court services. He also explained that judicial appointments to the Tokyo District Court would be prioritized in anticipation of an increase in nuclear-related litigation.[12] According to data we obtained through a freedom of information request, the number of judges assigned to the Tokyo District Court increased from 369 in 2011 to 383 in 2012.[13] However, in 2013 the Supreme Court reduced the number of judges allocated to the Tokyo District Court to 372. There was no discussion of the nuclear disaster in the Diet committees' consideration of court personnel in 2013. It seems that while the Supreme Court did allocate additional human resources to deal with an anticipated increase in disaster-related litigation, by 2013 it began to wind back these additional allocations.

In summary, the data suggests that while TEPCO and the ADR Center increased their capacity in response to an increase in the number of claims, there was relatively little disaster-related litigation, so the court system reduced its disaster preparedness. The three organizations that could provide nuclear damage compensation – TEPCO, the ADR Center and the court system – each adjusted their staffing levels in response to the changing number of claims they received at different times.

Evaluation of the compensation system

What explains the disparity between the larger number of claims received by TEPCO and the ADR Center compared with the courts? In this section we look at the progress of nuclear damage-related litigation through the courts, based on a survey of legal teams representing the plaintiffs in a number of nuclear damage lawsuits.[14] We consider the relationship between these lawsuits and the dispute resolution mechanism provided by the ADR Center. We also consider the impact of the 2009 amendments to the Compensation Act and the Justice System Reform on nuclear damage compensation after the Fukushima accident.

Legal defense, numbers of plaintiffs and adversaries

Practicing lawyers working on nuclear damage compensation claims generally serve two main functions. They represent claimants during the ADR Center mediation process and represent nuclear damage victims as plaintiffs if they choose to take TEPCO and/or the government to court. We surveyed 16 legal teams working on nuclear damage cases who represented victims in both ADR claims and litigation, seven who worked on ADR claims alone and two who were only involved in litigation. Based on the responses from the 18 groups that represented plaintiffs in nuclear damage-related litigation, we can make the following observations about the progress of these suits.

Most of the litigants represented by the respondents were located in the Kantō and Tōhoku regions. The legal teams were involved in seven suits that involved 60 or fewer plaintiffs, six involving between 61 and 300 plaintiffs, and five involving more than 300 plaintiffs. Of these, one team representing more than 300 plaintiffs was based in Kantō, one was in the Tōhoku/Hokkaidō area (excluding Fukushima) and three were in Fukushima Prefecture. The class actions in Fukushima were particularly large, with 3,865 plaintiffs, 872 plaintiffs and 473 plaintiffs involved in the three cases.

All the suits were brought against TEPCO, with 16 also naming the government as a defendant. All sought general damages, with eight suits also seeking specific damages for the cost of replacing lost residential property, buildings and household belongings, and to cover evacuation costs (not including court costs and lawyers' fees). Some suits also asked the defendants to restore the plaintiffs' home towns to their original states (that is, to the uncontaminated state before the accident).

Timing of the litigation

The timing of the litigation is significant. None of the plaintiffs filed suits in 2011. The first suit was filed in 2012, followed by 11 in 2013 and six in 2014. Clearly, 2013 was the peak year for nuclear damage litigation. Furthermore, if we look at how these suits progressed over the same period, while there were no suits before the courts in 2011, there was one in 2012, 12 in 2013 and 15 in 2014. The number of cases before the courts increased over time.

If we examine the timing of the suits by region, a pattern emerges. The first was filed in the Fukushima District Court in 2012. In 2013 suits were filed in nearby regions including the Sapporo District Court, the Yamagata District Court and the Sendai District Court; in the Yokohama and Chiba District Courts in the Kantō region; and in the Nagoya, Kyōto and Kōbe District Courts in the Kansai and Tōkai regions. Finally, in 2014, suits were filed in the Okayama, Hiroshima and Fukuoka District Courts in the Chūgoku and Kyūshū regions. As time passed, there was a steady geographical expansion radiating out from the affected prefectures.

The majority of lawsuits began in 2013. We found evidence of seven lawsuits related to the September 1999 Tōkaimura nuclear accident that had been filed by February 2003. Of these, three were filed in 2002 (CRIEPI 2003; MEXT 2008). The fact that there were still cases before the courts in 2014 suggests that there has been little improvement in terms of the smooth operation of the legal system since the Tōkaimura nuclear accident.

Complementarity of the courts and the ADR Center

Why did it take until 2013 for the majority of legal proceedings to appear before the courts despite the establishment of the ADR Center? In order to explore this question, we first examine the relationship between litigation and the ADR process.

In the survey, legal teams were asked about whether they had recommended their clients to pursue their claims through the ADR Center or through the court system in the first instance. Ten groups responded that they did not specifically advise either process. Eleven recommended the ADR Center process and three recommended the courts. It seems that most lawyers did not recommend that claimants pursue their claims through the court system in the first instance.

In that case, why did some claimants choose to go to court? One legal team gave the example of a client who had lost his wife to suicide in the aftermath of the disaster:

> [On first meeting him] I felt like he was completely lost. It was as if he could not grasp the reality of what had happened, the accident, the evacuation, and the death of his wife. I really felt that by fighting the matter in court, he wanted to reclaim a part of himself that was lost. The lawsuit performed what psychologists call 'the work of mourning'.

Litigation was connected with the victim's emotional recovery. Another team in western Japan that recommended legal action said that it hoped 'to further a national movement towards resolution of the compensation issue'. This suggests an understanding of legal action as a form of social movement activism.

These examples suggest that, from the perspective of emotional recovery and social movement activism, litigation had a particular significance for some victims. Nevertheless, the fact that most lawyers recommended that their clients pursue their claims for compensation though the ADR Center shows that, in other ways, the ADR process did function as a substitute for legal action in the courts. The court system and the ADR Center were complementary.

Delays at the ADR Center

Lawyers who had recommended the ADR Center process noted that 'in the beginning, it took more than a year to reach a settlement through the ADR Center. Furthermore, at times the mediators displayed little sympathy or understanding for the victims of nuclear damage. This made the process difficult.' Although the ADR Center was intended to speed up the provision of aid to disaster victims, some people were critical of the ADR Center's handling of claims in the immediate aftermath of the disaster. We would like to suggest two reasons for this.

First, the ADR Center had no power to compel the parties to a settlement. Therefore, if TEPCO did not accept the settlement, payments were delayed. On July 5, 2012, the general committee released a public statement in which it criticized TEPCO for failing to adhere to the spirit of the dispute resolution process, leading to lengthy delays for victims in receiving appropriate compensation. TEPCO responded publicly on July 6, promising to take the general committee's findings seriously and ensure that the settlements reached through the ADR process would be handled appropriately.

Second, the guidelines for determining the level of compensation were unclear. The ADR Center published the amount of compensation to which victims were entitled based on two sets of guidelines: the *Interim Guidelines on Determination of the Scope of Nuclear Damage Resulting from the Accident at the Tokyo Electric Power Company Fukushima Daiichi and Daini Nuclear Power Plants* issued by the Dispute Reconciliation Committee (2011a) and the *General Standards* established by the ADR Center's own general committee on February 14, 2012 (ADR Center 2012). Under these two sets of guidelines, the amount of general damages payable for deaths that took place after evacuation was calculated in two parts. A standard base amount was payable in all such cases. Mediators then added an attributable amount to the base amount based on an estimate of the degree of influence the nuclear accident had on the person's death. However, an internal document that was leaked to the media showed that mediators were routinely estimating the attributable amount at 50% of the total possible (*Mainichi Shimbun*, August 30, 2014, electronic edn). Tanaka Masa'aki, then deputy director-general of MEXT, claimed that the document was simply a summary of an internal discussion among some mediators.[15]

Both these problems originated in the power to establish guidelines for nuclear damage compensation that were given to the Dispute Reconciliation Committee as part of the 2009 amendments to the Compensation Act. These amendments were intended to mediate settlements under administrative law (Takahashi 2014: 93), to reduce the burden of proof on injured parties under the civil law and to give the state power (Advisory Committee on Nuclear Damage Compensation System 2008). However, there was confusion concerning the implementation of the guidelines in practice, contributing to delays in resolving claims for compensation through the ADR Center. This is why the majority of nuclear damage lawsuits only came before the courts beginning in 2013.

Influence of the Justice System Reform

Next we examine the influence the Justice System Reform had on nuclear damage compensation after Fukushima – in particular, whether it had anything to do with the delay in the commencement of legal proceedings over compensation claims. We examine two particular aspects: the number of personnel working in the court system and the expansion of ADR mechanisms.

The attempt to expand the justice system

First let us review the debate on staffing levels in the court system. The legal profession has three branches, practicing lawyers, prosecutors and judges. The Justice System Reform aimed to increase the size of the legal profession from approximately 20,000 in 1997 to 50,000 by 2018. A simple calculation reveals that this means an increase of 1,500 per year, or 15,000 over ten years. The proposed expansion had the support of business, consumer and union representatives and was supported unanimously by members of the Justice System Reform Council (established in 1999) (Kondō 2001). However, the Supreme Court calculated that, in order to reach the target, it would need to appoint at least 500 new judges over ten years.[16] In 1999 only 2,143 judges were working in the court system. If the Justice System Reform Council's ambitious goal for increasing the profession was taken into account, the Supreme Court's calculation seemed too low. Members of the Justice System Reform Council, including Yoshioka Hatsuko (then secretary-general of the Japan Housewives Association[17]), Nakabō Kōhei (former chair of the Japan Federation of Bar Associations) and Yamamoto Masaru (then executive vice-president of TEPCO), all described the Supreme Court's proposed increases as inadequate. Fujita Kōzō (former president of the Hiroshima High Court) also pointed out that, with the transfer to the family court of jurisdiction for legal actions related to personal status, there would also be a need to increase the number of family court judges.

Responding to these criticisms, Nakayama Takao (then director-general of the General Secretariat of the Supreme Court) pointed out that, given that the number of judges had only increased by 650 over the previous 50 years, an increase of 500 over ten years was 'a major personnel increase for the court system'. In fact, from the time the Justice System Reform Council issued its recommendations in 2001, the size of the judiciary grew over ten years from 2,243 to 2,850. The number of prosecutors also increased from 1,375 to 1,791 and the number of practicing lawyers from 18,246 to 30,518 over the same period (MOJ 2012). This amounts to an increase of 27% in the size of the judiciary, 30% in the number of prosecutors and 67% in the number of practicing lawyers. The rate of increase for judges and prosecutors was smaller than for practicing lawyers.

However, between 1990 and 2000 the Japanese government engaged in a round of public service cuts. Ōta Sei'ichi, director-general of the Bureau of Internal Affairs and Communications, recognized the need to increase the number of judges and asked for them to be exempt from the broader cuts. Yoshioka Hatsuko, of the Justice System Reform Council, pointed out that in order to realize the goals of the Justice System Reform to speed up legal proceedings, across-the-board cuts to the number of court clerks and secretaries would be undesirable. The Supreme Court publicly stated its intention to continue its existing recruitment program for new graduates.

However, court staff other than judges could not escape the tide of reform. On June 30, 2006, the Cabinet endorsed a planned reshuffle of public service staff and a hiring freeze.[18] The plan called on the courts to cooperate with the government's shake-up of the public service. The Supreme Court agreed to implement the efficiency increases expected of other government agencies in administrative agencies where the nature of the work performed was equivalent to that of other agencies.[19] The size of the administrative workforce in the court system only increased by 42 between 2001 (22,047) and 2011 (22,089) (Supreme Court of Japan 2002; General Secretariat of the Supreme Court of Japan 2013). In spite of calls for the courts to perform better in the administration of justice, the major staff increases that would have been necessary to do so did not eventuate.

Alternative dispute resolution

Another aspect of the Justice System Reform Council was the call for an expanded role for ADR. At the time, however, there was some criticism of the idea of expanding ADR without simultaneously expanding the court system (Sawafuji 2001). Nevertheless, the Justice System Reform proposal to expand ADR had universal support, with each interest group proposing a number of conditions (Satō et al. 2003).

The business community had been pushing for the expansion of ADR for some time (Irie 2013). During an inquiry into ADR, the Japan Business Federation (Keidanren) called for the expansion of ADR as a mechanism to quickly and easily resolve disputes between companies that involved highly specialized knowledge.[20] Although the Japanese Trade Union Confederation (Rengō) raised a number of outstanding issues with regard to speed and the problem of ADR mechanisms that lack the power to compel parties, it called for the development of an ADR process with the powers of a government agency.[21]

The quasi-legal professions, including judicial scriveners, patent agents and accountants, were also positive about the expansion of ADR. Among their specific claims was the call to relax Article 72 of the *Attorney Act*, which prohibits anyone other than practicing lawyers from operating for-profit law offices. The Japan Federation of Bar Associations, sensing potential competition from the quasi-legals, cautioned that only judicial scriveners and patent agents should be able to represent claimants in ADR processes. However, apart from insisting on its own central role in the ADR process, the federation did not oppose the idea of expanding the system.

Review of these discussions took place during the Judicial System Reform process, and we can conclude that the establishment of the ADR Center after the Fukushima accident was in keeping with the original intention of the reform. It was the legal profession that called for the establishment of the ADR Center after Fukushima. On June 24, 2011, the Japan Federation of Bar Associations put a proposal to the government arguing for the establishment of an ADR mechanism for nuclear compensation. It gave two reasons. First, direct negotiations between TEPCO and the victims might place the latter at a disadvantage. Second, due to the court system's limited physical and human resources, relying on the courts to resolve disputes would lead to long delays.[22] On the second point, Hayakawa Toshiaki, a senior bureaucrat with responsibility for nuclear damage compensation issues, hoped that 'we can resolve the claims rapidly, without having to go through the time-consuming court process'.[23]

However, in reality the ADR Center had 515 unresolved claims on its books in 2011, increasing to 3,201 by 2012 (ADR Center 2015). In 2012 Idei Naoki, then deputy chief of the secretariat of the ADR Center, made a personal statement about this issue:

> We are receiving more claims than we can handle. This is an indication of the public's trust in the ADR Center. As a Center, we need to meet these expectations by ensuring our operation is shipshape. However, we simply cannot hope to mediate settlements for the great majority of the tens of thousands of expected disputes over nuclear damages. (Idei 2012a)

Idei gave a number of reasons for the delays. First, mediators were initially extremely cautious in their approach. Second, the large number of claims meant that it was taking much longer than anticipated to check the facts of each case. Third, the mediators, research clerks and administrative staff at the ADR Center were ill prepared for the task they had been given (Idei 2012b). The ADR Center then recruited additional personnel. As a result, there was a significant decrease in the unresolved caseload for each research clerk, from 30 in 2012 to fewer than 15 cases in 2013.

While the expansion of the court system originally planned as part of the Justice System Reform was limited, out-of-court dispute resolution processes have become more widespread. As a result, in the compensation process after the Fukushima accident, the ADR Center received more claims than it could handle, making it necessary to hire additional staff.

Conflict in the judicial administration and the bureaucracy

In this chapter, we have examined the issue of nuclear damage compensation after the Fukushima nuclear accident. We have discussed the impact different compensation mechanisms had on the provision of aid to disaster victims. In particular, we looked at the different resources that were allocated to each of the three mechanisms: TEPCO, the ADR Center and the courts. We looked at how each of these mechanisms functioned and how this affected the aid that victims received based on a survey of legal teams that represented disaster victims after Fukushima.

The government's establishment of the ADR Center gave victims a choice of three methods for claiming compensation: direct application to TEPCO, applying for settlement mediation through the ADR Center, or suing TEPCO and/or the government through the courts. While TEPCO and the ADR Center had to increase staff numbers to cope with the growing number of applications, fewer litigants appeared before the court system than expected, and the Supreme Court, therefore, scaled back its disaster response.

A survey of legal teams who represented disaster victims showed that most lawsuits only commenced as of 2013. While litigation helped some victims to achieve psychological closure and served as a form of social movement activism, in other respects the ADR Center acted as a substitute for the courts. Legal action and ADR are mutually complementary. However, in the immediate aftermath of the nuclear accident, there were delays in the ADR Center process. There were a number of reasons for this. The settlements reached through the ADR Center were not legally binding, and the guidelines for deciding how much compensation to award in each case were unclear. Both these problems originated in the power to establish guidelines for nuclear damage compensation that were given to the Dispute Reconciliation Committee as part of the 2009 amendments to the Compensation Act. Confusion over these guidelines contributed to delays in mediating claims through the ADR Center.

We also discussed the Justice System Reform and its impact on nuclear damage compensation. Following the Justice System Reform, the number of judges did not increase in proportion to the increase in the number of practicing lawyers. Furthermore, public service cuts led to a decrease in the number of administrative staff in the court system. This resulted in a large number of applications for settlement mediation through the ADR Center, but the organization did not have the resources to deal with them. Staffing levels were only increased sufficiently to ease the bottleneck in April 2014.

Based on these findings, we can make a number of conclusions about the nuclear damage compensation system. The ADR Center, established as part of the Dispute Reconciliation Committee, handled many more claims than did the Dispute Reconciliation Committee established after the Tōkaimura nuclear accident and was much more successful in mediating settlements. This demonstrates that the government learned from the lessons of the Tōkaimura nuclear accident. It also shows the effectiveness of the Justice System Reform. However, the ADR Center's inability to meet the expectations placed on it led to delays in reaching settlements and, consequently, to delays in victims accessing the court system. The survey of legal teams that represented disaster victims found that litigation not only provided an avenue for victims to receive financial compensation, it also functioned as a means of psychological recovery and as a form of social movement activism. The delay in commencing legal proceedings for those seeking nuclear damage compensation may have also prolonged the psychological torment experienced by some victims and inhibited the use of the courts as a form of social movement activism.

The ADR Center, established as a consequence of the Justice System Reform, is located within MEXT. The delay in providing adequate staff to deal with the ADR Center's heavy case load is linked to its status as a unit within the government bureaucracy. The experience of nuclear damage compensation after Fukushima suggests that the extent to which the establishment of the quasi-judicial ADR Center within the bureaucracy limited its effectiveness will depend upon the degree of preparedness on the part of the judicial administration and the government. Of course, it is difficult to predict the legal disputes that may arise after a disaster. However, lack of preparation should not be an excuse for delays in providing aid to people who have experienced loss after a disaster. In order to prevent this from occurring in the future, both the government and the private sector need to anticipate the potential for disputes to arise after a disaster and ensure that they mobilize ample resources when they occur.

Acknowledgments

We are grateful for the suggestions we received from Bar Associations throughout the country and for the assistance of the legal teams who participated in the survey discussed in this chapter. We also received advice from Akashi Jun'ichi of the University of Tsukuba. We would like to thank everyone who assisted us in the preparation of this chapter. Both authors equally contributed to the creation of this paper. Any errors that remain are, of course, our own.

Part II
The Triumph of Network Governance

7 Design and Development of the Recovery Agencies

Itō Masatsugu

The political process associated with the formation of government administrative structures is two-pronged. On the one hand, those involved, the actors, debate the desired features of the new organization in terms of respective self-interest. On the other hand, innovation is to some extent constrained by the options available in terms of existing administrative structures. This account well fits the manner in which new Japanese government reconstruction and recovery organizational structures emerged in response to the unprecedented scale of damage caused by the March 11 (3.11) Great East Japan Earthquake disaster.

Japan's institutional response to earthquake reconstruction, nevertheless, included a number of specific elements not previously present in the country's disaster recovery structures. While the Reconstruction Agency, established on February 10, 2012, was headed by the Prime Minister as the head of Cabinet, a new Minister for Reconstruction position was also appointed. In addition, in each of the three disaster-affected prefectures – Fukushima, Miyagi and Iwate – regional reconstruction bureaus were set up to operate as one-stop liaison offices for reconstruction policy.

This chapter considers the theoretical framework of rational choice institutionalism as one means of understanding both the processes involved in the creation of post-3.11 governmental disaster reconstruction structures and the special characteristics that these structures displayed. Rational choice institutionalism is offered, first, to explain the process by which the Reconstruction Agency, the main government body overseeing reconstruction activities, was established and also to articulate its organizational characteristics. Consideration is then given to how the government structures created to support Great East Japan Earthquake recovery actually functioned following changes that occurred with the December 2012 defeat of the incumbent Democratic Party of Japan (DJP) government and the election of the Liberal Democratic Party (LDP) and the Komeito administration. Finally, the chapter profiles the special characteristics of these Japanese governmental disaster recovery structures and comments on future possible developments in this field.

Organizational options and the Reconstruction Agency

The establishment of new administrative structures has strategic political significance in terms of supporting the power needed both to acquire and to hold government office. According to rational choice institutionalism, decisions made by each actor concerning the configuration and authority of any new organizational structure aim to maximize controllability and the benefits that accrue from the policies that the new organization will administer. On the other hand, this decision making also attempts to minimize the various transaction costs that accompany the establishment of such an authority. While a number of elements are involved, two in particular are significant from the point of view of those with administrative responsibility.

Political officeholders are obliged to delegate the creation of a new structure to the bureaucracy. The first cost incurred is that of the policy distortion that can result from this delegation process and which can lead to outcomes that are different from those intended by elected representatives. Horn (1995) refers to this cost incurred by the principal, or political officeholder, in terms of its dealings with the bureaucracy as 'agency cost'. Agency costs can be minimized by elected political representatives restricting the decision-making parameters of the bureaucracy while also having some input into the appointment of personnel.

Creating a structure that favors government control, however, comes with its own risks. This is particularly the case if a change of government is likely. In the event of the government changing hands, there is a danger that the structural framework that an administration designs to limit the power of the bureaucracy will then favor the new incumbents who have policies and views opposed to the party that originally created the structure. In other words, an incumbent administration risks creating a structure that will eventually work to its own disadvantage. Horn (1995) refers to this cost, which is the negative aspect of a party that holds political power placing excessive controls on the bureaucratic apparatus, as 'commitment cost'.

To some extent, these two costs operate to balance out each other. Nevertheless, political officeholders must consider each when designing a new administrative structure. While institutions such as independent regulatory commissions that operate in relative isolation from prime ministerial or presidential executive branch control generate agency costs, their high level of political neutrality results in limited commitment costs. A structure that is directly attached, however, to the ministries that comprise an executive branch-type administration – one that is headed by a president or prime minister – will have high commitment costs. This is because if power is wrested away from the incumbents, control of the structure will shift to the opposition. Nevertheless, given that such a structure permits the implementation of policies that benefit the party holding political office, the agency costs are low.

In the case of the stable governance that occurs when a party has majority support in an elected assembly, those holding office tend to minimize agency costs by establishing new administrative structures that report directly to the executive. Lewis (2003), however, notes that, in the case of a minority or divided government (the latter referring to the situation in which one party has control of the executive while an opposing party has control of one or more of the houses of a legislature), a governing party must share political power with the opposition. In this instance, the party in power is obliged to incorporate opposition proposals into new administrative designs, making it highly likely that the preference will be for a structure with a high degree of independence – that is, with high agency costs.

Administrative structure options in Japan

Rational choice institutionalism uses the notion of transaction costs to provide an integrated theoretical framework for understanding the design of administrative organizations. In reality, however, there are various provisos and limitations that influence both the design and the administrative location of any new organization. Choices made regarding the form and power of any new structure must occur within the context of existing systems.

In the Japanese governance system, the Cabinet Office and the various ministries and their agencies – which operate as agents of head of state-style governance – and bodies such as independent regulatory commissions – which operate as agents of council-style governance – function according to legislation enacted by the Diet, while the various committees that act as advisory and consultative bodies function according to either legislation or Cabinet ordinance. The system also features headquarters and consultative groups, comprising Cabinet officials and expert advisors, which report to Cabinet and which form at the request of either Cabinet or the Office of the Prime Minister.

Headquarters and consultative groups, which report to Cabinet and whose members are drawn from Cabinet and can include the prime minister, make decisions that accord with government policy. As a result, their agency costs are low. On the other hand, agency costs rise when a new administrative body with a high level of independence is created. This is particularly evident in the case of regulatory commission-type bodies, which, because they are characterized by wide-ranging representation, are difficult for a governing party to control.

What sort of choices were made, then, in the formation of administrative structures charged with overseeing the reconstruction necessary in the wake of the Great East Japan Earthquake?

Earthquake reconstruction government structure options

On May 13, 2011, two months after the disaster, the Kan Naoto DPJ Cabinet tabled a bill in the Diet outlining the government's fundamental polices and related administrative structures for reconstruction following the Great East Japan Earthquake. The bill proposed the Reconstruction Headquarters in Response to the Great East Japan Earthquake, chaired by the prime minister and with membership of all Cabinet ministers. This headquarters was to have carriage of all planning and general coordinator matters related to reconstruction. The bill also clarified the role of the Reconstruction Design Council in response to the earthquake, which had commenced operation in April 2011, as a body with wide-ranging membership that would investigate and deliberate on substantive matters related to reconstruction.

The structure proposed by the Kan administration, which drew on a central headquarters reconstruction structure model with advisory committees made up of specialist and expert membership, replicated the recovery structures created at the time of the January 1995 Great Hanshin–Awaji Earthquake. In the case of that disaster, while an expert advisory group known as The Committee for Reconstruction of the Hanshin–Awaji Area was established in February, the bill tabled in the Diet concerning fundamental policies and administrative structures related to reconstruction also proposed a Headquarters for Reconstruction of the Hanshin–Awaji Area chaired by the Prime Minster and with membership of all members of Cabinet. This was in contrast to the response to the September 1923 Great Kantō Earthquake at which time an Imperial Capital Reconstruction Authority headed by the Minister for Home Affairs was established. Although this latter model was canvassed in the case of the 1995 disaster, the then Murayama Tomiichi Cabinet ultimately opted for the headquarters model. According to Ito Shigeru (2005:92), it was thought that the headquarters model involving the participation of all Cabinet ministers would facilitate the swift and flexible commencement of reconstruction ventures by each of the ministries and government agencies involved.

The Kan Cabinet thus based its earthquake recovery plan on the Hanshin–Awaji model. Drawing on the rational choice institutionalism theory discussed above, we can explain this decision as follows.

While Kan and his fellow ministers undoubtedly recognized the merit in establishing a discrete organizational entity through which the implementation of reconstruction policies could be overseen, in addition to incurring start-up costs such a move would have created a structure that stood in isolation from existing agencies such as the Ministry of Land, Infrastructure, Transport and Tourism. We can therefore assume that there was a preference for a headquarters model that facilitated planning activities and general regulatory functions to be performed in tandem with the activities of Cabinet. Kan's ruling DPJ had, however, lost its majority in the July 2010 upper house elections. Therefore, earthquake reconstruction plans and policies were only able to pass the Diet through compromises with the then opposition parties, the LDP and Komeito.

The LDP put forward draft legislation that, recalling the creation of an overseeing authority-type body in response to the 1923 Great Kantō disaster, proposed the establishment within Cabinet of a Great East Japan Disaster Reconstruction and Renewal Authority, with its own responsible minister. This body was to have central oversight of the formulation and implementation of reconstruction policy. Komeito also presented draft legislation proposing the creation of an Earthquake Reconstruction Agency with centralized responsibility for all matters related to construction, including policy implementation. In addition, Komeito proposed the creation of a Special Reconstruction Zone System. Put another way, we might say that rather than the central headquarters model of the Hanshin–Awaji reconstruction process, the opposition at the time of the Great East Japan Earthquake gave support to the reconstruction authority-type model, which, as noted above, operated following the Great Kantō disaster.

Ultimately, because it held a majority in the upper house, the opposition was able to block the formation of the reconstruction structures proposed by the Kan administration and instead insist upon the formation of a new centralized authority. If the government had granted a high level of independence when establishing such a body, agency costs would have been high because of the likelihood of a future change of government and the resulting drawback, as outlined above, of the formation of a new Cabinet by the opposition. Yet, in spite of criticism of its response to the disaster (which made the position of both Prime Minister Kan and his administration increasingly vulnerable) and the fact that Kan eventually stepped down (at which time the prime ministership passed to Noda Yoshihiko), the DPJ made it clear that it had no intention of relinquishing office. With the realization that an election – and the possibility of the LDP and Komeito assuming government – would not occur in the immediate future, the LDP and Komeito sought to impose agency costs on the government by demanding the establishment of a new structure similar to the 1923 standalone Imperial Capital Reconstruction Authority.

In the legislation initially tabled in the Diet that related to Great East Japan Earthquake fundamental reconstruction policies and implementation structures, the Kan administration did respond to some extent to the proposals from the opposition parties. This saw several important measures incorporated, including the creation within a year of the legislation being adopted of a new Reconstruction Agency with carriage for the planning and overall regulation of reconstruction policy.

With the opposition remaining tenacious in its demands, the Kan administration had no option but to accept the LDP and Komeito proposals more generally. Accordingly, the final reconstruction legislative draft that went before the Diet included amendments that saw the government's original proposal of a Reconstruction Headquarters in Response to the Great East Japan Earthquake headed by the prime minister become the Reconstruction Agency with central control of reconstruction and recovery policy. After each of the three parties – DPJ, LDP and Komeito – agreed that the chair of the Lower House Special Committee for Great East Japan Earthquake Recovery would have responsibility for the final form of the Great East Japan Earthquake Reconstruction Bill, the legislation passed an upper house plenary session and was enacted on June 20 (Iwasaki 2011).

Clause 3 of the bill stipulated that the Reconstruction Headquarters in Response to the Great East Japan Earthquake would make decisions in July regarding the 'fundamental policy direction for recovery from the Great East Japan Earthquake', while the following were prescribed in reference to the new Reconstruction Agency:

- national measures related to recovery will involve the installation of a system that surpasses the existing ministerial and agency framework and which permits a direct one-stop response to local community and municipality needs
- a system will be initiated that facilitates focused analysis of matters related to reconstruction and recovery.

Accordingly, a preparatory office for the Reconstruction Agency was set up in the Cabinet Secretariat, and on November 1 the Reconstruction Agency Establishment Bill was tabled in the lower house. Various amendments came from both side of the house relating to matters such as administrative jurisdiction and the role of advisors to the Minister for Reconstruction, in addition to the prefectural-based Reconstruction Bureaus. The bill was eventually enacted after winning a majority in a plenary session of the upper house on December 9 (Sakurai et al. 2012: 17–18).

The above discussion seeks to explain the creation of the Reconstruction Agency in terms of rational choice institutionalism. Detailed examination of the design of this organization can help understand the operation of government organizations in Japan and also to identify the specific characteristics of those organizations. The next section therefore discusses institutional characteristics of the Reconstruction Agency with a view to noting their meaning in the context of official Japanese administrative systems.

Institutional characteristics of the Reconstruction Agency

The Reconstruction Agency, established as the government structure to take carriage of core matters related to earthquake reconstruction and recovery, was granted a very particular position in relation to existing government organizations. First, in addition to being given a ten-year limited life, the agency was exempt from the provisions of the *Cabinet Office Establishment Act* and the *National Government Organization Act*, both of which stipulate the conditions under which government organizations function in Japan. Being located within – rather than merely reporting to – the structure of Cabinet, the Reconstruction Agency was accorded the same rank as the Cabinet Office, and thus had a higher status than other ministries. In addition, like the head of the Cabinet Office, the prime minister was named the head of the Reconstruction Agency. The location of the agency directly within Cabinet came from a suggestion by Komeito, which argued that the creation of a body so positioned and led by the prime minister would assist in alleviating the anxiety of disaster victims and would also promote the swift development of reconstruction policy.[1]

The second significant difference was that, in addition to the prime minister being named head of the Reconstruction Agency, a Minister for Reconstruction was appointed and attached to the agency. According to Clause 8, Paragraph 3, of the Reconstruction Agency Establishment Act, the role of the Minister for Reconstruction was to 'assist the Prime Minister, superintend the affairs of the Reconstruction Agency, and supervise the public service offered by its officials'. The appointment of this Minister for Reconstruction came in response to opposition demands. If the agency had been located in the Cabinet Office, instead of directly within the Cabinet, its status would have been one of the external bureaus of the Cabinet Office. In this case, it would have been headed by a director-general or secretary, who, as a general rule, is chosen from the ranks of career bureaucrats. Placing the agency directly within Cabinet and appointing a Minister for Reconstruction, however, ensured that reconstruction occurred in the context of political, rather than bureaucratic, leadership. Since there were few previous structures headed by the prime minister that also had a responsible minister, formalization of this arrangement (whereby the Minister for Reconstruction was not in fact the head of the entity for which the minister had ministerial responsibility) required intervention by the Cabinet Legislation Bureau, which oversees legislative drafting processes.[2]

The third point specific to the organization of the Reconstruction Agency concerned the authority and powers given to the minister. According to the legislation, the Minister for Reconstruction could request document details and explanations from the heads of related government agencies (Reconstruction Agency Establishment Law, Clause 8, Paragraph 4), provide advice (Paragraph 5), request the information necessary to provide advice (Paragraph 6) and offer opinions to the prime minister as head of Cabinet (Paragraph 7). These are the same powers as those stipulated for the Minister of State for Special Missions.[3] Furthermore, the Reconstruction Agency Establishment Law stipulated that the heads of relevant government agencies were obligated to comply with advice given by the Minister for Reconstruction. This stipulation, instituted through an amendment from the floor during Diet proceedings, can be read as an attempt to strengthen the ability of the Reconstruction Agency to guarantee the effectiveness of reconstruction policies.

Central control and one-stop service

As the discussion above implies, there was an expectation that, since the Reconstruction Agency was directly attached to the prime minister, it would function to coordinate reconstruction policy planning and the various agencies involved, and also to liaise with municipalities and local communities. During the consultative stages of the bill upon which the agency's creation was based, these functions were labelled as 'central control'. In the consultative stage, too, the LDP and Komeito insisted that the agency provide a one-stop service point of implementation of reconstruction policies.[4]

With the Japanese government administrative structural reforms of the early 2000s, there has been a growing sense of a need to grant a central control function to newly created government agencies as a means to enhance the regulation of government policy. For example, the central bureaucratic restructure that occurred under the Koizumi Cabinet saw the creation of the Council on Economic and Fiscal

Policy, a body intended to have control of such policy. Statistical system reforms proposed by that council led to the formation within the Cabinet Office of a National Commission for Statistics. This was to be the government organization with responsibility for the central control function of the country's statistical systems (Matsui 2012: 112–17). Furthermore, in September 2009 the Consumer Affairs Agency was established with central control over consumer governance.

What, then, was the concrete central control function expected of the Reconstruction Agency? Okamoto Masakatsu, who was appointed as administrative vice-minister of the agency, highlighted three points necessary for central control of reconstruction policies: integration of information and the coordination of administrative matters, collaboration between the relevant organizations, and organizational visibility and transparency (Okamoto 2013: 8). At the same time, however, there were calls for 'municipality-led reconstruction' (Reconstruction Design Council in Response to the Great East Japan Earthquake 2011: 16), a development that gave other relevant ministries substantive carriage of the planning of the majority of reconstruction projects. In reality, it thus became difficult for the Reconstruction Agency to exercise a top-down control function towards the other official entities involved.

Comparable problems have, in fact, been experienced by similarly constituted central control units. The role of the Council on Economic and Fiscal Policy, for example, differed according to the policies of the Cabinet in power at any given time. And even after the creation of the central control National Commission for Statistics, decentralized statistical collection processes continued. The Consumer Affairs Agency, too, has an uneasy and complicated power relationship with related ministries and agencies. Contrary to the discourse and desired image, Japanese central control organizations are therefore probably best understood as symbolic entities that function to supplement, rather than centrally control, the operation of existing governance organizations.

A second distinguishing feature of the Reconstruction Agency, namely its role as a one-stop service response point, can also be placed somewhere along the complementary function spectrum. In Iwate, Miyagi and Fukushima prefectures, Reconstruction Bureaus were established as provincial offices of the Reconstruction Agency. Rather than exercising strict controls over provincial matters in terms of relations with either the municipalities or ministries impacted upon by the disaster, these bureaus were expected to act as contact or advice centers, and as centers that would facilitate networking and cooperation between the various parties involved in reconstruction.

The Reconstruction Agency's particular functions of both control center and one-stop service response point, however, did not necessarily develop as planned. The following section discusses the radical review of disaster recovery activities that occurred as a result of the December 2012 change in government, and also seeks to clarify the development processes of both the Reconstruction Agency's control center function and one-stop service response.

Development of reconstruction support structures

The Reconstruction Agency was established on October 2, 2012, under Prime Minister Noda Yoshihiko. Hirano Tatsuo, who had served as Special Minister of State for Disaster Management, was appointed as the minister for the newly formed Reconstruction Agency. In addition, the initial agency structure included three senior vice-ministers and four parliamentary secretaries, each of whom had an existing role in another department. In order to ensure coordination between the relevant ministries involved in reconstruction policy implementation, a Reconstruction Promotion Council, with membership of each Cabinet minister, was located in the Reconstruction Agency. Furthermore, in March the first meeting was held of a Reconstruction Promotion Committee whose members included the governors of each of the three disaster-affected prefectures and relevant specialists and experts. The membership of this Reconstruction Promotion Committee, which was inaugurated as a consultative committee with strong links to the (Cabinet-based) Reconstruction Promotion Council, was mainly drawn from the same political scientists who had played a leading role in the Reconstruction Design Council, mentioned at the outset of the discussion, which had formed in the month immediately following the 2011 disaster.[5] Through these structures, formal recovery activities finally came into operation 11 months after the disaster occurred.

In the general election of December 16, 2012, however, the ruling DPJ suffered a crushing defeat and the second Abe Shinzō Cabinet, comprising members from both the LDP and Komeito, came into being. With this change of government, disaster recovery activities entered a new dimension.

In a general policy speech made in January 2013, Prime Minister Abe announced a significant change of direction regarding government earthquake reconstruction structures.[6] Nemoto Takumi, appointed as Minister for Reconstruction in the newly formed Cabinet, emphasized three points designed to accelerate recovery and to 'create a new Tōhoku':[7] (1) full overhaul and reformulation of policy measures, (2) full commitment to a hands-on approach and (3) strengthening of central control of the recovery effort. Criticizing delays in DPJ-initiated recovery efforts, the Abe Cabinet clearly flagged its intention to revise official disaster recovery systems policy.

One such change occurred in March 2013 when membership of the Reconstruction Promotion Committee was radically altered with the exception of the governors of the three disaster-affected prefectures. In addition, where in the past the Reconstruction Promotion Committee had comprised mainly political scientists with connections to the Reconstruction Design Council, membership under the new administration came mainly from fields such as economics, business and finance. Furthermore, in April 2014 the Reconstruction Promotion Committee released a document entitled 'Towards the creation of a "New Tōhoku" (Recommendations)'. Designed to strengthen the role of the Reconstruction Agency, these recommendations advocated policies that included a full commitment to a hands-on approach (that is, reconstruction of an interactive relationship between central and local government and relaxation of reconstruction financial grant processes), a strengthening of the central control function (that is, designating – as outlined below – Tokyo and Fukushima as the two main centers of reconstruction policy activity and also creating a cluster of new central control taskforces), and a review of reconstruction budgetary and finance frameworks (that is, ensuring the swift and flexible implementation of budgetary matters).[8] Both the membership changes of the Reconstruction Promotion Committee and the newly elected government's overhaul of the official disaster reconstruction support procedures sought to enhance and strengthen the central control function and one-stop response of the Reconstruction Agency.

Yet, how, in fact, did the central control function and one-stop response of the Reconstruction Agency develop? The following section elaborates upon official disaster recovery support changes in terms of the structure and personnel of the Reconstruction Agency and also with respect to the Reconstruction Bureaus.

Central control functions – structure and personnel

The Reconstruction Agency comprised the internal departmental Tokyo base office and the Reconstruction Bureaus located in Iwate, Miyagi and Fukushima prefectures. There were also offices in Aomori and Ibaraki prefectures. It was expected, however, that the Reconstruction Agency would exert its central control function by making maximum use of existing administrative resources. As a result, the following characteristics related to organizational structure and deployment of personnel emerged.

First, unlike other government agencies and also unlike the Imperial Capital Reconstruction Authority on which it was fundamentally modelled, the structure of the Reconstruction Agency's head office was not divided into bureaus or divisions. Rather, the office featured an administrative vice-minister, beneath who were director-general, councillor and counselor positions. In addition, the office was largely structured into teams with respective carriage of the following: general matters and planning, victim support health and daily life, recovery and reconstruction from earthquake and tsunami damage, and Fukushima and the reconstruction of nuclear energy damage. The adoption of a flexible, team-based organizational structure was intended to enable the agency to respond as required to the recovery agenda.[9]

A second characteristic of Reconstruction Agency staffing was the secondment of personnel from other ministries. Given its limited life, the Reconstruction Agency did not recruit new staff. Instead, bureaucrats were seconded for a period of two years to each Reconstruction Agency post, with staff rotating through the various agency tasks and projects. At the time of writing in July 2016, successive executive postings had largely been allocated as 'booked seats' to certain ministries. With the exception of the third incumbent, who had come from the Ministry of Internal Affairs and Communications (MIC), the post of administrative vice-minister, for example, has to date been monopolized by Ministry of Land, Infrastructure, Transport and Tourism officials (MLIT). Furthermore, following the February 2013 addition of a director-general post with specific responsibility for Fukushima, the agency operates with three directors-general under the now established practice of one each from MIC, MLIT, and the Ministry of Economy, Trade and Industry (METI).[10] While the two councillor positions are booked in advance for METI bureaucrats, the head of the Iwate Bureau comes from the Ministry of Agriculture, Forestry and Fisheries, the head of the Miyagi Bureau from MLIT, and the head of the Fukushima Bureau from MIC.

The third personnel point of note concerns the counselor positions. In order to enhance flexible communication with ministries and agencies, these staff members hold existing positions in other ministries. Thus, although there were only 197 permanent Reconstruction Agency staff at the end of the 2016 financial year, as of October 2015 the actual number of people employed was approximately 600 (around 240 in the Head Office, 80 in the Iwate Bureau, 130 in Miyagi and 140 in Fukushima).

Apart from the transient 400 or so from other ministries,[11] we might add here temporary staff and staff dispatched from private industry. The director-general division, for example, with the exception of those who work in other ministries, has nine designated staff. Nevertheless, at the end of July 2015 there were, in fact, 30, including those already with other ministries.[12] In Japan public service employee and bureaucrat numbers are restricted by law. We can nonetheless say that, in spite of the restrictions under which it operated, the Reconstruction Agency strived to effect 'maximum mobilization' (Muramatsu 1994) through a strategy of seconding from other ministries while also devising a staffing system that facilitated the flow of communication between the relevant ministries regarding the reconstruction agenda.

The Tokyo-Fukushima Dual Headquarters System and taskforces

With the calls for strengthening central control functions that followed the November 2012 change of government, a number of developments occurred in the organizational and personnel aspects of the Reconstruction Agency. First, there was the formulation of the so-called Tokyo-Fukushima Dual Headquarters System. This system arose following a directive made by Abe Shinzō at the fifth meeting of the Reconstruction Promotion Council, held on January 10, 2013, to develop a structure that would permit decision making at the local level, including by the local Reconstruction Bureau, regarding Fukushima reconstruction.[13] Accordingly, on February 1, 2013, a Fukushima Headquarters for Reconstruction and Revitalization was established in Fukushima City with a Tokyo Headquarters for Fukushima Reconstruction and Revitalization established within the Tokyo office of the Reconstruction Agency. The Fukushima Headquarters, headed by the Minister for Reconstruction, encompassed the Fukushima Reconstruction Bureau, the Fukushima Regional Office for Revitalizing Environment, the Ministry of the Environment, and the Local Nuclear Emergency Response Headquarters. A previous administrative vice-minister of the Reconstruction Agency was given the post of Head of the Secretariat of the Fukushima Headquarters for Fukushima Reconstruction and Revitalization, and, as a means of creating an integrated operation of local structures, a secretariat was formed from the current Reconstruction Agency administrative vice-minister, the councillor with carriage of Fukushima issues, the head of the Fukushima Reconstruction Bureau, the head of the Local Nuclear Emergency Response Headquarters (a METI appointment) and the head of the Fukushima Regional Office for Revitalizing Environment (a Ministry of the Environment appointment).

The Tokyo Headquarters for Fukushima Reconstruction and Revitalization, on the other hand, was established in order to strengthen centralized government functions related to Fukushima reconstruction. The head of the Tokyo Headquarters was also the Minister for Reconstruction, with personnel drawn from bureau chief-level staff from other ministries seconded to the Reconstruction Agency. This was intended to permit the minister to oversee directly each of the Tokyo Headquarters staff (Yanase 2013: 1–4).

A second point regarding organizational and staffing changes to the agency was the formation of a series of taskforces centered on the Minister for Reconstruction. Each taskforce assembled a collection of bureau chief-level staff with the aim of providing targeted problem solving by personnel from a cross-section of the relevant ministries and agencies. By the end of October 2014 taskforces had been established to address decontamination and the acceleration of reconstruction, acceleration of housing and community reconstruction, strategies to contain reputational damage incurred through the nuclear accident, health/welfare and daily life support for disaster victims, and the promotion of industry reconstruction.

The Reconstruction Agency thus attempted to liaise with associated ministries mainly through the people-to-people contact that came by both deploying bureaucrats from other ministries and engaging staff who had existing positions with other ministries. However, in spite of attempts to implement the February 2013 directions to tighten central control, it is difficult to judge the extent to which this has actually been possible.

For example, the taskforce charged with accelerating housing and community reconstruction has convened seven times and has progressed planned measures in this field to the fifth stage. We might therefore say that in this case the taskforce approach has achieved a degree of success. On the other hand, at the time of writing in July 2016, no activity could be confirmed for the Local Nuclear Emergency Response Headquarters, apart from its first meeting on February 15, 2013, and the compilation of an Early Stage Return and Stabilization Plan on March 7 of the same year. This may reflect the specific reconstruction and recovery problems of Fukushima. Nevertheless, it also suggests that, at least in the current situation, rather than operating top-down through the office of the minister in charge, the oversight function of the Reconstruction Agency is better achieved by maintaining collaborative contact with other ministries through appointing staff currently holding positions with those ministries.

The development of the one-stop response

The Reconstruction Bureaus act as provincial offices of the Reconstruction Agency. It was therefore expected that they would administer relevant matters in the areas designated as Special Reconstruction Zones, including the distribution of grants. There were also expectations that the bureaus would act as a one-stop shop consultation apparatus for the cities, towns and villages impacted by the disaster (Figure 7.1). Local communities, however, criticized the Reconstruction Bureaus for failing to perform this role adequately. Criticism was particularly harsh in the case of the distribution of reconstruction grant monies, with Miyagi governor Murai Yoshihirō expressing his strong dissatisfaction regarding the first grant distribution round made on March 2, 2012 (*Mainichi Newspaper*, 4 March, 2014). Murai pointed out that the Miyagi Bureau had not provided funds in accordance with the financial reconstruction plans of local disaster-affected communities. Rather, he declared, because it seemed more interested in complying with the wishes of the Treasury and other finance-related ministries, the Reconstruction Agency had made a largely token response to local community needs.

The later introduction of flexibility through, for example, the expansion of projects eligible for relief monies, certainly alleviated some degree of local dissatisfaction. Even so, it was not necessarily the case that the local Reconstruction Bureaus were, in fact, effectively discharging their intended function as one-stop points of response to community needs.

First, concerns were persistently voiced regarding matters such as delays in the response of the bureaus and the non-user friendly nature of the reconstruction grant process. For example, while on the one hand local people appreciated that the February 2013 establishment of the Fukushima Headquarters indeed permitted local issues to be addressed by ministries in tandem at one point, there were concerns regarding the response delay to issues such as decontamination (*Fukushima Minpo News*, January 30, 2014). Dissatisfaction continued, too, regarding reconstruction grants. At the end of 2013, for example, Minamisanriku made an application for funds to create 14 municipal roads with a distance of 9.5 kilometers. This was necessary because some communities had been moved to higher ground following the disaster and roads were needed to connect these relocated communities to the fishing harbor. It was reported, however, that after an 'advance survey' conducted by the Reconstruction Agency during the application stage, the distance had been cut back to 1.4 kilometers (*Mainichi Newspaper*, March 9, 2014).

A second issue concerned the Special Reconstruction Zone System formed according to the Law on the Special Reconstruction Zones in Response to the Great East Japan Earthquake. When this came into force, the involvement of related ministries through their local offices continued as in the past.

The Special Reconstruction Zone System instituted a scheme that required decisions relating to the identification of land for reconstruction purposes to occur through deliberations of the

Figure 7.1: One-stop response and general coordination at the local level by the Reconstruction Bureaus

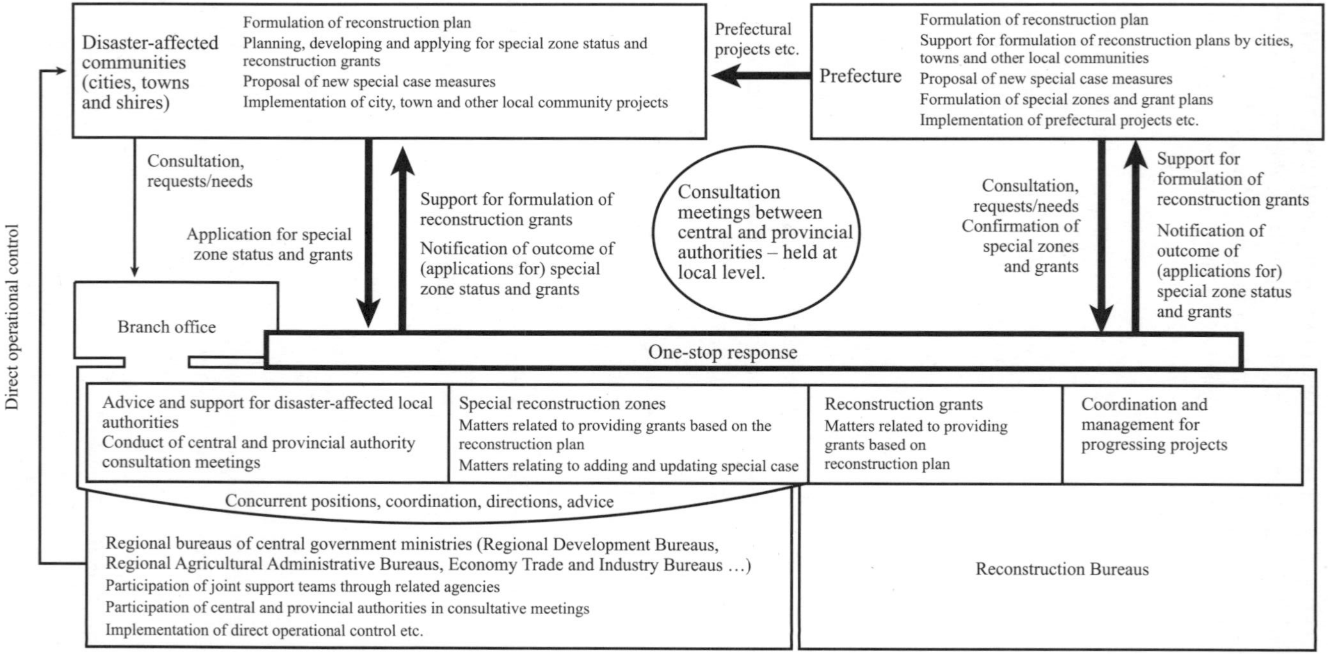

Source: Reconstruction Agency homepage (https:/ /www.reconstruction.go.jp/topics/recon_onestop.pdf).

Reconstruction Development Council, established in municipalities, on which all related agencies had representation (Harada 2012: 181–2). Because of a separate ordinance stating the need for all parties legally involved in such transactions to agree, however, matters could not proceed unless prefectural governors were of the same opinion as the minister. With local authorities following the ordinance, the one-stop review and reassignment of land use in the Special Reconstruction Zones only had meaning in terms of simplifying and speeding up those processes that involved consultation between the prefecture and the onsite agencies, and regarding complicated processes upon which the agreement of each authority was reached as necessary at Reconstruction Development Council gatherings. There was no demand for the Reconstruction Bureaus to assume the powers previously held by the various ministries or to convene formal meetings with the local communities. What was expected, however, was a preparedness to consult with and respond to the desires of these communities regarding both the planning of reconstruction projects and the timing of Reconstruction Development Council meetings and a willingness to provide indirect support in planning reconstruction projects.

Third, the prefectures began to engage in their own one-stop response activities, even to the extent of appearing to formalize these in their own systems. For instance, to speed up the multiple authorization procedures pertaining to reconstruction projects, in January 2014 Iwate Prefecture declared its intention to form a one-stop response-type Team for Facilitating Reconstruction Projects in the relevant prefectural agencies.[14] As reconstruction proceeded, what had initially been a simmering undercurrent of competition between prefectural administrations and the Reconstruction Agency became more overt. One result has been the Reconstruction Agency's ultimate inability to monopolize the provision of a one-stop response to local community reconstruction.

Reconstruction acceleration and outreach

One measure for accelerating reconstruction was the development by the Reconstruction Agency and Reconstruction Bureaus of what might be called a one-stop response outreach strategy.[15] A first step in this process occurred in February 2014 when the Reconstruction Agency created a dedicated Support for

Accelerating Land Acquisition team. In October 2013 the agency had initiated a Land Acquisition Acceleration Program with a view to facilitating the site acquisition of, for example, land for public housing for disaster victims. A range of specific factors operated, however, to make acquiring such land problematic. As a result it was difficult to devise successful acceleration measures at the level of the local community. The newly formed Support for Accelerating Land Acquisition team therefore took advice from local communities regarding the specifics of issues relating to the land being sought. Following this, the Reconstruction Agency and relevant Reconstruction Bureau worked in tandem with the Legal Affairs Bureau of the Ministry of Justice and the Tōhoku Regional Development Bureau, MLIT, to solve the procedural issues involved. They did this by conferring, for example, with agencies such as the Solicitor Association. From January 2014, in fact, the Reconstruction Agency made a practice of appointing notaries to its staff and locating these personnel in disaster-affected local communities.

Second, in August 2014 a similar Support for Construction Acceleration team was formed to solve the problems related to the construction of public housing. This team was staffed by personnel such as Reconstruction Agency ministerial secretaries responsible for local communities and staff of the Reconstruction Bureaus, with connections also to MLIT. Members of this Support for Construction Acceleration team, moreover, made it their mission to go directly to regional areas and to listen to and provide support for local communities.

Third, the Reconstruction Bureaus also advanced their outreach one-stop response into the area of daily life support for disaster victims. For example, one successful project, referred to as the Project for Promoting Group Relocation for Disaster Mitigation, sought to ensure greater availability of land on which to build residential dwellings and growth in the construction of individual residential dwellings. Because of the success of the project, however, it became apparent that there was also a need to provide disaster victims with a one-stop point of information on related procedural matters. Accordingly, in May 2014, and in liaison with other agencies, the Miyagi Reconstruction Bureau held information and consultative meetings throughout the prefecture on the subject of residential rebuilding in disaster-affected precincts.

Although subject to system and government resource restrictions, local-level outreach strategies (such as those discussed above) helped transform the one-stop response implemented by the Reconstruction Bureaus from a token program to a policy with initiative. In other words, as the reconstruction process progressed, the Reconstruction Bureaus went from being passive to active partners. This willingness to make adjustments can, of course, be seen as an attempt to better meet the needs of the disaster-affected communities. Given the sense of competition between the various levels of administration – particularly national and prefectural – that marked reconstruction and recovery processes, we might also understand such moves, however, as a survival strategy on the part of both the Reconstruction Agency and the Reconstruction Bureaus.

Concluding remarks and discussion

The Reconstruction Agency, unlike the central headquarters organizational model proposed by Cabinet at the time of the earthquake, was formed as a special government entity situated directly within Cabinet. Furthermore, the design and development of government organizations set up to promote reconstruction following the Great East Japan Earthquake came about from ongoing input from both sides of Japanese politics. The choice of a system that did not accord with the preference of the then Cabinet was a function of the governing DPJ's loss of the upper house in the July 2010 general election. This left the government with no choice but to accede to the demands of the LDP and Komeito opposition parties in order to pass legislation.[16] With the change of government that occurred in 2012, the Reconstruction Agency – already formed as a result of opposition demands during the time of the DPJ's time in government – was pressured to strengthen its central control function and also to provide a one-stop response center in the prefectures impacted by the disaster.

As this discussion clearly demonstrates, however, there remains a gap between the expectations that were held for the Reconstruction Agency and the agency's actual performance. The central control function that the Reconstruction Agency ultimately adopted was not a command function that took a lead in reconstruction matters and accordingly exercised strong regulation and coordination powers over the relevant ministries. Rather, in keeping with its responsibility to oversee a flexible administrative network grounded in personnel exchange between ministries, the agency developed a more collaborative approach in its promotion of reconstruction. There is no sense of the agency's one-stop response involving on-the-ground stakeholder consultation with a view to meeting all requests made by local communities affected by the disaster. Instead, this central entity operates as a point of indirect support, with particular attention given to communities facing reconstruction challenges.

Some might interpret the gap between what was expected of the Reconstruction Agency and what the agency delivered in reality as grounds for criticism in terms of the delays that occurred in the reconstruction process. However, if the central control function and the one-stop response had been pursued in the literal meaning of these terms, little attention would have been given to local recovery initiatives or to providing coordination and support for reconstruction at the prefectural level. In the final instance, the Reconstruction Agency was required to respect provincial independence while also accelerating the rate of recovery. This it did while responding to calls to demonstrate leadership through liaising with the relevant organizations, including local authorities in disaster-affected areas, and through linking up relevant agencies and organizations. Such a model profiles the agency as what might be called 'a hub of interagency collaboration'. Consideration of previous disaster response structures in Japan indicates that reconstruction after the Great East Japan Earthquake is the first occasion in which such a collaborative hub structure has been considered necessary or desirable.[17]

Until 2020, when its functions will come to an end, the Reconstruction Agency will in all likelihood continue to be expected to realize organizational objectives that improve and strengthen its control center function and its one-stop response. At the same time, there will be calls for this body to work closely in tandem with disaster-affected communities and to steadily accumulate a track record of collaboration through a flexible approach to problem-solving. In this way, the Reconstruction Agency will be able to give concrete meaning to the rather ambiguous discourses surrounding the notions of central control and one-stop response service. In doing so, the agency will be given the chance to redefine its own mission.

8 Deployment of Local Government Personnel: Autonomy and Cooperation

Inatsugu Hiroaki

The Great East Japan Earthquake caused considerable disruptions to the functioning of local government. Fourteen municipal government offices were damaged or destroyed and many local government personnel lost their lives. These losses paralyzed local governments, preventing them from taking their rightful place at the head of reconstruction efforts. In many regions, neighboring municipalities were similarly affected. Local governments from outside the disaster zone, most of them located far from the disaster zone, responded by dispatching their own personnel to assist. By July 1, 2011 (15 weeks after the earthquake struck), approximately 57,000 local government personnel or one in 40 of the total local government workforce of 2.4 million (neither figure includes police or firefighters) had been dispatched to the disaster zone.

Why did local governments outside the affected areas choose to dispatch their own personnel to assist in disaster relief and reconstruction efforts? Local government employees have a responsibility to serve the local government by which they are employed (*Local Public Service Act*, Article 35). Most local government employees in Japan first obtain employment in their twenties and continue working for the same organization until they reach the mandatory retirement age at 60. This pattern is typical of employment relations in Japan more generally. The *Local Autonomy Act* stipulates that local government employees must work for the benefit of the residents of the jurisdiction in which they are employed (Article 1-2).[1] However, after the Great East Japan Earthquake, many local governments located far from the disaster zone dispatched workers to help those affected.[2]

Japan has a relatively small number of public service employees compared with other countries.[3] Yet these local governments are responsible for providing education, welfare and a host of other services to local residents. Expenditure on local government services accounts for some 60% of total government spending. However, in recent years there have been extensive and ongoing cuts to the local government workforce resulting in a decrease in its overall numbers from 3.2 million in 2000 to 2.7 million in 2015. Many municipalities face budgetary pressure. They have undertaken a series of administrative reforms to cut staff and today they operate with a bare minimum of employees. Under normal circumstances, local governments are unable to even consider the possibility of providing services beyond those that are directly connected to the welfare of local residents.

In these difficult circumstances, why did so many local governments dispatch personnel to the disaster-affected areas? What kind of local government personnel were sent to the disaster zone and how did the deployment take place? This chapter explores these questions.

Local government and the transfer of personnel

During the Edo period, local administration was carried out by the 300 clans, each of which ruled over their own domains. Following the Meiji Restoration, the new government set out to build a unified modern state. The Meiji reformers created a centralized administration under the Ministry of the Interior and its regional offices, which were termed ‘prefectures’. Within this system, local municipalities below the prefecture level (cities, towns and villages) were granted limited autonomy. While the municipalities were nominally the smallest ‘autonomous’ units within the Meiji state, the central government also saw them as the smallest unit of government administration. The authority of the state was divided up between the regionally based local governments, which acted on its delegated authority. This system was called Agency Delegated Function. Municipal mayors had the authority to carry out local administration on behalf of the state. Regional administration was carried out through the prefectures, large regional government bodies and the smaller, local municipalities. For the municipalities, therefore, their function as an organ of the national state and as an autonomous local government became indivisible. Some scholars named this status as the Fusion model (Amakawa 1983).

After the end of the Second World War, Japan enacted a new constitution. Chapter Eight of this constitution guarantees local autonomy and provides for the direct election of local government leaders (Article 93, paragraph 2). Prefectural governors also became subject to election, and the prefectures, which had previously been simply administrative units, gained legal status as autonomous regional government bodies. However, the existing system of delegated authority, whereby the municipalities were responsible for implementing national government policies, was also expanded to include the prefectures. The national government continued to use the prefectures and the municipalities as its agency to implement its policies at a regional and local level. The existing system for the administration of local finance, which enabled central government control, also remained intact. Many former officials of the Ministry of the Interior who had worked in the prefectural governments before the Second World War remained in place in prefectural governments and became senior administrators. It also became customary for career public servants from the newly established Local Government Agency to undertake regular secondments in local government, particularly in the prefectures.

Critics have traditionally seen this situation as a continuation of the central government's pre-war administrative and financial dominance, despite the provisions in the new constitution for electing prefectural governors. In the 1980s, however, some theorists proposed an alternative to this traditional view. They argued that by paying attention to the politics of regional government relations with the center, rather than focusing exclusively on the way the center exercises power through controlling the administrative system, a different understanding of regional government relations was possible. Contrasting the new Horizontal Political Competition Model (HPCM) with the earlier Vertical Administration Control Model (VACM), the new theorists emphasized the political influence that popularly elected mayors, as representatives of local interests, could have on the central government. They carried out empirical studies to try and demonstrate the superiority of their model.

Muramatsu Michio (1997: 28–31) argues that both perspectives are necessary in order to understand the relationship between central and regional governments. VACM and HPCM approaches produce different ways of understanding the secondment of central government officials to local government bodies. The traditional critique argued that the secondment of young public servants from central government agencies to prefectural and designated city governments was a means of maintaining central government control over the regions. These theorists argued that this custom had a negative impact on local autonomy and that it ought to be abolished. However, since the 1960s, when prefectural governments gradually began to recruit staff from among the local population, the data suggests that a major shift occurred in the prefectural governments. Although senior posts of the prefectural governments had previously been staffed primarily through dispatches from the center, they have been gradually displaced by local staff. The ministries that dispatch to local governments shifted from traditional ministries to other ministries. The number of local government personnel who were seconded to central government agencies also increased. Such data suggests that prefectures, under the leadership of elected governors, adopt a strategic approach to obtaining resources from the center by exercising a strong influence on which central government agencies personnel are seconded from (Inatsugu 2000). Personnel exchanges between central and local governments are no longer one-sided and involve local government personnel spending time in the central government as well. It is now seen as a form of networking between the central state and local governments. Both levels of government see the system as having advantages for them and it has continued to expand.

The development of the welfare state has given citizens greater access to public services. This has increased the resources and negotiating power of local governments and increased local autonomy. Central and local governments depend upon one another in order to deliver these services. The development of the welfare state has thus forced the two levels of government to recognize their interdependence (Muramatsu 1997: 130; Rhodes 1999). The two levels of government have come to depend on one another in terms of both personnel and policy. Muramatsu (1997: 140) argues 'that central-local relations are relations of overlapping authority, that the key to them is an understanding of the bargaining process, and that many local government strategies in bargaining can be explained by the principle of competition between local governments'. The implication of this argument is that when local governments decided to dispatch personnel (excluding firefighters and police) outside of their jurisdiction after the Great East Japan Earthquake, they were

Table 8.1: Number of local government personnel dispatched

			Unit: no. of people						
	Cumulative total	Currently dispatched	By prefecture				By profession		
			Iwate	Miyagi	Fukushima	Other	General	Civil engineering etc.	Other
2011. 7. 1	56,923	2,460	501	1,517	404	38			
2011.10. 1	73,802	1,211	250	644	290	27			
2012. 1. 4	79,107	804	186	373	230	15			
2012. 4.16	81,544	1,407	379	669	349	10	491	773	143
2012.10. 1		1,682	450	842	381	9	561	939	182
2013. 5.14	85,096	2,056	552	1,096	404	4	796	906	354
2013.10. 1		2,084	574	1,103	403	4	814	903	367

Note: Until the end of 2011, the cumulative total is available in the MIC data but from 2012, only the number currently deployed is publicly available on the website. In the cumulative totals given above, 81,544 is as at March 31, 2012, and 85,096 is as at March 31, 2013.
Source: Constructed by the author based on data from the Ministry of Internal Affairs and Communication website.

very aware that other local governments were doing the same. This is a prime example of the positive aspects of a system based on interdependence.[4] I explore this argument further in the conclusion to this chapter.

Local government personnel dispatches after 3.11

In this section I begin with an overview of the scale of the dispatch of local government personnel that took place after the earthquake. The statistical data provided by the Ministry of Internal Affairs and Communications (MIC) does not include the number of firefighters and police who took part in emergency assistance activities. I therefore discuss the fire and police data in a later section.

Scale

By July 1, 2011, 56,923 local government employees had been dispatched to the disaster zone. This was two-thirds of the total number dispatched in the two years following the earthquake.

Looking at the way the relief effort developed over time, 2,460 dispatched personnel were working in the disaster zone on July 1, 2011, compared with only 1,211 three months later on October 1. On January 4, 2012, there were only 804 dispatched personnel working in the disaster zone, demonstrating a decrease in the number of dispatched personnel over time. Dispatched personnel also performed different types of work and stayed for different lengths of time. On July 1, 2011, 484 people were helping manage evacuation shelters, 442 were assisting in the distribution of financial aid to disaster victims, and 348 were providing medical, health and hygiene services. By October 1, the total number of dispatched workers had shrunk by half. Only 25 were helping to manage evacuation shelters compared with the 484 who were working in this area just three months earlier. However, more of the dispatched personnel were assisting with rebuilding infrastructure (283→382) and reconstruction measures (129→146). In terms of occupation, civil engineers, builders and other technical specialists tended to be dispatched for longer periods. From the latter part of 2011, more of the dispatched workers were taking part in reconstruction efforts. From April 2012 there was a small but stable increase in the overall number of dispatched personnel working

Table 8.2: Schemes for providing local government assistance after the Great East Japan Earthquake

	Centralized	Decentralized
Coordinated approaches		
In advance	Fire, police	Disaster Medical Assistance Team (DMAT)
Afterwards		National Governors Association (NGA) scheme MIC/Japan Association of City Mayors (JACM) scheme
Dispersed approaches		Disaster assistance and sister-city agreements

in the disaster zone. In terms of profession, almost half were civil engineers and other technical specialists, suggesting that the primary focus of the dispatch was on providing technical support for disaster reconstruction.

Mechanisms for dispatching local government personnel

There are a number of ways of categorizing the different methods of dispatch for local government personnel after the earthquake. Some of the dispatches were directed by the national government while others were organized on a voluntary basis. Firefighters were dispatched to the disaster zone at the direction of the national government. The dispatch of police followed a similar scheme. I refer to these as centralized schemes.

Unlike police and firefighters, however, the central government does not have the power to direct other local government personnel to assist with disaster relief. The dispatch of local government administrators and technical specialists invariably occurred following requests by local governments in the disaster-affected areas or by national local government associations. I refer to these as decentralized schemes. The decentralized schemes can, in turn, be further classified depending on how they were coordinated. In order for the dispatch of local government personnel from outside the disaster zone to take place, there had to be a mechanism for accepting requests for assistance from the affected local governments and matching them with offers of assistance from outside the affected area. In this section, I discuss each of these approaches in turn.

The first approach, which I refer to as the 'in-advance' approach, is typified by the dispatch of the Disaster Medical Assistance Teams (DMATs). The Ministry of Health, Labour and Welfare (MHLW) coordinates the DMAT program and maintains a register of the teams that are available for deployment in a disaster situation. The fact that information about the DMATs was already available in a single location meant that deployments could proceed quickly in response to identified needs.

The second approach, which I refer to as the 'dispersed' approach, is typified by local governments in the disaster zone that had pre-existing disaster assistance or sister-city agreements with a local government outside the zone. Examples of disaster-affected local governments that received one-on-one assistance based on this type of agreement were widely reported in the press. During the initial stages of the disaster, while the national government (including MIC and other agencies) still lacked an overall grasp of the situation, these agreements provided for the dispatch of personnel to the disaster zone almost on a 'guerrilla' basis.

A number of mechanisms whereby local government personnel were dispatched to help with disaster relief cannot be classified under either of these approaches. These schemes were not based on pre-existing agreements or systems. At the prefectural level, the Union of Kansai Governments led the way in developing a scheme to provide support to the affected prefectures that was mainly administered by the National Governors Association (NGA). At the municipal level, a scheme was devised by MIC and the Japan Association of City Mayors (JACM) to match the needs of affected municipalities with offers of help from outside. These schemes do fit with a centralized, national government approach but they differed from the in-advance approach exemplified by DMAT. Information about the resources available was only collected incrementally after the disaster had occurred, so I refer to this as an 'afterwards' approach.

Of the decentralized schemes, the in-advance approach taken by DMAT was very highly regarded. The dispersed approach, which involved the dispatch of personnel by individual municipalities on the basis of existing agreements, was generally rated highly in terms of the speed of the response, but critics have pointed out that it resulted in an uneven distribution of human resources. This approach depended on the existence of individual connections between local government, making it incapable of meeting the needs of all of the affected areas. Furthermore, the number of personnel dispatched via this approach was not as high as might be suggested by the degree of publicity it received.[5] The majority of personnel were dispatched to the disaster zone based on schemes that used the afterwards approach.

In the next two sections I examine the dispatch of human resources for disaster relief after the Great East Japan Earthquake under each type of scheme.

Centralized schemes

Firefighters

In Japan firefighting is provided at the municipal level. There is generally one brigade stationed in each municipality, though some smaller municipalities have banded together to establish joint fire brigades with their neighbors. Article 44 of the *Fire and Disaster Management Organization Act 1947* provides for the establishment of Emergency Fire Response Teams (EFRTs) for deployment during large-scale disasters. The EFRTs were established in June 1995, after the experience of the Great Hanshin–Awaji Earthquake. The purpose of the teams is to enable effective and rapid deployments on a national scale in case of major disasters that exceed the capacity of fire brigades within the affected prefecture. EFRTs include members of fire brigades throughout Japan. In June 2003 the *Fire and Disaster Management Organization Act* was amended to provide a legislative basis for the EFRT system and to empower the commissioner of the Fire and Disaster Management Agency (FDMA) to direct operations during times of major disaster. Under the amendment, the Minister for Internal Affairs and Communications is responsible for producing a basic plan for the EFRTs and the commissioner of the FDMA is responsible for registering individual EFRTs. As at June 1, 2012, 781 fire brigades across Japan (98% of all fire brigades) were taking part in the EFRT program, which had a total of 4,431 teams (FDMA 2013: 379). The FDMA maintains a central register of all available EFRTs.

At 3.40 pm on March 11, the commissioner issued a mobilization directive to land-based EFRTs from 20 prefectures under Article 44, paragraph 5, of the *Fire and Disaster Management Organization Act*. Additional teams were mobilized as the scale of the disaster was revealed. Firefighting helicopters across Japan were also put into a state of readiness. Those helicopters that could not make it to the affected area by sunset were ordered to forward-operating bases in Kantō in preparation for rapid deployment to the affected areas. The EFRTs conducted firefighting and search and rescue operations in the disaster zone in conjunction with local fire brigades and other agencies in the extremely difficult conditions that prevailed immediately after the disaster. The peak of the mobilization was on March 18, one week after the disaster, when 1,870 EFRTs containing 6,835 individual firefighters were deployed. The deployment continued for 88 days, from March 11 until June 6. During this period, 8,854 units from the EFRTs were deployed at any one time (for a total of 31,166 units over the course of the mobilization), involving 30,684 individual firefighters (109,919 over the course of the mobilization) (FDMA 2013: 20–3, 379–83).

The mobilization of the EFRTs was directed by the commissioner under the legislation. This is an example of a centralized disaster response scheme, enshrined in legislation.

Police

The police in Japan are organized on a prefectural basis, with a police department located in each of the 47 prefectures. A central government agency, the National Police Agency (NPA), is responsible for coordinating the police force as a whole. The Inter-prefectural Emergency Rescue Unit (IERU) was established in June 1995, after the Great Hanshin–Awaji Earthquake, to provide a rapid, wide-area response by police to large-scale disasters. The IERU's specialist personnel conduct high-level emergency rescue operations. There are a total of 4,800 personnel attached to the IERU across Japan. Their activities are coordinated by the NPA. The IERU includes members of Prefectural Riot Police Units, Regional Police Riot Units, the Traffic Police Force and Expressway Traffic Police Units that have the relevant skills and experience for disaster relief work.

After the Great East Japan Earthquake, the Public Safety Commissions of Iwate, Miyagi and Fukushima prefectures requested assistance from the NPA, which responded by mobilizing a total of 389,000 members of the IERU (by June 20, 2011), up to 4,800 per day. In coordination with the Self-Defense Forces, local governments and fire brigades, the IERUs took part in a range of measures to support the safety and security of people in the disaster-affected areas. These included managing evacuations, search and rescue operations, and emergency traffic management (NPA 2012: 1). In addition to the 8,000 prefectural police in the three affected prefectures, the 4,800 police from across the country who were mobilized through the IERUs on a daily basis put a total of 12,800 police on the ground. Like the EFRTs, the police mobilization was based on a pre-existing, centralized scheme.

Decentralized schemes

The in-advance approach

Emergency disaster medical relief was organized through the DMATs, which include medical staff from both public and private hospitals. The DMAT scheme was first developed in 2005, under the leadership of the MHLW. They are mobile medical teams trained in acute disaster response and are administered by a secretariat located inside the National Disaster Medical Center. In the context of disaster medicine, the term 'acute' refers to the initial stages of a disaster, when large numbers of casualties suddenly require treatment. Acute disaster response involves the ability to carry out medical care while assessing the severity of symptoms and making assessments as to their urgency so as to determine the priority for treatment (triage). Each team is made up of approximately five members, including doctors, nurses and ancillary staff (healthcare and other professionals). As at June 2011, there were 882 DMATs spread across Japan (with 5,357 individual staff). When the Great East Japan Earthquake struck, the MHLW immediately requested the DMAT teams to deploy to the disaster zone, where they provided medical support at local hospitals, wide-area medical transport and other essential medical aid. The peak of the DMAT deployment was reached on March 13, when 193 teams were operating in the disaster zone (MHLW 2011: 149). By March 22, when the deployment ended, approximately 380 teams (1,800 personnel) had taken part (*Kahoku Shinpō*, February 18, 2012).

The DMAT system was established by regulation rather than legislation. Article 36, paragraph 1, of the *Disaster Countermeasures Basic Act* requires the MHLW to formulate a disaster management plan. DMAT operations are specified in Chapter 3, Section 2, of this plan. However, as it is not based on legislation, the establishment of the DMAT teams and the issuing of a deployment request by the ministry do not carry the force of law. Although it is a decentralized scheme in this sense, DMAT is managed by a corporate entity under the jurisdiction of the MHLW, which maintains a central register of the teams that are available when a disaster takes place. Therefore, the management of the deployment proceeded relatively smoothly.

However, after the deployment came to an end, many chronic patients with extensive healthcare needs remained in the disaster zone. In order to continue to provide medical care after the conclusion of the acute stage provided by DMAT, the MHLW asked the Japan Medical Association, hospital providers and other health services to cooperate with it to send doctors and other healthcare providers to the zone on request of affected prefectures. The Japan Medical Association Teams organized by the Japan Medical Association deployed along with other medical teams to the zone. Up to 156 teams (including approximately 706 personnel) primarily providing services to chronic patients were operating in the disaster zone on a daily basis (as at April 15). By July 12, a total of 2,438 teams involving 11,549 personnel had taken part (MHLW 2011: 150). By the end of June, 1,807 doctors, 2,560 nurses and 2,104 other staff from local government hospitals alone had contributed to the disaster medical effort. Of the 936 hospitals that are members of the Japan Municipal Hospitals Association, 53% (493 hospitals) sent medical teams to the disaster zone (Japan Municipal Hospitals Association 2011).

The afterwards approach

In 2010 prefectural governments in the Kansai area established the Union of Kansai Governments (UKG). The UKG played an active role in responding to the Great East Japan Earthquake. On March 13 its regional committee held an emergency meeting. The committee issued a statement indicating that it would begin providing aid to the disaster-affected prefectures. The system was based on pairing UKG member prefectures with the three prefectures in the disaster zone. Kyoto and Shiga were to take responsibility for Fukushima; Osaka and Wakayama would provide assistance to Iwate; and Hyōgo, Tokushima and Tottori would support Miyagi. On March 14 the Iwate and Miyagi prefectural governments established offices to coordinate the scheme. The UKG created a system for identifying the needs of the affected areas and matching them with appropriate support. The UKG's scheme was already getting underway when the NGA also took action. Local governments such as Hyōgo Prefecture, which is a member of the UKG, and its capital city of Kōbe, sustained severe damage during the Great Hanshin–Awaji Earthquake. They were therefore able to make use of their past experiences to grasp the needs of governments in the disaster zone and determine the best ways in which to support them.

The NGA established an emergency regional disaster response headquarters the day after the disaster and created a system whereby prefectures throughout Japan could provide material and personnel support to the affected prefectures. The NGA's role would be to coordinate the system (NGA 2013: 2–12). Soon after the disaster, the affected prefectures asked that additional personnel be sent to help collect and distribute material disaster aid and manage evacuation shelters. They also asked for caseworkers who could address victims' psychological needs. The NGA's disaster response headquarters tried to supplement the existing national government response by carefully taking into account the specific requests it received from the affected prefectures. In the first weeks after the disaster, most of the additional personnel who were dispatched to help in the disaster zone were sent on by each prefecture individually. After March 23 the NGA's disaster response headquarters was able to begin collecting requests for assistance from the affected prefectures. They then tried to match these requests to the available resources. The NGA used the same matching system as the UKG, enabling the coordination of the dispatch to proceed relatively smoothly.

On March 12 MIC established a contact center within the Local Administration Bureau, where municipal governments in the disaster zone could access support in order to maintain local government services. By March 22 the ministry had created a system whereby personnel from municipalities around the country could be dispatched to the disaster zone on a short-term basis. Prefectural governments in the disaster zone were responsible for coordinating requests for assistance from their local municipalities and forwarding them to MIC. With the cooperation of the JACM and the National Association of Towns and Villages (NATV), MIC matched the requests for assistance from local governments to the available resources.

Table 8.3: Personnel dispatch under the NGA scheme

	General	Engineering	Other	Total
Requested by affected prefectures	417	210	75	702
Matched by NGA	402	204	26	632
Available for dispatch	516	434	151	1,101

Source: NGA 2013.

As I discuss further below, some one-to-one assistance was already taking place between those municipalities that already had disaster assistance or sister-city agreements. MIC's main focus was on providing support to those municipalities that did not have access to such networks. Figure 8.1 shows how the dispatches under the scheme were coordinated. The JACM and the NATV asked municipal mayors to provide personnel to take part in the scheme. The municipal affairs departments in each of the three affected prefectures collected requests for assistance from the municipalities. MIC's Local Public Service Personnel Department stood in the middle, coordinating the scheme and allocating the available resources to the affected prefectures.

The JACM, the NATV, the prefectural municipal affairs departments and MIC cooperated in order to coordinate the municipal employee dispatch scheme. However, the complexity of the process meant that multiple inquiries were often needed to ensure that the appropriate resources were allocated to each municipality. As a result, it took some time to organize the dispatch of municipal employees for medium- to long-term assignments.[6]

The process for dispatching water and sewerage workers to the affected municipalities was somewhat different. The Japan Water Works Association and Japan Sewage Works Association worked in conjunction with the Water Supply Division in MHLW's Health Service Bureau and the Sewerage and Wastewater Management Department in the Ministry of Land, Infrastructure, Transport and Tourism respectively to determine the availability of municipal water and sewage workers for dispatch to the disaster zone and to request their assistance. For example, in the afternoon of March 11, Neyagawa, in Osaka Prefecture, received an inquiry from the prefectural office of the Japan Water Works Association in Osaka

Figure 8.1: The Municipal Employees Dispatch Scheme

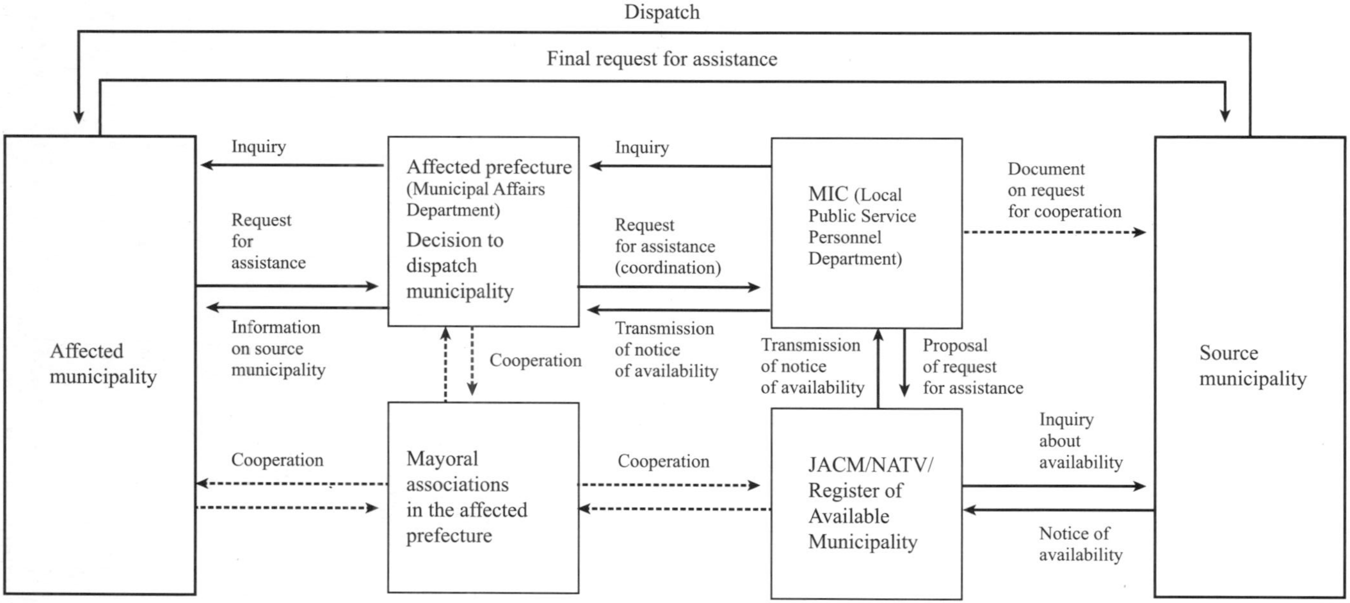

Source: Ministry of Internal Affairs and Communications homepage, '*Sōmushō ni okeru hisai chihōo kōkyō dantai ni taisuru jinteki shien no torikumi*' (*System for handling requests to the Ministry for the dispatch of municipal employees*).

about the availability of personnel and water supply vehicles. On March 16 the city sent a team of four water workers, one water supply vehicle and one support vehicle to Miyako in Iwate Prefecture. On March 19 Neyagawa sent a second team of two workers to supplement the first.[7] The Japan Water Works Association (and the MHLW) coordinated the matching. Due to the specialized nature of the work, the task of matching requests with available resources was probably easier than was the case for the more general MIC/JACM scheme.

The dispersed approach

There were numerous reports in the mass media about individual municipalities that received assistance based on existing disaster assistance and sister-city agreements. In the early stages of the disaster, MIC failed to keep abreast of these dispersed assistance efforts.

Sendai, the largest city in the Tōhoku region and the prefectural capital of Miyagi, is a signatory to a disaster agreement with Tokyo and 18 other designated cities throughout Japan known as the Twenty Major Cities Disaster Cooperation Agreement. After the Great East Japan Earthquake, Sendai received assistance under this agreement. Yokohama is also a signatory to the agreement. Both cities have female mayors (which is still very rare in Japan), which has been the basis of longstanding cooperation between the two. After the disaster, Yokohama dispatched a considerable number of personnel to Sendai. On March 13 an advance party arrived to start work in Sendai. By May 21 they were joined by the main force of 366 personnel to form a strong administrative team of 57 personnel, a logistics team of 93 and an evacuation center management team of 266 (FDMA 2013: 563).

A number of other municipalities also had individual disaster assistance agreements. After witnessing the Great Hanshin–Awaji Earthquake, Shiroishi, in Miyagi Prefecture, concluded a disaster assistance agreement with its sister city, Ebina, in Kanagawa Prefecture. After the 2011 earthquake Ebina immediately began sending material aid to Shiroishi and has continued to provide personnel to assist in activities such as inspecting damaged homes (FDMA 2013: 562).

The municipality of Date, in Hokkaidō, has sister-city agreements with Yamamoto and Watari, both located in Miyagi. Three days after the disaster, Date established a sister-city support office to coordinate its disaster response. After discussions with the affected municipalities and verifying the situation on the ground, Date decided to focus its disaster relief effort on Yamamoto, which was badly damaged. Beginning on March 20, Date sent support personnel to Yamamoto on a number of occasions (FDMA 2013: 563). Minamisōma, in Fukushima Prefecture, has a mutual assistance agreement with Tokyo's Suginami Ward. Along with Nayoro in Hokkaidō and a number of other municipalities that have similar agreements with Suginami, the ward established a Local Government Support Committee (*Hokkaidō Shimbun*, April 27, 2011). Suginami owns a residential facility in Higashiagatsuma, Gunma Prefecture, which was formerly a recreation center. After verifying its suitability and that of a number of similar facilities that existed in the area, on March 16 Suginami bussed 200 evacuees from Minamisōma to Higashiagatsuma. Every year, Kesennuma in Miyagi Prefecture sends fish to Meguro Ward in Tokyo for the Meguro Pacific Saury Festival. On the basis of this connection, the two municipalities concluded a sister-city agreement in 2010. After the disaster, Meguro sent six to ten staff to Kesennuma each week to provide assistance (*Asahi Shimbun*, November 12, 2011).

These dispatches, based on mutual assistance and sister-city agreements, did not depend on mediation by the national government. They involved one-to-one or one-to-many support and facilitated a quick response. However, the dispersed approach depended on pre-existing sister-city relationships and personal connections between municipal mayors. It was not a systematic solution. Furthermore, this type of ad hoc assistance did not constitute a large proportion of the overall number of local government personnel that were dispatched to the disaster zone. The example of Miyako, in Iwate Prefecture, demonstrates this point.

Miyako is located at the easternmost point of Honshū, Japan's main island. It formed an alliance with three other municipalities located at the extremes of Honshū, including Ōma in Aomori Prefecture, the northernmost point, and Shimonoseki in Yamaguchi Prefecture, the westernmost point. In 2009 the four cities also concluded a mutual assistance agreement. After the disaster, Miyako's partner cities sent large quantities of material support to Miyako. Shimonoseki dispatched personnel on an ongoing basis.

Miyako also has a sister-city agreement with Kuroishi in Aomori Prefecture, which also led to the dispatch of personnel after the disaster. Miyako had also been sending fish to Tokyo for the Meguro Pacific Saury Festival since it first began in neighboring Shinagawa in 1996. After the disaster, Shinagawa dispatched municipal personnel to Miyako to provide disaster relief.

However, the total number of personnel dispatched to Miyako based on these pre-existing agreements and relationships did not make up a large proportion of the overall number. On November 1, 2012, 18 months after the earthquake, 25 personnel from other municipalities were stationed in Miyako on a medium- to long-term basis. Only 20% of these came from cities that had an existing relationship with Miyako (four from Shinagawa and one from Shimonoseki). The remaining personnel were allocated under the JACM and NGA schemes. Some of the additional personnel who came to Miyako under a Ministry of Land, Infrastructure, Transport and Tourism scheme for technical specialists happened to come from Okayama. Town planners from Okayama were also sent to Miyako at a later date (three as at November 1, 2012) but there was no connection with the earlier dispatch of technical specialists.[8] Eighty per cent of the personnel who were dispatched to Miyako came through schemes that were devised after the disaster.

Twenty designated cities signed up to the mutual assistance agreement with Sendai mentioned above. Of these, Yokohama dispatched the largest number of personnel to Sendai. The other designated cities sent personnel elsewhere. For example, Kitakyūshū sent a significant number of personnel to Kamaishi, as did Nagoya to Rikuzentakata. These dispatches were not made under the Twenty Major Cities agreement but were based on relationships established after the disaster.

Rikuzentakata lost 68 of its 295 full-time municipal employees in the disaster. Many of these were core staff in their thirties and forties who were not far off becoming section managers. The municipal government offices also sustained heavy damage, forcing the municipal government to work out of a temporary office housed in a prefabricated building. As the reconstruction process got underway, the administrative workload increased rapidly. The disaster created debris that was equivalent to 200 times the amount of waste generated by the municipality in an average year. This gives an indication of the scale of the challenge facing the municipality. In 2010 the city's initial budget was 11.32 billion yen. In 2013 it was 101.91 billion yen, nine times that amount. As at March 1, 2012, 60 personnel from other municipalities were stationed in the local government offices on a medium- to long-term basis to help Rikuzentakata deal with this enormous workload; 32 were from Nagoya.[9]

The relationship between Rikuzentakata and Nagoya was only established after the disaster. Having sent a survey team to the disaster zone, Nagoya learned that Rikuzentakata was suffering from a massive shortage of personnel, leading to a breakdown in the functioning of government. The city decided to focus its entire disaster relief effort on Rikuzentakata.[10] Of the other outside personnel working in Rikuzentakata, eight were from the Iwate prefectural government and 11 were from Ichinoseki. Thirteen of the Nagoya workers returned home in 2012 and were replaced by personnel from other local governments. Rikuzentakata's mayor had a personal connection with the mayor of Takeo, in Saga Prefecture. Based on this relationship, Takeo sent two workers to assist Rikuzentakata. The remaining personnel were dispatched through the JACM/MIC scheme. Most of the personnel who were dispatched to Rikuzentakata came either through relationships that were established after the disaster or through the JACM/MIC scheme. Only a few were dispatched based on existing agreements.

Seventeen people were dispatched in 2011 to the municipality of Kamaishi, in Iwate Prefecture, from other municipalities (five from Gifu Prefecture and four from Kitakyūshū). In 2012 this number had grown to 37 (ten from Kitakyūshū, five from the Gifu Mayors Association and three from Osaka municipality).[11] Kamaishi did not have a specific agreement with Kitakyūshū before the disaster. Initially, Kitakyūshū dispatched local government workers to a number of municipalities in Tōhoku at the request of governments and other agencies. Later, the Kitakyūshū municipal government adopted the idea of focusing the city's relief effort on a particular municipality, based on Kitakyūshū's own characteristics and strengths. This led to the decision to provide active and ongoing support to Kamaishi.

Both municipalities had a connection with steelmaking. When nurses from Kitakyūshū were sent to work in Kamaishi at the request of the MHLW three days after the disaster, Kitakyūshū was prompted to offer its assistance to Kamaishi, to which it received a positive reply. Kitakyūshū began by helping Kamaishi with managing its evacuation shelters and continued to provide ongoing support. The other local government workers dispatched to Kamaishi came through the JACM/MIC scheme and through connections forged after the disaster.

The interdependence of local government

In this chapter I have discussed the different mechanisms by which local governments across Japan provided assistance to their counterparts in the disaster zone and the different types of assistance they provided. The Emergency Fire Response Teams and the Inter-prefectural Emergency Rescue Units were two centralized schemes for providing disaster relief that were provided for in legislation. Under these schemes, the central government had the power to direct firefighters and police stationed in local government areas throughout Japan to deploy to the disaster zone.

The decentralized schemes, on the other hand, did not rely on central government directives. The deployment of the Disaster Medical Assistance Teams, which were formed before the disaster, adopted an in-advance approach. The DMATs are coordinated by a corporate entity that is under the jurisdiction of the MHLW. This enabled a rapid deployment. While this is a decentralized scheme, in the sense that it was not directed by the government, the existence of a coordinating body meant that provisions for a rapid response were already in place.

However, fire, police and medical assistance were not the only services required in the disaster zone. Local governments in the affected areas suffered from a severe shortage of personnel. Local governments had to survey damage, issue certificates to disaster victims to ensure they could access aid and certify disaster damage to homes, deal with debris, provide financial aid, construct temporary accommodation and make plans for reconstruction. Their ability to carry out these functions, not to mention the massive reconstruction effort that would become necessary later on, was hampered by an overwhelming lack of human resources. General expenditure for some local governments increased by more than ten times the level before the disaster, while investment increased by upwards of one hundred times. Local governments faced a rapid increase in their administrative workload. They needed more personnel to carry out this work.

Many municipalities had to tackle projects in which they had no experience. Many urban municipalities employ people with extensive town planning experience. On the Sanriku Coast, few local government employees have such experience. The small municipalities along the Sanriku Coast also lacked experience in carrying out reconstruction and revitalization works, such as assisting with the relocation of large numbers of disaster-affected households or constructing public housing. These municipalities needed the assistance of experienced local government employees from outside.

Local government employees from outside the area have the necessary knowledge. Their contribution to reconstruction efforts is vital. In a survey carried out by the Subcommittee on Public Administration and Local Government of the Japan Society for the Promotion of Science's Social Scientific Research Committee, 705 respondents were asked to name the 'kind of personnel… required in disaster situations where the administration is not functioning adequately'. Fifty-six per cent replied that 'public servants with the requisite technical knowledge and skill' were necessary. Local government employees, particularly municipal employees, perform administrative tasks for local residents every day. They can carry out tasks such as issuing disaster certificates or managing evacuation shelters without additional training. Local government employees also find it easier to gain the trust of disaster victims.

Those local governments that were already party to disaster assistance agreements received additional personnel via these agreements. Local governments that are linked through sister-city agreements already interact with one another. In many cases, individual citizens also have ties. Their children often take part in exchanges and similar activities, as do their mayors and municipal councillors. Some of these agreements include provisions for local government personnel exchanges lasting one to two years. The spirit of cooperation in times of crisis is already well established. These relationships provided the basis for the dispatch of personnel after the disaster.

However, only a limited number of personnel were dispatched via these agreements and only a limited number of local governments were covered by them. Those local governments that were unable to take advantage of these agreements could turn to the MIC/JACM scheme and the NGA scheme that preceded it.

The NGA scheme was based on the UKG program. Under this scheme, Osaka and Wakayama prefectures agreed to assist Iwate Prefecture. After Osaka Prefecture decided to help Iwate, its employees were dispatched immediately, but they were unable to meet the enormous need for assistance. Osaka responded by calling on its municipalities for help. The Osaka prefectural government asked the municipalities, which were represented in the Osaka Mayors Association, to supply personnel to take part in the relief effort. The mayors, upon receiving this request, decided how to respond. In doing so they paid close attention to the number and profession of the personnel being sent by other municipalities.[12] The NGA took decisions about the size of each prefecture's contribution based on the size of each prefecture and other factors. In order to carry out the scheme, MIC was completely dependent on the coordinating role played by NGA.

The dispatch of personnel through the MIC/JACM scheme was decided upon with the cooperation of JACM, the municipal affairs departments in each prefecture and MIC.

The central government did not play a leading role under the decentralized schemes, nor did it direct the regions to carry out these schemes. The central government followed the example of the NGA and obtained the full cooperation of the JACM in designing a scheme to reach those local governments in the disaster zone that had not yet received relief. Organizations such as the NGA and JACM did not have the resources to implement a disaster relief scheme on their own. They needed the government to collect information, distribute funds and issue the appropriate legislative orders and notifications.

In a discussion on inter-governmental relations, Rhodes (1999: 78) argues that:

1. Any organization is dependent upon other organizations for resources.
2. In order to achieve their goals, the organizations have to exchange resources.
3. Although decision making within the organization is constrained by other organizations, the dominant coalition retains some discretion. The appreciative system of the dominant coalition influences which relationships are seen as a problem and which resources will be sought.

In the example discussed in this chapter, Rhodes' notion of a 'dominant coalition' applies to both the NGA and the JACM. They were completely dependent on the national government for human resources and power.

Municipalities that dispatched personnel to the disaster zone did so for a number of reasons, in spite of the shortage of human resources they faced already. First, they had to go along with what other municipalities in their prefecture were doing. When municipalities received a request for assistance from the municipal affairs department in their prefectural government or from the prefectural mayors association, they had to think about their future working relationships with these organizations. They therefore had little choice but to send a similar number of personnel to other municipalities. Their own residents also urged them to do so. When reports of other cities sending personnel to the disaster zone appeared in the mass media, many city government offices received telephone calls from residents asking for confirmation that their local municipality would be participating in the relief effort. Despite the squeeze on their existing human resources, city governments could hardly ignore such calls. Reports also appeared in the media about local governments that were dispatching personnel to the disaster zone based on disaster assistance agreements and sister-city agreements. Examples of governments that simply dispatched their staff without such prior agreements also appeared. This generated a kind of competition between local governments.[13] Another factor that motivated municipalities to participate in the relief effort was the desire to build their own municipal workforce capacity to respond to disasters. This would be useful if they were faced with a similar disaster in the future.[14]

It seems reasonable to conclude that because the advantages of sending staff to help in the disaster zone (relationships with other agencies, responding to the urging of local residents, maintaining the municipality's reputation and developing its skills to cope with disasters at home in the future) outweighed the potential disadvantages (criticism from local residents, relationships with other organizations, cost of dispatching personnel), local governments decided to dispatch their own personnel to participate in the relief effort.

9 Service and Support by Local Governments outside the Disaster Zone

Wada Akiko

A basic principle of local government in Japan is the necessity for a resident to register with a local authority in order to be eligible for the broad range of administrative services offered by that authority. Enforcement of this principle during the Great East Japan Earthquake, however, would have jeopardized access to administrative services for disaster victims who were forced to leave their homes at the time of the emergency and evacuate to other parts of Japan. While disaster victim support is a feature of government services everywhere, the provision of such aid in the Japanese system is generally a function of local government. While local government provides evacuee support to disaster victims, some support requires proof of disaster victim status, which is another basic principle in Japan. Nevertheless, a mismatch between reality and basic principles after the earthquake made exceptions to these principles necessary. Decisions relating to disaster victims made outside the policy guidelines could generate monetary costs that could lead to questions being asked at local assemblies. In many cases, however, central government initiatives ensured financial support. While altruistic motives were certainly at play, this chapter investigates the types of incentives that encouraged local authorities to offer administrative services and evacuee support to disaster victims who relocated outside disaster-affected prefectures. Particular consideration is given to Yamagata Prefecture and the cities of Yamagata and Yonezawa within that prefecture. These authorities offered a range of administrative services and disaster victim support to all evacuees, including those who fled Fukushima for fear of nuclear contamination. The chapter seeks to suggest reasons for these authorities offering such services and support, regardless of whether or not evacuees held resident registration or proof of disaster victim status.

Characteristics of Japanese local governments

Japanese local governance is delivered by 47 prefectural governments and approximately 1,700 municipalities, comprising cities, towns and other small communities.[1] It is stipulated in law that the local government closest to the resident has responsibility, as far as possible, for the delivery of administrative services (Local Government Law, Article 1-2). In concrete terms, Japanese local government provides residents with an extremely wide range of administrative services that include health and welfare provisions, education, policing, fire and rescue services, garbage collection, maintenance of roads and water reticulation infrastructure, business and commerce support, environmental protection and the administration of electoral services. These local services are supported by three main sources of funding, namely local taxes, national taxes tagged for disbursal to local authorities (local allocation tax) and national government subsidies (national Treasury disbursements). Local taxes largely comprise personal income tax[2] and fixed asset and property taxes. Local allocation tax is the portion of national taxes distributed to ensure income parity between different local administrations and thereby to guarantee that each has sufficient funds to offer a defined range of services. As with local taxes, no parameters limit spending. In contrast to local taxes, however, the funds are provided by the central authorities. Local allocation tax consists of the usual tax grants distributed according to a common national standard and special grants that take account of particular circumstances such as disaster. The national Treasury disbursements include multiple disbursals for matters such as central government block grants for the delivery of compulsory education. In the 2014 financial year, all local government in Japan was funded by 36.0% local taxes, 17.1% local allocation tax and 15.2% national Treasury disbursements (White Paper on Local Public Finance 2016). Since it is difficult to offer the extraordinarily wide range of services provided by Japanese local government on local taxes alone, many local administrations rely on central government funding.[3]

In the wake of a disaster, the provision of support for those who are evacuated from their homes (hereafter, evacuee support) is also the responsibility of local governments.[4] It is stipulated by law (*Disaster Relief Act*, Articles 2 and 13) that prefectural governors or heads of local authorities will enact the provision of services such as evacuation center facilities and temporary accommodation for persons who require such support. The evacuation center is a place where those whose homes are no longer habitable as a result of disaster receive short-term provision of food and shelter. Schools and community centers are often used for this purpose. Temporary accommodation, on the other hand, provides evacuees with provisional housing, often in prefabricated structures.[5] Once temporary accommodation becomes available, evacuees move from evacuation centers to this type of housing, where they wait for permanent accommodation to be built. Based on the law, the cost of providing temporary accommodation is the responsibility of the prefectural authority in which the disaster occurred. Evacuees do not have to pay for it. Depending on the budgetary circumstances of the authority concerned, however, it is not unusual for between 50% and 90% of this funding to come from the central Treasury.[6]

In addition to providing regular administrative services, then, Japanese local government also administers disaster aid and relief. Compared with other countries, however, two points are particular to the system in Japan. The first is the manner in which local governments provide services for residents based on resident registration (Ōnishi 1993: 64–105, 1994: 137–69). This resident register is maintained as a list of all who reside in a particular local area. In addition to the name of the person, the register lists the household in which the person resides. In the event of a change of address, the resident is obligated to inform the local authority concerned. While other countries draw on similar resident registration systems, Japan is the only country that in principle bases its delivery of local administrative services on this resident information source. Authorities also use the register to provide information to residents. Details extracted from the register regarding school-age children, for example, are used to make contact with parents regarding available primary school options. At election time, including national elections, it is the responsibility of local authorities to compile electoral rolls and to inform people of the location of polling booths. Local authorities compile these rolls based on resident registry information.

It might be noted that, notwithstanding the importance of resident registration in local government service delivery, a number of systematic exceptions are offered.[7] Some user-pay services are also available to unregistered residents.[8] However, the provision of services based on registry information is very much the accepted practice.

The second point concerning practices in Japan that differ from those in other countries relates to the provision of disaster aid. In order for an evacuee to receive aid of some kind, she or he is required to provide proof of disaster victim status. Disaster support is only available to disaster victims and cannot be accessed by those who do not have victim status. As a rule, for example, temporary accommodation is only provided to those with proof that they have been forced to evacuate.[9] The concrete form of this proof is generally a Disaster Evidence Certificate that is issued by the head of the relevant disaster-affected local authority. The purpose of these certificates is to prevent dishonest claims being made on temporary accommodation services.

The response of local and prefectural authorities

At the time of the Great East Japan Earthquake, the combination of earthquake, tsunami and nuclear power station accident created large numbers of disaster victims.[10] A particular aspect of that event was the fact that many people evacuated the local area or prefecture in which they lived at the time of the disaster to move to other parts of Japan. Japan has a high incidence of natural disasters and the Great East Japan Earthquake was not the first time that disaster evacuees had crossed local or prefectural authority borders.[11] However, at the time of the earthquake, a number of the local and prefectural authorities that accepted disaster victim evacuees (hereafter, receiving authorities) undertook policy initiatives that had not been seen before.

First, a number of receiving authorities offered certain administrative services to evacuees regardless of whether or not they held the relevant resident registration. As noted above, a fundamental principle of access to local and prefectural government services in Japan is registration on the resident register of the jurisdiction in which one lives. Yet some authorities provided certain services to evacuees regardless of their registration status. These services included infant health checks and vaccination, in addition to the provision of school attendance allowances for the children of families in financial distress. In other words, administrative services to disaster victims were provided in a manner that deviated from the resident register principle.

The second point concerned the relaxation of the condition for those accessing temporary accommodation to provide a Disaster Evidence Certificate. As a rule; the provision of temporary accommodation is the responsibility of prefectural or metropolitan authorities. Such provision generally only occurs when a Disaster Evidence Certificate has been issued by the head of the disaster-affected local authority in which the evacuee originally resided. Following the Great East Japan Earthquake, however, some receiving authorities made temporary accommodation available even to evacuees without these certificates. Persons eligible for Disaster Evidence Certificates at the time were those whose homes had been lost or damaged in either the earthquake or tsunami. Also eligible were those forced from their homes following the evacuation orders issued by the central authorities in the wake of the nuclear accident. There were, however, people living outside the evacuation order zone who left their homes because of concerns about radiation contamination related to the nuclear reactor meltdown. While there were many people in this category, known as voluntary evacuees, who could not access Disaster Evidence Certificates, a number of local authorities nevertheless provided these people with temporary accommodation.[12]

Why did some local authorities provide partial administrative services for evacuees who were not registered on the local resident registry? And why did some prefectural authorities provide temporary accommodation even for evacuees without the necessary Disaster Evidence Certificates? In providing answers to these questions, we might consider the cases of Yamagata Prefecture and the cities of Yamagata and Yonezawa in that prefecture.

Yamagata Prefecture is adjacent to Miyagi and Fukushima prefectures. After the disaster-affected prefectures of Iwate, Miyagi and Fukushima themselves, Yamagata was the prefecture that took the largest number of disaster victim evacuees.[13] Yamagata was particularly distinctive in that it accepted large numbers of voluntary evacuees (Wada 2015: 194). Of all the local administrations in Japan, the Yamagata Prefecture city authorities of Yamagata, where the prefectural offices are located, and Yonezawa, which directly adjoins Fukushima Prefecture, accepted the largest numbers of disaster victim evacuees.[14] Yamagata Prefecture provided temporary accommodation for voluntary evacuees (that is, for evacuees without the usual evidentiary documentation), while the cities of both Yamagata and Yonezawa offered some administrative services for evacuees even without resident registration. Looking more closely at these examples will help provide answers to the questions posed above.

Disaster victim evacuee support[15]

Yamagata Prefecture borders Miyagi Prefecture in the east and Fukushima Prefecture in the south. In March 2014 the population was approximately 1,137,000 (Yamagata Prefecture 2014). Yamagata City, where the prefectural head office is located, is in the east of the prefecture and, in addition to sharing a common border with Miyagi Prefecture, is about a one-hour ride to Fukushima by *shinkansen* (bullet train). In March 2014 the city had a population of 254,000 (Yamagata City 2014). Yonezawa City, in Yamagata Prefecture's south, borders Fukushima Prefecture. Its population in March 2014 was 87,000 (Yonezawa City 2014).

After the Fukushima Daiichi Nuclear Power Station accident that followed the March 11, 2011, earthquake, the government ordered the evacuation of residents within a three-kilometer radius of the plant.[16] The following day, the evacuation order was expanded to a ten-kilometer, and eventually a 20-kilometer, radius. Many of those affected evacuated to places that were not merely outside the zone but outside the prefecture.

As two main locations that accepted evacuees from the nuclear evacuation zone, Yamagata City and Yonezawa City opened evacuation centers on March 13 and March 14 respectively.[17] Soon after this, both public housing and properties rented from private landlords were made available as temporary accommodation (hereafter, designated temporary accommodation). With the movement of evacuees into this accommodation, the evacuation centers closed on May 18 in Yonezawa City and June 30 in Yamagata City. Following these closures, both cities established Disaster Evacuee Support Centers[18] as a means of continuing evacuee support.

On June 15 Yamagata Prefecture expanded the definition of those eligible for entry to the designated temporary accommodation to include voluntary evacuees.[19] As a result, there was an increase in the numbers of voluntary Fukushima evacuees seeking residence in designated temporary accommodation in both Yamagata City and Yonezawa City. On October 31, however, a shortage of available properties in these cities forced a cessation in the receipt of temporary accommodation applications. Evacuee numbers peaked on December 1 at 5,844 in Yamagata City (Yamagata Prefecture 2011a) (approximately 2.3% of the total city population) and on November 17, 2011, at 3,895 in Yonezawa (approximately 4.4% of the city population). After these dates, the numbers in each city began to fall.

On 18 March, 2014, among the 2,140 evacuees in Yamagata City (approximately 0.8% of the population), 1,998 came from Fukushima Prefecture. Among these, 532 evacuees (24.9% of all evacuees in Yamagata City) came from the 13 municipalities within Fukushima designated in the Nuclear Power Disaster Victim Evacuee Special Law.[20] Another 1,466 (68.5%) were from other municipalities in Fukushima Prefecture. At that time, there were 127 evacuees (5.9%) from Miyagi Prefecture and 15 (0.7%) from Iwate Prefecture. While most of the 13 local Fukushima authorities referred to in the special law were in the evacuation order zone, some were outside or partially outside. Evacuees who had formerly been resident in other municipalities in Fukushima Prefecture, however, were regarded basically as voluntary evacuees who did not hold Disaster Evidence Certificates. The exceptions were a small number who had received such documentation by virtue of tsunami damage to their properties.

Because Yonezawa City directly borders Fukushima Prefecture, there was a considerably larger ratio of voluntary evacuees here than in Yamagata City. On March 20, 2014, Yonezawa was home to 1,560 evacuees (approximately 4.4% of the city population). Among these, 353 (22.6% of all evacuees in the city) were from the 13 designated local authorities in Fukushima Prefecture, while 1,196 (76.7%)[21] were from other municipalities.[22]

Administrative service provision

As a rule, local authorities in Japan provided relief for disaster victim evacuees. From September 2011, however, in addition to regular relief, Yamagata City initiated the provision of 23 administrative services even to evacuees who were not on the resident register. These services included various health and education services such as offering infant health checks and vaccination and the provision of funds for school attendance to households in straitened circumstances.

From October of the same year, Yonezawa City also began to offer partial administrative services to non-registered evacuees. While there was an overlap in the nature of some of these services offered in both places, others were specific to either Yamagata or Yonezawa.

From January 1, 2012, evacuees from the 13 designated municipalities in Fukushima Prefecture who had come to either Yamagata or Yonezawa City became eligible for a range of services under the Nuclear Power Disaster Victim Evacuee Special Law (enacted in August 2011). However, since many of these services had been offered from the previous year, other than benefit aid for children with disabilities, few new services were actually introduced at that time. From January 1, 2012, in accord with the stipulations of the special law, local administrations throughout Japan began to offer designated services to evacuees without resident registration. Why, however, had this practice commenced several months earlier in the cities of Yamagata and Yonezawa? An explanation for this is considered in the next section.

Although it was difficult to precisely track unregistered disaster victim evacuees, responses to a series of annual surveys of evacuees conducted by Yamagata Prefecture indicated that the numbers of those who had not transferred their resident registration diminished each year.[23] In percentage terms these were 71.3% of evacuees in October 2012, 65.6% in October 2013, and 57.2% in October 2014. In addition to the fact that some did, indeed, register with the local Yamagata authorities that accepted them, the return of other evacuees to Fukushima is likely to have been a factor in this reduction.

Analysis

As explained above, in terms of the provision of temporary accommodation, Yamagata Prefecture provided this service for all disaster victim evacuees, even those who had left their homes on a 'voluntary' basis. With respect to administrative services, however, the specifics depended on whether the evacuee fell into one of three categories – those who had transferred their resident registration to the records of the local authority that received them, those from the 13 designated Fukushima authorities who did not transfer their registration, and those from outside those 13 authorities who did not

Figure 9.1: Administrative services and disaster victim evacuee aid provided by local and prefectural authorities that accepted evacuees

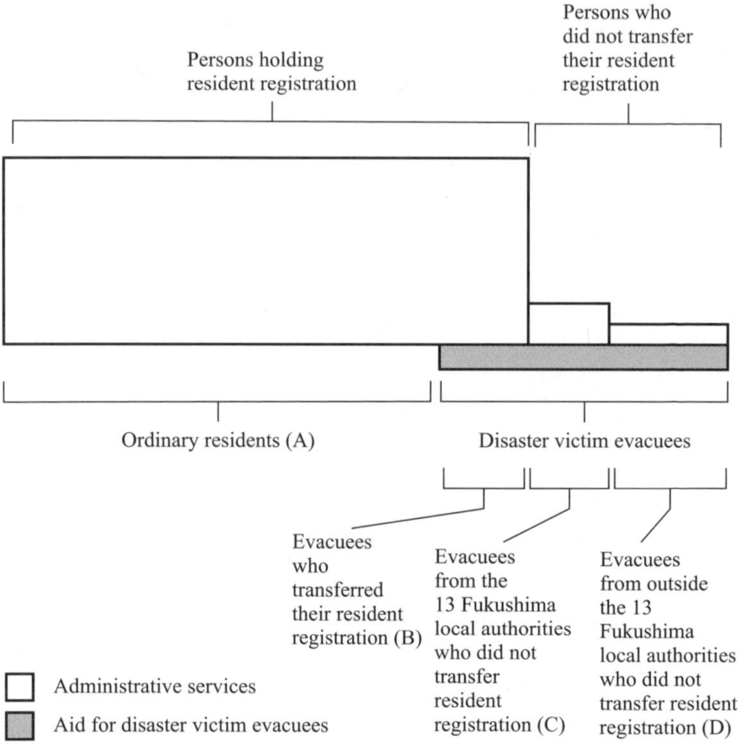

Source: Compiled by the author.

transfer their registration. We might consider the circumstances of each of these groups in comparison also with regular non-evacuee residents (Figure 9.1).

First, administrative services for ordinary residents (A) were provided by the local authority with whom they were registered. This practice derives from the governance principle of service delivery according to resident registration information.

Second, disaster evacuees who transferred their resident registration (B) were entitled on the same grounds to the service provision given to ordinary residents. They were also entitled to the supplementary support and aid provided for disaster victim evacuees. Among these were some who, although they had transferred their resident registration, did not have Disaster Evidence Certificates.

Third, evacuees from the 13 designated Fukushima local authorities who did not transfer their resident registration (C) received support and aid due to disaster victim evacuees. This group also received the partial delivery of administrative services given to evacuees who did not register as residents. From January 2012 evacuees from the 13 designated authorities became eligible for certain services under the Nuclear Power Disaster Victim Evacuee Special Law. While the cities of Yamagata and Yonezawa had already commenced delivery of many services, these were expanded from January 2012 in response to the special law.

Fourth, there were evacuees from outside the 13 designated Fukushima local authorities who did not transfer their resident registration (D). In addition to receiving aid as disaster victim evacuees, this group also received the partial administrative service provision that ordinarily would not have been provided to those without resident registration. This category of evacuee benefited from the judgment made by both Yamagata and Yonezawa cities to deviate from the usual practice of providing services based only on resident register information.

At this point it is useful to restate the two guiding questions of this discussion. First, why did the prefectural authorities supply temporary accommodation to all categories of disaster evacuee, including the voluntary evacuees who did not hold Disaster Evidence Certificates (B, C and D in Figure 9.1)? And, second, why did Yamagata City and Yonezawa City deviate from the usual local government principle of service provision to registered residents only and offer partial administrative services, ahead of central initiatives in this respect, to disaster evacuees not holding resident registration (C and D in Figure 9.1)?

The provision of temporary accommodation

In Yamagata Prefecture, temporary accommodation was provided in the form of public housing and private rental accommodation. Access to this type of designated temporary accommodation would generally only be possible if an applicant held a Disaster Evidence Certificate. Yamagata Prefecture, however, waived this requirement.[24] There were reports of complaints from registered residents who pay for public housing or private rental accommodation regarding the Yamagata offer of free designated temporary accommodation to evacuees. Why did Yamagata Prefecture adopt these temporary accommodation policies regardless of resident dissatisfaction?

Immediately after the disaster, many Fukushima residents contacted authorities in Yamagata Prefecture with inquiries concerning the possibility of voluntary evacuation and access to rental accommodation in that prefecture (Yamagata Prefecture-wide Support Measures Headquarters 2011). In May 2011, as these inquiries were being received, Fukushima Prefecture notified all other prefectural authorities in Japan to confirm that private rentals accessed by Fukushima evacuees would be considered as temporary accommodation (Fukushima Prefectural Governor 2011). That means that the costs of this type of designated temporary accommodation would be borne by Fukushima Prefecture in conjunction with the central authorities.[25] The Fukushima notice in fact cited a Ministry of Health, Labour and Welfare statement to this effect (Head of the Social Support Bureau 2011). In other words, it was clear that the central government had given systematic financial approval to the provision of designated temporary accommodation for evacuees, including voluntary evacuees. This was undoubtedly a key factor both in the Fukushima notice and then in the Yamagata Prefecture temporary housing policy decisions.[26]

The provision of administrative services

The discussion now turns to the issue of why both Yamagata City and Yonezawa City deviated from the usual governance principles regarding resident registration in relation to the provision of administration services to evacuees. On 26 August, 2011, the Yamagata City Promotion of a Stable Lifestyle for Citizens Headquarters, chaired by the mayor, decided to provide partial administrative services even to disaster victim evacuees who had not transferred their resident registration. Prior to this, on August 19, the Minister for Internal Affairs and Communications had contacted all prefectural and metropolitan governors to request that they inform all heads of local authorities under their jurisdiction of the promulgation and enforcement on August 12 of the Nuclear Power Disaster Victim Evacuee Special Law (Minister for Internal

Affairs and Communication 2011). In addition to stipulating that the country would bear the financial cost of administrative services provided by receiving local administrations for 'disaster victim evacuee residents' (Article 9), the law indicated similar central government support for the delivery of services over and above these to such evacuee residents (Article 10). There was also an indication of support, based on the same law, for a similar level of service provision to evacuees who were not disaster victim evacuee residents (Addendum Article 3).

At the time of the August 19 ministerial statement, no definition of 'disaster victim evacuee resident' was provided, nor were the administrative services to be offered to disaster victim evacuee residents under the special law set out. Nevertheless, the principle of central government funding support in the case of service provision for non-registered evacuees was to some extent indicated. In the case of Yamagata City, the Minister of Internal Affairs and Communications communique seems to have been interpreted as confirming support for the provision of services to non-registered evacuees. In other words, Yamagata City's decision seems undoubtedly to have been influenced by the promulgation of the special law.

The provision of administrative services to non-registered evacuees began in Yamagata City in September 2011. Media reports of this initiative clearly influenced the direction taken by other local authorities. Since Yonezawa City bordered Fukushima Prefecture, large numbers of voluntary evacuees had relocated there. Yonezawa City initially hesitated, however, to deliver services outside the usual legislative parameters. Nevertheless, news that Yamagata City, the site of the prefectural headquarters, was offering services to evacuees, coupled with dissatisfaction expressed by evacuees in Yonezawa (often with the support of local residents regarding the limited support there), appears to have motivated that city to undertake similar evacuee policy directions. The influence of the Yamagata City decision on Yonezawa City's October 2011 commencement of administrative services to non-registered disaster victim evacuees is an example of what is known in local government discourse as 'cross-referencing'. This is the term used when one local authority considers the decision of a peer authority in setting its own policy directions.[27]

In November 2011 a definition of 'disaster victim evacuee residents' was made public (that is, evacuees from the 13 Fukushima nuclear accident-impacted local authorities) (Head of the Local Government Administration Bureau 2011). The types of administrative services approved under the special law were also stipulated, as was the level of support to be provided by disaster reconstruction special grants of local allocation tax. At this time, central authorities committed to covering 100% of costs incurred in the provision of administrative services based on the special law, and 80% of costs for services delivered over and above the special law. This left only 20% of the cost involved to be borne by the receiving local authority. Since close to 70% of the disaster victim evacuees in Yamagata City and Yonezawa City were evacuees from outside the 13 Fukushima local authorities, many of the administrative services provided incurred a 20% outlay from the city's own funds. Yet few registered residents appeared to be concerned. This is likely to have been due to the fact that the outlay was not all that significant in overall budgetary terms. There was also, undoubtedly, widespread sympathy for evacuees who desired to flee the possible effects of the nuclear accident. Furthermore, the numbers of evacuees in both Yamagata City and Yonezawa City gradually decreased, a trend that is probably explained by the eventual return to Fukushima by those who had departed on a voluntary basis.[28] Each of these factors is a matter for separate investigation.

Conclusion

The discussion above sought to explain why Yamagata Prefecture, which accepted large numbers of voluntary disaster victim evacuees, gave aid entitlements to voluntary evacuees, and why Yamagata City and Yonezawa City offered administrative services to disaster victim evacuees who did not hold the usual resident registration. Clearly, a major reason was the systemized financial support of the central government. In Japan, although the provision of administrative services and disaster victim evacuee support occurs at the local level, the resources that support these activities come from the central government. It is not surprising, therefore, that systematic financial support from the state became a key factor in the decisions made by the local authorities concerned.

The above fact indicates a possibility that special consideration for voluntary disaster victim evacuees and evacuees who did not hold the relevant registration has been made not only in Yamagata Prefecture but throughout Japan.[29] Given the nature of the support required, the absence of central government financial backing may, in the final analysis, have meant an inability to deliver these services.

In June 2015 Yamagata Prefecture decided that from April 2017 it would no longer offer designated temporary accommodation to voluntary evacuees (Yamagata Prefecture 2015). This decision came after notification to prefectural administrations throughout Japan to the effect that, as of April of that year, Fukushima Prefecture itself would no longer make temporary accommodation available to voluntary evacuees (Fukushima Prefecture 2015). Although there is generally a limit of two years of access to temporary accommodation, prefectural and metropolitan governors had received Office of the Prime Minister and Cabinet approval to extend this. The Fukushima decision indicated the withdrawal of this central support. Thus we can conclude that the Yamagata Prefecture decision to cease providing designated temporary evacuation to voluntary evacuees also resulted from the central government trend to withdraw systematic financial support.[30]

On August 29, 2016, Yamagata Prefecture announced that it would investigate the possibility of providing free accommodation to low-income voluntary evacuees in the form of the accommodation originally built for prefectural employees (*Yamagata Shimbun*, August 30, 2016). This announcement, which is specific to Yamagata Prefecture, came in response to pressure from groups established by volunteer evacuees and their supporters in order to agitate for the continued provision of designated temporary accommodation. Petitions advocating this had also been adopted by various local authorities throughout the prefecture. Such a provision would be yet another deviation from the usual practice. Nevertheless, the accommodation in question was not occupied, and unlike public housing or private rental accommodation, is not available to ordinary citizens. These factors, together with a largely zero monetary cost to the prefecture, suggest that complaints from registered residents would be minimal. The governor of the prefecture has tried to consider all stakeholders. These include those registered residents who feel a sense of injustice when the usual practice is overlooked for voluntary evacuees, and the voluntary evacuees and their supporters who seek the continuation of housing support. There has been support for the evacuees, too, from various local assemblies throughout the prefecture. In light of these developments, we might see the announcement as an expression of the governor's desire to provide support for low-income evacuees (something that few would protest) without generating added costs for the prefectural administration.

In principle, local government in Japan operates according to various legislative provisions and associated regulations. In times of major earthquakes and similar emergencies, however, authorities seek the flexible application of these in a manner that considers the viewpoint of all stakeholders, including the disaster victims for whom support is necessary, ordinary residents and representative assemblies.

10 Local Government Response to 3.11: Staff Perceptions

Matsui Nozomi

When the Great East Japan Earthquake struck, local government services were temporarily suspended. There were, in fact, municipalities that lost physical infrastructure, including town and city halls, to the disaster. In spite of this, given that local government officials had responsibility for the implementation of onsite disaster response activities, the role played by municipal employees in tsunami and earthquake recovery and reconstruction was immense (Imai 2014: 95). Yet, with some fragmentary exceptions, there are few accounts of the disaster-related activities of local government officials. While we have support staff notes (Tokyo Metropolitan Office 2012, 2013, 2014) and the records of professional workers, including doctors, nurses and educators, and also of labor union associates (Nihon kango kyōkai shuppankai henshūbu hen 2011; Kokumin kyōiku bunka sōgō kenkyūjō 2013), there is currently no systematic data collected from the ordinary municipal staff who were directly involved in on-the-ground reconstruction. What, then, did these staff actually feel about recovery and reconstruction processes and what contribution did they perceive themselves to have made?

With the above in mind, three years after the disaster in 2014 we conducted a survey (distributed by post) of 1,325 government employees of section or department head status.[1] Respondents were employed either in the office of one of the three disaster-affected prefectures – Iwate, Miyagi and Fukushima – or in one of 37 cities, towns or other municipal communities located along the coastline of those prefectures. Based on the results of that survey, this chapter examines the nature of Japanese local government response to disaster recovery and reconstruction. The chapter particularly addresses three themes. The first is municipal staff awareness of and response to disaster-related duties. Information is provided regarding the measures taken by public officials during recovery and reconstruction, and about respondents' perceptions of the duration of the recovery phase. The second theme concerns the degree of contact that occurred between the various parties involved as reconstruction progressed. This section discusses employee perceptions regarding the nature of the contact between the 37 communities and their residents, in addition to contact between the three prefectural offices and prefectural residents. Attention will also be given to contact between the 37 communities and national and prefectural governments, and between a given prefecture and other levels of government (fourth part below). The third theme concerns the various factors and structures regarded as necessary to progress recovery and reconstruction. The chapter concludes by collating, in terms of these three themes, the findings for provincial government that have emerged from Great East Japan Earthquake recovery and reconstruction processes.

Response to changes in tasks performed

Government administration operates through both division of labor and task collaboration. After the Great East Japan Earthquake, however, many government employees were confronted with the need to undertake tasks that went well beyond their previous job descriptions. Although some officials could accommodate these new duties into their existing work, others were given new tasks that differed from previous duties. What processes operated, then, to see workers take on the various duties they performed during the disaster recovery and reconstruction process?

First, we will consider workers in the 37 municipalities surveyed. Here, 55.2% of respondents indicated that additional disaster-related tasks were added to their regular duties; 36.6%, however, were diverted from their usual work to undertake disaster-specific tasks. From this it is apparent that recovery and reconstruction process tasks were accomplished by local government employees carrying the double burden of undertaking pre-disaster tasks in addition to tasks related to the disaster, and by staff having new disaster-related tasks added to their existing work. Such a trend is apparent also in responses given to questions concerning the relationship between disaster-related tasks undertaken by respondents and the tasks for which the department to which the respondent was attached had responsibility (Table 10.1). Combining the 31.2% response rate given to tasks largely related to the disaster and the 21.8% response to 'All work was related', we can see that there was an allocation of tasks that saw 53.0% of municipal employees undertake disaster recovery and reconstruction tasks that had some relation to their usual work. On the other hand, there was a 46.3% rate response to the categories 'partially related' or 'not related at all'. This indicates that there were staff who were required to carry out disaster recovery and reconstruction tasks that had no relationship whatsoever to work they had previously performed.

In the case of the three prefectures surveyed, there was a similar situation. In prefectural offices, more than half of all staff members were required to add earthquake-related tasks to their usual work (Table 10.2). Data from this element of the survey indicates, however, that there was a greater likelihood here than in the 37 municipalities of these additional tasks being related to the respondent's usual work ('All work was related', 30.2%; 'Most work was related', 40.4%). Furthermore, the allocation of (new) duties was largely connected to matters for which an employee's department already had responsibility.

How, then, did some staff find themselves undertaking tasks that were unrelated to their existing work? Among the 37 municipalities, the highest number of respondents (44.3%) indicated that they had been instructed to do by a supervisor. This suggests that there were limits to the degree to which such decisions were the result of individual staff judgments or staff peer discussion. In the prefectures, too, instruction by supervising staff was the main reason for new duties being undertaken.

From the above results, the following becomes evident. If we classify tasks that are related to the disaster as non-routine and tasks that existed prior to the earthquake as routine, we can see that a double responsibility was assumed by staff for both. Even immediately following the earthquake, few staff members were able to focus on non-routine tasks only. Further, the allocation of new work came as a result of workplace directives. This means that employees, who were often also disaster victims, were seen as single employees – organizational cogs in the machinery of government. That is to say, the response to the disaster occurred within the organizational structures and staffing patterns as they existed before the earthquake. It is therefore apparent that any support for workers, or assistance for the increased work demands imposed on them as a

Table 10.1: Post-disaster tasks in the 37 municipalities surveyed

	Was disaster-related work undertaken related to the division to which you were attached?					Total	
	All work was related	Most work was related	Work was partially related	Work was not at all related	No response	%	*N*
Iwate Prefecture	21.1	36.0	24.2	17.4	1.2	100	161
Miyagi Prefecture	26.0	32.0	23.7	17.7	0.6	100	350
Fukushima Prefecture	11.8	23.5	30.9	33.8	–	100	136
Total	21.8	31.2	25.3	21.0	0.6	100	647

Table 10.2: Post-disaster tasks in the three prefectures surveyed

	Was disaster-related work undertaken related to the division to which you were attached?					Total	
	All work was related	Most work was related	Work was partially related	Work was not at all related	No response	%	*N*
Iwate Prefecture	32.0	49.3	14.7	4.0	–	100	75
Miyagi Prefecture	46.6	36.2	11.7	9.6	–	100	94
Fukushima Prefecture	19.0	37.9	24.1	19.0	–	100	116
Total	30.2	40.4	17.5	11.9	–	100	285

result of the disaster, could only be provided within the parameters of previous staffing structures and existing job descriptions.

Perceptions of the recovery duration

A feature of the Great East Japan Earthquake recovery and reconstruction has been the periodical release of data, mainly relating to Reconstruction Agency and prefectural activity, giving information on task progress. This commitment to what we might call 'project transparency' has characterized the government's disaster response. Official figures indicate that, in the three years since the disaster, recovery has occurred in terms of the restoration of social infrastructure and that daily life for residents is also gradually returning to normal.

At what point, then, did the public employees who actually carried out on-the-ground recovery and reconstruction tasks perceive that recovery might have occurred? First, we will consider data from the 37 municipalities (Table 10.3). Of respondents employed by these municipalities, 20.4% thought that pre-earthquake conditions had been restored within a year. In contrast, 20.1% believed that the situation had not normalized even three years after the disaster. A prefectural breakup of municipalities, however, provides specific patterns of response. While 23.5% and 21.7% of municipal officials from Iwate and Miyagi prefectures respectively thought that the situation had normalized within a year, the Fukushima figure was only 13.1%. The sense of the time at which recovery had occurred or was likely to occur differed between prefectures. Neither was there any common understanding across the 37 municipalities. Differences were evident, too, depending on the agency or division involved. Those who indicated a return to regular tasks within a year were connected to agriculture, forestry and fishing, to the legislative assembly, to commerce and tourism, and to health and welfare. On the other hand, many who worked in fields such as urban development, reconstruction and education felt that things had not yet returned to pre-disaster conditions even three years after the incident. These differing responses perhaps reflect the differing degree of direct control exercised by the various groups over actual recovery and reconstruction operations.

Data were gathered also from each of the three prefectures (Table 10.4). Here it was evident that, in comparison to the 37 municipalities, any sense of a return to normal in the workplace was low. For example, the highest response of any category was 'Not yet returned [to normal]' (35.8%). Looking at the results for specific prefectures, 39.7% of Fukushima employees felt that even three years after the disaster normal conditions had yet not been restored. This is a significant rate of difference to the 28.3% response given by Fukushima Prefecture's municipal employees. Further investigation of the data reveals other acute differences. For example, even after one year, there was a sense among prefectural employees that recovery – as opposed to the more active phase of reconstruction – remained ongoing. Among employees of the 37 municipalities surveyed, however, there was a feeling that the one year post-disaster point had marked a shift from recovery to reconstruction. The reason for the differing prefectural and municipality responses – and also for the difference between the prefectures – perhaps relates (as elaborated upon below in the section 'Perceptions of contact between government bodies') to the fact that prefectures and cities have different degrees of on-the-ground contact. Even among the 37 municipalities, there were different specific rates of recovery and reconstruction progress. It is for this reasons that the mode of recovery and reconstruction progress also differed between the various cities, towns and other local areas surveyed. Most prefectural offices, furthermore, mainly had contact with municipalities in which reconstruction was ongoing. It is for reasons such as this that the prefectures may have had less of a sense that things were returning to normal.

Although survey results indicated a frequent perception that the work situation had stabilized within a year, there was a group of respondents who felt that this was not the case. What reasons did these respondents give to explain their feelings that pre-disaster conditions did not yet prevail? A number of options were suggested to respondents, including the attitudes of residents, employee numbers, employee sphere of authority and available funds. Staffing shortfalls emerged as a key factor at both levels of administration. This was apparent from the 56.3% and 43.8% response to that issue recorded in the 37 municipalities and the three prefectures respectively. On the other hand, neither funding matters nor employee sphere of authority were regarded as especially crucial. In other words, there was a clear perception that shortage of staff was the main factor preventing a return to previous work conditions.

Response to contact and request for assistance

Municipal administration is the level of government that has the greatest direct contact with residents. Both under ordinary conditions and during times of crisis, its core business is to respond to resident needs. It is not surprising then that, regardless of size, all divisions surveyed from the 37 municipalities had direct resident contact. Since most, however, dealt only with residents whose problems related to the issues for which their department had responsibility, there was a limit to the number of residents with whom any particular employee came into contact

Table 10.3: Time taken to return to pre-disaster work conditions in the 37 municipalities surveyed

	From about when did you resume your pre-disaster work conditions?								Total	
	Within 1 month	Within 3 months	Within 6 months	Within 1 year	In excess of 2 years	Not yet returned	Without returning to previous work, assigned to a new position	No response	%	*N*
Iwate Prefecture	7.0	11.2	16.6	23.5	5.9	19.8	15.5	0.5	100	187
Miyagi Prefecture	2.9	15.0	18.8	21.7	10.2	17.2	13.7	0.5	100	373
Fukushima Prefecture	2.8	11.7	15.2	13.1	10.3	28.3	17.9	0.7	100	145
Total	4.0	13.3	17.4	20.4	9.1	20.1	15.0	0.6	100	705

Table 10.4: Time taken to return to pre-disaster work conditions in the three prefectures surveyed

	From about when did you resume your pre-disaster work conditions?								Total	
	Within 1 month	Within 3 months	Within 6 months	Within 1 year	In excess of 2 years	Not yet returned	Without returning to previous work, assigned to a new position	No response	%	*N*
Iwate Prefecture	4.8	8.4	18.1	12.0	8.4	33.7	14.5	—	100	83
Miyagi Prefecture	2.9	6.7	12.5	15.4	5.8	32.7	24.0	—	100	104
Fukushima Prefecture	3.2	6.3	7.9	17.5	4.8	39.7	19.8	0.8	100	126
Total	3.5	7.0	12.1	15.3	6.1	35.8	19.8	0.3	100	313

on a day-to-day basis. This meant that a specialized response to resident needs became the norm.

We might consider the scope of inquiries that came from residents and the nature of the responses made at the three-year point after the disaster (Tables 10.5 to 10.8). Seventy per cent of respondents from the 37 municipalities had received some sort of inquiry from residents (Table 10.5). Only 23.8% responded that they had not received very many and a mere 6.0% had received none at all; in fact, 40.7% indicated that they had received inquiries 'quite often', suggesting that inquiry frequency was generally high. However, among the municipalities, there was a big difference depending on prefecture. In Fukushima 80% of municipal employees had either received inquiries 'very often' or 'quite often'. Since the combined percentage of these two response categories for Iwate Prefecture and Miyagi Prefecture was only 65.2% and 68.1% respectively, we should interpret the Fukushima situation as a special case. Regarding the nature of inquiries,[2] the highest response figure indicated that matters raised were 'partly unexpected' (59.7%). We can deduce, therefore, that inquiries made by residents more often than not went beyond what employees had anticipated. In all, 83.5% of requests related to matters that were unexpected to some extent. In the three prefectures, on the other hand, survey results indicated that only 53.7% of respondents had received inquiries from residents. The fact that this figure is low in comparison to municipalities is perhaps explained by the difference in the frequency of day-to-day contact with residents.

In terms of providing a concrete response to residents, 56.2% of respondents from the 37 municipalities indicated that they were 'Fully able to respond' or 'Able to respond fairly well' to accommodate requests made (Table 10.7). When broken down by prefecture, however, specific patterns emerge. In Fukushima municipalities, 55.3% of respondents indicated an inability to meet requests adequately, while 4.3% indicated they were unable to do so at all. This is a total of 59.6% that Fukushima municipal officials felt unable to respond to adequately, a figure that is high in comparison to Iwate and Miyagi prefectures, where the combined response to these two survey options was 33.6% and 37.3% respectively. When asked to explain the inability to meet resident requests,[3] 40.0% noted that, even if the request involved tasks related to their job description, staff shortages prevented them taking action, while 40.0% indicated

Table 10.5: Inquiries from residents in the 37 municipalities surveyed

	Rate at which direct inquiries, requests and calls for assistance have been received from residents in the three years since the disaster					Total	
	Received very often	**Received quite often**	**Not received very often**	**Not received at all**	**No response**	**%**	***N***
Iwate Prefecture	21.9	43.3	27.3	5.9	1.6	100	187
Miyagi Prefecture	26.5	41.6	24.7	7.2	_	100	373
Fukushima Prefecture	44.8	35.2	17.2	2.8	_	100	145
Total	29.1	40.7	23.8	6.0	0.4	100	705

Table 10.6: Inquiries from residents in the three prefectures surveyed

	Rate at which direct inquiries, requests and calls for assistance have been received from residents in the three years since the disaster					Total	
	Received very often	**Received quite often**	**Not received very often**	**Not received at all**	**No response**	**%**	***N***
Iwate Prefecture	10.8	34.9	41.0	12.0	1.2	100	83
Miyagi Prefecture	17.3	34.6	33.7	12.5	1.9	100	104
Fukushima Prefecture	17.5	39.7	32.5	9.5	0.8	100	126
Total	15.7	36.7	35.1	11.2	1.3	100	313

Table 10.7: Responses to residents in the 37 municipalities surveyed

	Were you able to respond to the concrete requests that came from residents?					Total	
	Fully able to respond	**Able to respond fairly well**	**Unable to respond very well**	**Unable to respond at all**	**No response**	**%**	***N***
Iwate Prefecture	1.7	63.0	32.4	1.2	1.7	100	173
Miyagi Prefecture	4.6	54.6	36.1	1.2	3.5	100	346
Fukushima Prefecture	3.5	34.8	55.3	4.3	2.1	100	141
Total	3.6	52.6	39.2	1.8	2.7	100	660

Table 10.8: Responses to residents in the three prefectures surveyed

	Were you able to respond to the concrete requests that came from residents?					Total	
	Fully able to respond	Able to respond fairly well	Unable to respond very well	Unable to respond at all	No response	%	*N*
Iwate Prefecture	6.9	62.5	30.6	–	–	100	72
Miyagi Prefecture	3.4	46.1	47.2	2.2	1.1	100	89
Fukushima Prefecture	1.8	37.2	54.9	1.8	4.4	100	113
Total	3.6	46.7	46.0	1.5	2.2	100	274

that unmet requests involved matters that went beyond what the system was equipped to deal with. In addition, 31.0% responded that requests fell outside the activity parameters of the agency involved. Thus, in addition to the staff shortages highlighted in the discussion regarding the point at which staff felt recovery had occurred, the inadequacy of the system itself also limited the capacity of staff to assist residents.

In the case of the three prefectures, 50.3% of survey respondents indicated that they were 'fully' or 'fairly well' able to meet resident requests (Table 10.8), a rate that is a little lower than that of the 37 municipalities. Among reasons given for prefectural inability to assist, a majority of 46.9% of respondents indicated that the request fell outside the work jurisdiction of their agency. However, while the 37 municipalities gave lack of staff as the highest-ranked restricting factor, only 23.8% of respondents from the three prefectures saw this as a problem.[4] This difference is also perhaps explained by the fact that, in comparison with municipalities, contact with residents at the prefectural level is restricted. This could be the reason that the staffing system that operated at non-disaster times was not regarded as a particular impediment when responding to resident disaster inquiries.

Perceptions of contact between government bodies

In the recovery and reconstruction process, extensive cooperation with organizations outside local government was vital. As Itō (2014:

79) points out, a particular feature of the current reconstruction process was the breadth and depth of such collaboration, which saw cooperative links forged between government and private organizations, government and residents' associations, and between different levels of government themselves. An argument has been made that the central government played a key role in these kinds of multi-organizational partnerships (Makihara 2014: 57).

Certainly, in the case of Self-Defense Force recovery activity and the central role played by the Reconstruction Agency in policy design, there was considerable input from the central government. Furthermore, the implementation by municipalities of various reconstruction systems was only possible with central government support. The practicalities of funding applications, for example, involved direct consultation and negotiation between the municipalities involved and the central authorities. Yet, while the center may have had a decisive role in terms of systems planning and revenue, implementation could only proceed when municipalities and the central authorities worked together in tandem. Nishio (2013: 208) notes a perception, however, of limited involvement by prefectural authorities in the various recovery and reconstruction activities. This section seeks to ascertain the actual rate of contact during the implementation of the multi-organizational reconstruction process between the three levels of government – disaster-affected municipalities, the prefectures and the central authorities.

In the survey on which this discussion is based, respondents were requested to consider the frequency of contact between governments at two points – within one month and within three years of the disaster. First, we will examine the situation in the 37 municipalities one month after the earthquake (Table 10.9). Results here indicate a 60.5% rate of contact between municipalities and each of the prefectures. On the other hand, the rate of municipal contact with the central authorities was 19.8% in the case of ministry offices and 25.7% for locally based national government agencies. In other words, while the rate of municipality contact with the prefectures was high, contact with the central government was limited. In contrast to previously stated theories and projections, therefore, we can say that, at least as perceived by municipal authority employees, there was a high rate of municipal contact with the prefecture compared with the central authorities. This result accorded with that of the three prefectures one month on from the disaster. First, contact with municipalities within each prefecture was the same for all three at 70.0% (Table 10.10). In the case of central government offices, however, the results were 37.6% for ministries and 32.6% for locally based national agencies. One month after the disaster, therefore, we can see that frequency of contact between prefectural governance and the municipalities was higher than the frequency of prefectural contact with central governance bodies.

Three years after the disaster it was apparent that little change had occurred. In the 37 municipalities, the partner with whom respondents had the highest rate of contact was the prefecture (64.6%). In the three years that elapsed following the disaster, however, changes did occur in the pattern of this frequency. As opposed to the 'every day' or 'once every 2–3 days' responses that were evident one month after the disaster, frequency of contact three years later fell to 'About once every month'. In the case of the rate of contact between the 37 municipalities and the central government three years after the disaster, there was a rise to a 30.5% response for ministries and 36.2% for local agencies. Thus, although there was little change between the one month and three year points in the comparatively high rate of contact with the prefectures, rates of contact with the central government did increase during that time. This rise in central government contact is likely to be due to the increased need, as discussed above, for direct negotiation between these two government levels on such issues as reconstruction funding. In terms of frequency, the response of 'About once every month' indicates that municipal contact with the central authorities at the three year point was, as in the past, limited. Furthermore, in the case of the three prefectures, the 69.1% rate of contact with municipalities three years after the disaster remained at much the same rate as it had been in the month following the earthquake (Table 10.10). Prefectural contact with the central authorities at the three year point was 58.1% for the various ministries and 44.8% for local agencies.

It is clear, then, that the perception of those with onsite responsibility for implementing disaster recovery and reconstruction processes directly contradicts any suggestion that the prefecture's role in these processes was limited. The government structure with which disaster-affected cities, towns and other communities had the most frequent contact was, in fact, the prefecture. This is confirmed by the responses given by prefectural employees regarding contact with the municipalities under their jurisdiction.

Frequency of contact, however, did not necessarily result in harmonious relations between the levels of government involved. Rather, as was evident from the survey responses, friction arose. One month after the disaster, 35.4% of respondents from the 37 municipalities ('Felt strongly', 11.6%; 'Felt somewhat', 23.8%) indicated a difference in opinion and perceptions between themselves and the prefecture. The figures for respondents who 'strongly' or 'somewhat' felt differences relating to the central authorities were 23.7% for the various ministries and 20.8% for the local agencies, which clearly was less than perceived differences between municipal employees and the prefectures (Table 10.11). Respondents from the three prefectures had a 27.8% perception of differences ('Felt strongly', 3.5%; 'Felt somewhat', 24.3%) in opinion and perception from that of municipalities (Table 10.12). The combined figures were 31.6% for the various ministries and 19.5% for local agencies associated with the central government (Table 10.12).

The lapse of time saw little change in these differences. Three years after the disaster, in a result that largely replicated that of the one month point, employees in the 37 municipalities recorded a 36.1% rate of difference of opinion and perception from that of the prefecture ('Felt strongly', 8.2%; 'Felt somewhat', 27.9%). In the case of the central government, the rate of difference of opinion ('Felt strongly' and 'Felt somewhat') rose at a similar rate for both the ministries and agencies over the three years to become 35.9% for the various ministries and 28.4% for local agencies (Table 10.11). On the other hand, the rate of difference between the three prefectures surveyed and the municipalities for which they were responsible increased incrementally in the three years after the disaster to 29.4% (Table 10.12). Differences of opinion and understanding between the prefecture and the central government, however, rose approximately 10% to figures of 44.4% in the case of the various ministries and 28.8% for local agencies. Clearly, at the three year point more prefectural staff perceived differences of opinion and perception between themselves and central government head offices.

Recovery and reconstruction factors and the staffing system

Finally we consider the factors regarded by the public officials surveyed as necessary for recovery and reconstruction progress.

Table 10.9: Rate of contact with other organizational units by disaster-affected communities in the 37 municipalities surveyed – frequency of contact one month and three years after the disaster

		Breakdown (%)						Total	
	Contact occurred	**About every day**	**About once every 2–3 days**	**About once every week**	**About once every month**	**Almost no contact**	**No response**	**%**	***N***
				One month after disaster					
National Emergency Response Headquarters	F 9.3	3.0	2.0	2.0	2.3	84.7	6.1	100	705
Various national ministries	F 19.8	2.6	3.5	5.0	8.7	73.9	6.4	100	705
National local agencies	F 25.7	4.8	6.2	6.8	7.9	67.9	6.2	100	705
Fire service	F 40.3	20.3	5.4	8.2	6.4	53.8	6.0	100	705
Police	F 38.4	17.4	6.0	7.9	7.1	56.3	5.1	100	705
Self-Defense Forces	F 40.1	24.4	5.8	7.1	2.8	55.0	4.8	100	705
Prefecture	F 60.5	18.9	15.3	16.2	10.1	35.3	4.3	100	705
Other municipalities within the prefecture	F 42.9	5.8	7.9	15.2	14.0	51.8	5.2	100	705
Municipalities outside the prefecture	F 27.0	3.3	4.8	9.4	9.5	66.8	6.2	100	705
Residents' associations (community and neighborhood associations etc.)	F 46.9	19.9	11.1	8.7	7.2	47.2	6.0	100	705
Social Welfare Council	F 38.5	11.5	10.5	9.5	7.0	55.5	6.1	100	705
Volunteer organizations, non-profit organizations	F 41.5	12.5	10.8	11.2	7.0	52.8	5.8	100	705
Healthcare agencies	F 32.4	11.8	7.9	7.0	5.7	62.1	5.5	100	705

Electric power corporations	F 29.7	6.5	7.0	8.5	7.7	64.8	5.5	100	705
Telecommunications corporations	F 26.4	5.8	5.4	7.8	7.4	67.9	5.7	100	705
Others	F 17.0	11.2	3.5	2.0	0.3	42.7	40.3	100	705
Three years after the disaster									
National Emergency Response Headquarters	F 28.2	1.1	2.8	4.3	20.0	65.7	6.1	100	705
Various national ministries	F 30.5	0.4	2.4	5.4	22.3	63.7	5.8	100	705
National local agencies	F 36.2	0.9	2.8	6.5	26.0	57.6	6.2	100	705
Fire service	F 21.7	1.8	1.8	4.5	13.6	71.6	6.5	100	705
Police	F 18.4	1.0	0.4	4.1	12.9	75.3	6.2	100	705
Self-Defense Forces	F 8.4	1.0	1.1	1.6	4.7	84.7	7.0	100	705
Prefecture	F 64.6	2.7	8.7	18.6	34.6	30.9	4.5	100	705
Other municipalities within the prefecture	F 46.6	0.6	2.1	9.1	34.8	47.9	5.5	100	705
Municipalities outside the prefecture	F 22.0	0.1	0.9	3.4	17.6	72.1	6.0	100	705
Residents' associations (community and neighborhood associations etc.)	F 38.4	2.0	3.4	10.6	22.4	55.2	6.4	100	705
Social Welfare Council	F 23.4	1.3	2.8	4.8	14.5	70.4	6.2	100	705
Volunteer organizations, non-profit organizations	F 28.0	0.9	2.3	7.1	17.7	66.0	6.1	100	705
Healthcare agencies	F 14.9	0.3	1.4	2.7	10.5	79.1	6.0	100	705
Electric power corporations	F 15.5	0.3	1.1	1.8	12.3	78.0	6.4	100	705
Telecommunications corporations	F 11.6	0.1	0.9	1.7	8.9	81.8	6.5	100	705
Others	F 8.6	1.8	1.4	1.7	3.7	50.9	40.4	100	705

Table 10.10: Rate of contact with other organizations by the three disaster-affected prefectures – frequency of contact one month and three years after the disaster

	Contact occurred	Breakdown (%)						Total	
		About every day	About once every 2–3 days	About once every week	About once every month	Almost no contact	No response	%	*N*
				One month after disaster					
National Emergency Response Headquarters	F 13.8	6.4	2.6	1.3	3.5	80.2	6.1	100	313
Various national ministries	F 37.6	8.9	12.1	9.6	7.0	58.1	4.2	100	313
National local agencies	F 32.6	9.9	4.8	10.9	7.0	62.3	5.1	100	313
Fire service	F 19.7	7.0	2.2	5.4	5.1	74.8	5.4	100	313
Police	F 26.8	9.3	3.8	6.4	7.3	68.1	5.1	100	313
Self-Defense Forces	F 24.6	11.8	5.1	4.5	3.2	70.6	4.8	100	313
Other prefectures	F 38.0	6.7	8.3	11.2	11.8	57.8	4.2	100	313
Municipalities within the prefecture	F 70.0	34.2	17.6	12.1	6.1	26.8	3.2	100	313
Municipalities outside the prefecture	F 11.2	1.9	1.9	2.6	4.8	82.7	6.1	100	313
Residents' associations (community and neighborhood associations etc.)	F 21.5	5.1	2.9	5.8	7.7	72.8	5.8	100	313
Social Welfare Council	F 16.4	2.6	2.6	5.4	5.8	78.0	5.8	100	313
Volunteer organizations, non-profit organizations	F 25.3	4.2	5.8	8.6	6.7	69.6	5.1	100	313
Healthcare agencies	F 15.7	6.1	2.9	2.9	3.8	78.6	5.8	100	313

Electric power corporations	F 18.0	2.9	3.5	5.8	5.8	77.0	5.1	100	313
Telecommunications corporations	F 12.8	2.9	2.2	4.2	3.5	82.1	5.1	100	313
Others	F 29.1	17.6	7.3	3.2	1.0	42.5	28.4	100	313
				Three years after the disaster					
National Emergency Response Headquarters	F 31.9	0.6	1.0	5.4	24.9	62.0	6.1	100	313
Various national ministries	F 58.1	1.9	8.6	12.8	34.8	37.4	4.5	100	313
National local agencies	F 44.8	1.9	4.5	10.9	27.5	49.2	6.1	100	313
Fire service	F 4.1	0.3	0.0	0.3	3.5	89.8	6.1	100	313
Police	F 8.3	0.0	0.0	1.0	7.3	85.6	6.1	100	313
Self-Defense Forces	F 4.2	0.0	0.3	1.0	2.9	89.8	6.1	100	313
Other prefectures	F 49.9	1.0	1.3	6.1	41.5	44.7	5.4	100	313
Municipalities within the prefecture	F 69.1	5.8	11.8	19.2	32.3	27.5	3.5	100	313
Municipalities outside the prefecture	F 7.7	0.6	0.0	1.3	5.8	8.6	5.8	100	313
Residents' associations (community and neighborhood associations etc.)	F 15.3	0.6	1.0	2.2	11.5	78.6	6.1	100	313
Social Welfare Council	F 7.1	0.0	0.3	1.0	5.8	87.2	5.8	100	313
Volunteer organizations, non-profit organizations	F 19.9	0.0	1.0	3.2	15.7	75.1	5.1	100	313
Healthcare agencies	F 9.2	0.3	0.6	1.3	7.0	85.0	5.8	100	313
Electric power corporations	F 11.2	0.3	1.0	1.0	8.9	83.1	5.8	100	313
Telecommunications corporations	F 4.8	0.0	0.3	0.0	4.5	89.5	5.8	100	313
Others	F 19.0	5.8	2.6	5.8	4.8	52.1	29.1	100	313

Table 10.11: Differences of opinion and understandings between the 37 disaster-affected communities and other organizations one month and three years after the disaster

	Contact occurred	Breakdown (%)				Almost no contact	No response	Total	
		Felt strongly	Felt somewhat	Not felt much	Not felt at all			%	*N*
				One month after disaster					
National Emergency Response Headquarters	F 24.8	6.7	10.4	4.3	3.4	68.8	6.5	100	705
Various national ministries	F 35.2	7.4	16.3	6.5	5.0	58.7	6.1	100	705
National local agencies	F 38.0	5.8	15.0	8.5	8.7	55.3	6.7	100	705
Fire service	F 47.1	0.4	2.7	14.8	29.2	45.8	7.1	100	705
Police	F 45.2	0.9	4.3	15.5	24.5	48.7	6.2	100	705
Self-Defense Forces	F 48.1	1.3	3.5	13.9	29.4	46.2	5.7	100	705
Prefecture	F 69.1	11.6	23.8	18.9	14.8	26.8	4.1	100	705
Other municipalities within the prefecture	F 51.5	1.0	7.4	18.6	24.5	43.0	5.5	100	705
Municipalities outside the prefecture	F 38.3	0.7	5.7	13.0	18.9	55.0	6.7	100	705
Residents' associations (community and neighborhood associations etc.)	F 54.2	4.5	17.9	14.6	17.2	39.3	6.5	100	705
Social Welfare Council	F 46.1	0.9	5.4	19.4	20.4	47.2	6.7	100	705
Volunteer organizations, non-profit organizations	F 49.0	2.8	11.9	16.6	17.7	45.0	6.0	100	705
Healthcare agencies	F 39.7	1.3	4.7	14.0	19.7	54.2	6.1	100	705

Electric power corporations	F 38.0	3.0	4.8	11.2	19.0	55.9	6.1	100	705
Telecommunications corporations	F 34.2	1.3	4.1	12.2	16.6	59.4	6.4	100	705
Others	F 17.3	0.9	2.8	5.4	8.2	41.8	40.9	100	705
Three years after the disaster									
National Emergency Response Headquarters	F 48.8	14.0	24.0	6.1	4.7	45.1	6.1	100	705
Various national ministries	F 52.8	10.5	25.4	10.5	6.4	4.1	6.2	100	705
National local agencies	F 53.5	6.0	22.4	15.0	10.1	40.3	6.2	100	705
Fire service	F 38.3	0.1	1.7	13.8	22.7	54.6	7.1	100	705
Police	F 36.3	0.1	2.4	13.9	19.9	57.2	6.5	100	705
Self-Defense Forces	F 29.6	0.1	1.0	9.4	19.1	63.0	7.4	100	705
Prefecture	F 73.1	8.2	27.9	21.0	16.0	22.3	4.5	100	705
Other municipalities within the prefecture	F 58.1	0.6	8.1	23.7	25.7	35.7	6.2	100	705
Municipalities outside the prefecture	F 39.7	0.4	6.5	14.9	17.9	53.2	7.1	100	705
Residents' associations (community and neighborhood associations etc.)	F 51.7	1.8	18.0	18.0	13.9	41.0	7.2	100	705
Social Welfare Council	F 38.3	0.1	3.1	18.2	16.9	55.2	6.5	100	705
Volunteer organizations, non-profit organizations	F 44.4	1.4	9.8	17.2	16.0	48.9	6.7	100	705
Healthcare agencies	F 31.0	0.4	2.8	13.3	14.5	62.4	6.5	100	705
Electric power corporations	F 30.2	2.3	3.4	11.2	13.3	63.0	6.8	100	705
Telecommunications corporations	F 27.5	0.4	3.1	11.5	12.5	65.4	7.1	100	705
Others	F 12.7	0.7	1.8	4.1	6.1	45.5	41.7	100	705

Table 10.12: Differences of opinion and understandings between the three disaster-affected prefectures and other organizations one month and three years after the disaster

		Breakdown (%)						Total	
	Contact occurred	**Felt strongly**	**Felt somewhat**	**Not felt much**	**Not felt at all**	**Almost no contact**	**No response**	**%**	***N***
				One month after disaster					
National Emergency Response Headquarters	F 28.1	8.9	7.7	6.4	5.1	66.8	5.1	100	313
Various national ministries	F 51.7	11.8	19.8	11.8	8.3	44.4	3.8	100	313
National local agencies	F 42.2	4.8	14.7	13.1	9.6	52.4	5.4	100	313
Fire Service	F 25.2	0.0	0.3	9.6	15.3	69.6	5.1	100	313
Police	F 32.0	0.0	1.0	12.8	18.2	62.6	5.4	100	313
Self-Defense Forces	F 29.3	0.0	2.2	10.2	16.9	65.5	5.1	100	313
Other prefectures	F 47.4	1.3	8.0	17.3	20.8	48.2	4.5	100	313
Municipalities within the prefecture	F 76.1	3.5	24.3	25.6	22.7	21.4	2.6	100	313
Municipalities outside the prefecture	F 16.2	0.3	3.5	5.1	7.3	78.3	5.4	100	313
Residents' Associations (community and neighborhood associations etc.)	F 28.1	1.9	6.7	11.2	8.3	66.5	5.4	100	313
Social Welfare Council	F 18.5	0.0	2.2	8.6	7.7	76.0	5.4	100	313
Volunteer organizations, non-profit organizations	F 31.1	1.6	11.8	8.3	9.6	64.5	4.2	100	313
Healthcare agencies	F 20.1	0.3	5.1	8.3	6.4	74.8	5.1	100	313

Electric power corporations	F 23.4	2.6	3.5	6.4	10.9	71.6	5.1	100	313
Telecommunications corporations	F 16.3	0.0	1.0	6.4	8.9	78.6	5.1	100	313
Others	F 26.8	2.2	6.4	8.3	9.9	43.5	29.7	100	313
Three years after the disaster									
National Emergency Response Headquarters	F 54.5	16.6	25.2	7.3	5.4	40.6	4.8	100	313
Various national ministries	F 69.0	13.7	30.7	15.7	8.9	27.8	3.2	100	313
National local agencies	F 56.6	5.8	23.0	18.5	9.3	38.7	4.8	100	313
Fire service	F 12.8	0.0	0.3	5.8	6.7	81.5	5.8	100	313
Police	F 17.8	0.0	0.6	8.3	8.9	76.4	5.8	100	313
Self-Defense Forces	F 13.8	0.0	1.3	5.8	6.7	80.5	5.8	100	313
Other prefectures	F 57.4	0.6	15.3	18.8	22.7	37.7	4.8	100	313
Municipalities within the prefecture	F 75.7	2.2	27.2	26.2	20.1	20.4	3.8	100	313
Municipalities outside the prefecture	F 14.1	0.0	3.2	4.2	6.7	80.2	5.8	100	313
Residents' associations (community and neighborhood associations etc.)	F 23.7	1.3	11.5	5.1	5.8	70.0	6.4	100	313
Social Welfare Council	F 13.1	0.0	1.6	6.1	5.4	81.2	5.8	100	313
Volunteer organizations, non-profit organizations	F 29.7	0.6	9.3	9.9	9.9	65.2	5.1	100	313
Healthcare agencies	F 14.8	0.0	2.6	6.1	6.1	79.6	5.8	100	313
Electric power corporations	F 18.9	4.2	4.8	5.1	4.8	75.4	5.8	100	313
Telecommunications corporations	F 9.3	0.0	1.3	3.8	4.2	85.0	5.8	100	313
Others	F 18.6	1.6	4.2	5.8	7.0	50.8	30.7	100	313

In the 37 municipalities, 92.8% of respondents indicated that staff motivation was the key ('Think that this has strong influence' and 'Think this has influence'; Table 10.13). Other factors seen as significant were staff numbers (89.8%), financial resources (88.5%), leadership qualities of those in charge (86.6%), dependability (85.6%), and clear objectives and timelines (76.7%). There was a sense, in other words, that recovery and reconstruction were contingent on the staff themselves. This result was largely confirmed in responses given by officials from the three prefectures. In that instance, 94.8% nominated staff motivation as a factor promoting recovery and reconstruction, 90.5% nominated leadership qualities of those in charge, 89.1% nominated staff numbers, 89.8% nominated financial resources, 87.9% nominated dependability, and 75.7% nominated clear objectives and timelines (Table 10.14).

A special feature of the Great East Japan Earthquake recovery and reconstruction process was the manner in which municipalities supported each other. In tandem with other back-up systems, many municipalities adopted the practice of dispatching staff to assist colleagues in jurisdictions experiencing staff shortages. What type of back-up municipal staffing, however, was regarded as useful in this process? Among the 37 municipalities surveyed,[5] the highest perceived need was for other public servants who were equipped with the necessary skills and knowledge (56.2%). In the event of these staff being unavailable, the next highest figure was for any individual who had the skills and knowledge needed (32.3%). From these results we can see that the most important quality sought in back-up staff was the possession of essential skills and knowledge. This was replicated in data collected from the three prefectures, where the highest response rate was for other public servants who were equipped with the necessary skills and knowledge (60.7%). In the prefectures, too, the second highest response (29.7%) was for any backup staff with those qualities.

Conclusion

This chapter has presented a range of results from a survey of the perceptions of public officials conducted three years after the Great East Japan Earthquake. When we collate survey findings, three key points emerge.

The first point relates to the implementation of tasks by local authorities. Local authorities were the bodies responsible for taking on the ever-changing tasks of recovery and reconstruction that they were required to absorb into the everyday job descriptions that operated prior to the disaster. From the initial recovery stage, the central government progressed reconstruction by distributing tasks between the relevant ministries. Local government, too, dealt with new tasks in terms of existing work allocation schedules. In other words, the response to non-routine matters had to be dealt with within the bounds of the routine system. We might question, however, the extent to which the kind of prior preparation or reserve could have provided an adequate response to such a disaster.

Second are the findings concerning relations between the various arms of government administration. There is a line of argument pointing out the significance of the role of the central authorities at the time of the Great East Japan Earthquake. Certainly, this level of administration played a major part in the design of systems to support recovery and the guarantee of financial support. From this it was assumed that the role of the prefecture was weak, an assumption that did not extend to other levels of non-central administration such as cities. The findings of this survey make it clear, however, that contrary to popular perception there was in fact a high level of contact between disaster-affected municipalities and the prefectures. This frequent contact did not necessarily lead to harmonious relations between the two, with survey data indicating, as both sides were clearly aware, that significant differences of opinion occurred in comparison with other levels of government. In the final analysis, links between related levels of provincial government, namely the prefectures and the municipalities, probably continued to function after the earthquake much as they had before.

The third point that emerges is the picture of the staff surveyed. Survey results particularly help us understand just what the system expected of municipal workers during disaster recovery and reconstruction. Staff themselves saw psychological factors such the leadership exercised by those in charge and individual motivation as key. And in instances where they were provided with back-up support in times of staff shortage, they prioritized specialized attributes such as essential skills and knowledge in those providing assistance. That is to say, while system

Table 10.13: Perceived level of influence of relevant factors on the rate of recovery and reconstruction progress (37 municipalities)

	Think that this has strong influence	Think that this has influence	Does not think this has too much influence	Does not think this has influence	Has no relationship to the relevant issue	No response	Total	
							%	*N*
Dependability towards the government felt by residents	40.4	45.2	6.1	0.4	7.1	0.7	100	705
Provision of (adequate) staff numbers	55.2	34.6	3.8	0.6	5.7	0.1	100	705
Provision of funds for project implementation	60.4	28.1	4.1	1.0	5.8	0.6	100	705
Clear objectives and timelines related to recovery plans	25.1	51.6	13.9	1.6	6.8	1.0	100	705
Leadership qualities of those in charge	53.0	33.6	5.7	0.7	6.4	0.6	100	705
Staff motivation	51.8	41.0	2.0	0.0	4.8	0.4	100	705

Table 10.14: Perceived level of influence of relevant factors on the rate of recovery and reconstruction progress (three prefectures)

	Think that this has strong influence	Think that this has influence	Does not think this has too much influence	Does not think this has influence	Has no relationship to the relevant issue	No response	Total	
							%	*N*
Dependability towards the government felt by residents	40.6	47.3	5.4	0.6	5.1	1.0	100	313
Provision of (adequate) staff numbers	54.3	34.8	6.4	1.0	2.2	1.3	100	313
Provision of funds for project implementation	60.4	29.4	5.1	0.6	3.2	1.3	100	313
Clear objectives and timelines related to recovery plans	13.7	62.0	17.3	2.2	3.5	1.3	100	313
Leadership qualities of those in charge	51.8	38.7	5.4	0.3	2.2	1.6	100	313
Staff motivation	46.6	48.2	2.2	0.3	1.6	1.0	100	313

efficiency and funding support were undoubtedly vital, a sense of psychological wellbeing and access to specialist skills and knowledge were the most important factors for municipal workers implementing reconstruction.

Among survey respondents, there were many whose homes were damaged and who were therefore forced to evacuate during the disaster (28.4% of municipal and 10.6% of prefectural employees). There were also those who tragically lost family members (4.8% of municipal respondents) or whose family members suffered injury (2.4% of municipal respondents). In other words, some respondents were disaster victims in addition to being local government employees. Survey results suggest that those staff members who were both disaster victims and public officials responsible for the progress of recovery and reconstruction faced a particular dilemma – 73% of respondents indicated an inability to shake off concerns about the safety of immediate or even extended family members while undertaking disaster response activities; 44.8% nevertheless indicated that they were fully committed to the recovery and reconstruction duties related to their work. This is a much greater number – 20% greater, in fact – than the 23.1% who indicated that family and other worries impacted on their work. Thus, three years after the disaster, a figure ultimately emerges from the survey of the lone public official who, although a disaster victim herself or himself, strived to prioritize public duty over family concerns.

11 The Practical Realities of Volunteer Activity in a Time of Disaster

NISHIDE Junro

Since the January 1995 Great Hanshin–Awaji Earthquake (known also outside Japan as the Kōbe Earthquake), widespread recognition has been given to the volunteers, including those working with non-governmental organizations (NGOs) and non-profit organizations (NPOs), who contribute to disaster recovery and reconstruction. Volunteers played a key role in events such as the 1997 Sea of Japan oil spill by the Russian tanker *Nakhodka*[1] and the 2004 Chūetsu Earthquake that struck Niigata Prefecture.[2] At the time of the Great East Japan Earthquake, also, people offering assistance flocked spontaneously to the three disaster-affected prefectures – Iwate, Miyagi and Fukushima. In an online survey conducted by this author,[3] 3.2% of respondents had offered onsite assistance in one of those sites. Of these, 47.9% were first-time volunteers, while 78.3% came from outside the Tōhoku region.[4] In other words, many Great East Japan Earthquake volunteers had no previous experience or in-depth knowledge of the disaster-affected area. In addition to the aid given by individuals, various organizations (including NGOs/NPOs staffed by accredited professionals and private sector groups) undertook volunteer activity throughout the three affected prefectures.

Recent trends have seen individuals and groups offering volunteer disaster recovery and reconstruction services, as well as the nature of these services, become both more diversified and more specialized. In the following discussion, any service voluntarily offered to victims of the Great East Japan Earthquake disaster is referred to as 'volunteer service' and any persons or organizations offering these services as 'volunteer actors'. The diversification and specialization of these services and actors has seen a corresponding improvement in the quality and range of disaster volunteer activities.

At the same time, the overall volunteering picture has become much more complicated.

Following the Great East Japan Earthquake, then, who were the volunteer actors, what sorts of volunteer services did they provide and what were the problems that emerged? This chapter aims to provide answers to these questions.

In addition to the individuals and NGOs/NPOs referred to above, volunteer community actors include the Disaster Volunteer Centers, hereafter, Disaster VCs, which acted as clearinghouses to process and coordinate individuals and volunteer groups, and private sector groups. In order to understand developments in these various groups, the chapter proceeds as follows. After an initial outline of issues to be addressed, it provides background to the emergence of Disaster VCs as a disaster response entity. Attention is given to the scope of activity and the kinds of services provided by these centers, and also to the numbers of people who provided volunteer support following the Tōhoku disaster. The chapter then investigates the kinds of back-up support provided for Disaster VCs, volunteer groups and individuals. It will be apparent that Disaster VCs may have been unable to operate without this support. The next section examines developments in organizations such as NGOs/NPOs, particularly the specialist nature of much NGO/NPO work and cooperation between various groups. The contribution of occupation-based and professional organizations is also considered. The final section draws on survey feedback from volunteer actors to note problems that emerged in Tōhoku and to make suggestions for future policy directions.

Disaster VC developments

The Disaster VC is a central crisis-time organization established in affected local areas following a large-scale disaster. In order to understand how these centers work, the historical background of Disaster VCs is outlined and consideration given to the operation of these centers in the wake of the Tōhoku disaster.

In Japan the presence of individual and NGO/NPO volunteer actors first emerged following the 1995 Great Hanshin–Awaji Earthquake that struck the Kōbe area. Most had no knowledge of or experience in disaster aid. One year after the Hanshin disaster, 1,380,000 volunteers had assisted the recovery effort and large numbers of volunteer actor groups, such as NPOs providing victim support, had formed (Great Hanshin–Awaji Earthquake Reconstruction Follow-up Committee Editorial Board 2009).[5]

At that time, however, there was no organized volunteer reception process or coordination of task allocation. Initially, therefore, the situation was completely chaotic. In order to remedy this, Japanese national and local governments sought as a matter of urgency to create a disaster volunteer coordination system. In July 1995 the notion of environmental maintenance was stipulated in the national Fundamentals of Disaster Prevention Plan.[6] By 2000 municipalities in regions throughout Japan drew on the idea of environmental maintenance to formulate Regional Disaster Prevention Plans. These plans, regarded as particularly crucial following the 2004 Niigata Prefecture Chūetsu Earthquake (Atsumi 2014),[7] clearly spelled out the role of both volunteers and Disaster VCs.[8]

In Regional Disaster Prevention Plans, the body with responsibility for both the establishment and operation of Disaster VCs is the local Social Welfare Council. These councils are government-initiated NPOs designed to promote social welfare activity within a region. Under Articles 109 and 110, Clause 10, of the Social Welfare Law, which governs the primary social welfare provision in Japan, a Social Welfare Council was established in each of the country's prefectural, metropolitan and municipal jurisdictions. As stipulated in Articles 109 and 110 of the legislation, a function of these councils at the local area level is to provide 'support for citizen participation in social welfare activity'. This, in fact, became the specific basis on which each local Social Welfare Council oversaw the establishment of the local Disaster VC and was thereby included as a stakeholder in Regional Disaster Prevention Plans.[9]

3.11 and the Disaster VC contribution

At the time of the Great East Japan Earthquake, Disaster VCs appeared in almost every municipality along the coastline of the three affected prefectures. In the year that followed, these centers processed more than a million volunteers. The onsite work of Disaster VCs and volunteers was not merely confined to directly physical tasks such as clearing away rubble or undertaking clean-up operations. Various services were also offered to support the psychological and emotional needs of those affected by the disaster.[10]

The nature of facilities

There are 37 coastal municipalities along the coast of the three affected prefectures. Disaster VCs were established in 29 of these. (Sites without Disaster VCs were the towns of Namie, Futaba, Ōkuma, Tomioka, Naraha and Hirono in Fukushima Prefecture, where the *Act on Special Measures Concerning Nuclear Emergency Preparedness*, enacted to deal with the special circumstances created by the Fukushima Daiichi reactor explosion and subsequent nuclear disaster, required residents in designated areas to evacuate. Neither did the small communities of Tanohata or Fudai in Iwate Prefecture form Disaster VCs.[11] Each of the 29 centers commenced operation at different times; of the 29 Disaster VCs, centers in seven cities and towns were operating within three days of the disaster, and centers in a further three cities opened by March 15, four days after the event. Of the remaining 19 jurisdictions, 16 had centers functioning before the end of March. For most municipalities, establishing a Disaster VC was a massive challenge. In both Kamaishi City and Ōfunato City in Iwate Prefecture, the Social Welfare Council premises had been inundated during the tsunami and were either totally or, in the case of Ōfunato, partially inaccessible. In Kesennuma City and Minamisanriku Town in Miyagi Prefecture, and in Rikuzentakata City, Ōtsuchi Town and the small community of Noda Village in Iwate Prefecture, council offices had been flooded, with the loss of life of senior Ōtsuchi officials (Japan National Council of Social Welfare 2013).

Other factors also came into play. In Fukushima's Minamisōma City, for example, different district evacuation orders saw a different rate of Disaster VC creation. In Kashima District, the processing of volunteers began on March 27 without any particular task specification. In Haramachi District, processing began on March 12 and, after a temporary suspension in response to evacuation orders, recommenced on March 23. Odaka District was declared a restricted zone until April 16, 2012, so the reception of volunteers did not commence there until May 18.[12]

The reception of volunteers

Table 11.1 provides figures on the numbers of volunteers received at the Disaster VCs in the three affected prefectures. By March 2012

Table 11.1: Volunteer numbers processed at Disaster Volunteer Centers in the three affected prefectures (as at the end of March 2014, total numbers)

	Iwate Prefecture	Miyagi Prefecture	Fukushima Prefecture	Total
March 2011 to end of May 2011	92,808	231,963	83,778	408,549
June 2011 to end of March 2012	254,691	293,778	64,371	612,840
April 2012 to end of March 2014	142,528	154,752	32,257	329,537
Total	490,027	680,493	180,406	1,350,926

Note: Because the role of these Social Welfare Councils has changed from the initial focus of recovery to reconstruction, each of the three affected prefectures has discontinued the operation of Disaster VCs and replaced them with new support structures.
Source: Calculations by the author based on figures from the homepages of Iwate Prefecture Social Welfare Council Volunteer and Citizen Activity Center, Miyagi Prefecture Disaster VC, and Fukushima Prefecture Everyday Life Reconstruction VC.

more than 1,000,000 volunteers had been processed, a figure that rose to 1,350,000 by March 2014, two years later. The peak of activity occurred on May 3 during the late April/early May Golden Week public holiday, when approximately 12,000 people participated.[13]

At the time of the disaster, however, a processing backlog was created when many Disaster VCs limited volunteer processing and reception to either local people or residents from inside the prefecture. Reports suggested that of the 65 Great East Japan Earthquake Disaster VCs (a figure that includes inland centers established throughout the three affected prefectures), 56 placed restrictions on the volunteer numbers processed (*Tokyo Shimbun*, April 26, 2011; *Kyōdō News* report). There was some criticism of this as an example of system inadequacy and inflexibility.[14]

The nature of volunteer service

Many volunteer services involved general tasks that any person could perform. Given concern around missing person searches and dangerous disaster site conditions, however, there were some limits to the sphere of suitable activities. As a result, volunteers were often involved in hard, physically onsite work that included repairing damaged houses and assisting residents to relocate, clearing away rubble, and removing accumulated mud from beneath floors or from gutters and drains. As restricted-entry zones gradually re-opened, permitting disaster-affected residents to return to their homes, appeals for help to clean out and wash down property became common (Japan National Council of Social Welfare 2013).

Requests that came from evacuation centers, on the other hand, were often for less physically demanding, or 'soft', volunteer assistance. This included distributing meals and foodstuffs, cleaning facilities, distributing drinking water, and taking people to and from bathing facilities and hospitals, in addition to helping the elderly and the disabled. When those who had lost their homes began to move into temporary accommodation, there was heavy demand for the provision of psychological support services for these people.[15] Important tasks included assisting with shopping, reading picture books and organizing gatherings that provided opportunities for interaction between those who had been affected by the disaster. This was especially the case in the areas in Fukushima Prefecture where the nuclear accident had forced evacuees out of their homes. Other services offered to meet the multiple needs of disaster victims were sorting materials in supply warehouses, assisting with government-related tasks such as filling in documentation, cleaning personal effects and photographs found in mud-filled drains and similar places, and arranging exhibitions to help owners reclaim these. Volunteers even assisted with the removal of number plates from deregistered vehicles. Some requests, such as assistance to clean hotel interiors or car parks, were regarded as the responsibility of individual or corporate interests and, as a result, were declined.[16]

Table 11.2 provides a breakdown of the specific types of volunteer services undertaken through the Kamaishi City Disaster VC. In the year after the disaster, 60% of volunteer activity was devoted solely to physical tasks such as clearing away and removing rubble, assisting victims to move residence and clearing away mud.

Support for Disaster VCs and volunteers

Contributions on the part of Disaster VCs and the volunteers that they coordinated, however, did not merely result from efforts by the centers and individuals involved. Ongoing assistance, in fact, came from local residents and private sector concerns, and also from professional volunteer actors such as NGOs/NPOs. What

Table 11.2: A percentage breakdown on types of services provided by volunteers through Kamaishi City Volunteer Center (now the Center for a Secure Everyday Life)

	No.	**%**
Removal of rubble	483	18.7
Couriering goods	452	17.5
Moving house	341	13.2
Removing mud	243	9.4
Support for official processes	210	8.1
Others	206	8.0
Organizing and cleaning	183	7.1
High-pressure cleaning	152	5.9
Reception substitute	127	4.9
Advertising events	44	1.7
Temporary matters (e.g. accommodation)	35	1.4
Cutting back grass	34	1.3
Stripping floors and walls	29	1.1
Searching for lost items	18	0.7
Disinfecting	12	0.5
Helping at evacuation centers	6	0.2
Distributing goods	5	0.2
Serving meals to victims	3	0.1
Total	2,583	100

Note: Data collected between March 2011 and March 2012. The same activity on different days counted as one activity. It might be noted that data recorded in the period immediately after the disaster is not necessarily totally accurate.
Source: Calculations by the author based on internal documents of the Kamaishi City Social Welfare Council.

sort of back-up support, then, was provided for Disaster VCs and volunteers?

Management (operational) support for Disaster VCs

Initial back-up support efforts concerned the establishment and operation of Disaster VCs. The fact that the Tōhoku disaster struck many small-scale municipalities meant that there were limits to the personnel available even to the Social Welfare Council charged with setting up Disaster VCs. There were also, as noted, cases where council offices had been inundated or senior staff had been lost. Cognizant of all this, a range of volunteer actors initiated on-the-ground support for Disaster VCs. Some local businesses dispatched staff to work at the centers or provided free fuel to support activities such as the distribution of rescue goods and the operation of snowplows. Others drew on specific business expertise to, for example, convert disaster victim information into online databases (Japan Business Federation Corporate Social Responsibility Committee 1% Club 2012). The support provided to set up and operate the centers largely came from Social Welfare Councils from other places throughout Japan, in conjunction with the Joint Committee for Coordinating and Supporting Voluntary Disaster Relief Activities (abbreviated to Joint Committee for Voluntary Relief) and from NGOs/NPOs that specialized in disaster aid.

Throughout Japan, prefectural and municipal Social Welfare Councils came together from designated regions to form single teams, or blocks, to deploy large-scale human resource support to Disaster VCs.[17] On March 25 social welfare teams across Japan, not including Tōhoku, initiated the processes necessary to dispatch staff to the three affected prefectures. The Kantō block (Kanagawa, Niigata, Yamanashi, Nagano and Shizuoka prefectures) and the Tōkai/Hokuriku block took responsibility for Iwate, and the Kinki, Chūbu and Shikoku block took responsibility for Miyagi, while a different Kantō block (Ibaraki, Tochigi, Gunma, Saitama and Chiba prefectures, with Tokyo Metropolis) and the Kyūshū block took responsibility for Fukushima Prefecture. From the time of the disaster until August 2011, a total of 30,685 staff members were dispatched to the designated sites through this system.

The Joint Committee for Voluntary Relief was the body that provided volunteers with material support and specialist know-how. Formed in January 2005 as a structure to both review and gain knowledge from volunteer activity associated with the October 2004 Niigata Chūetsu Earthquake, it was networked around the Community Chest of Japan,[18] the national social welfare funding body, with participants drawn from business, NPOs and Social Welfare Councils. While the Joint Committee's activities in non-crisis times center on survey research and staff training in disaster support, its main role during times of crisis is operational support for Disaster VCs. During the Tōhoku disaster, the Joint Committee dispatched staff it had trained in disaster support to various disaster sites. It also collaborated with (and, as needed, dispatched staff to) Disaster VCs and onsite NPOs, the Japan NGO Center for International Cooperation (JANIC), which promotes networking between Japan's international NGOs, and the Japan Volunteers Coordinators Association, which conducts staff training designed to promote collaboration between volunteers and volunteer organizations. These operational support staff members travelled from site to site and were therefore able to provide onsite operational know-how to the Disaster VCs and to supply the materials needed for relief activity. Between March 2011 and March 2014 a total of 1,498 workers were dispatched in this way to provide Disaster VC operational support in the three affected prefectures (Japan National Council of Social Welfare 2012; Joint Committee for Coordinating and Supporting Voluntary Disaster Relief Activities 2014).

Some NGOs/NPOs provided intensive disaster effort support by dispatching permanent staff to designated Disaster VCs. These NGOs/NPOs had an abundance of disaster relief experience and participated in-depth in Disaster VC decision making. Various methods were used to identify sites requiring support. For example, the international NGO Adra Japan assessed the situation across a range of locations, eventually deciding in terms of perceived need to support the Disaster VC in Miyagi Prefecture's Yamamoto Town. The disaster specialty NPO Rescue Stock Yard, however, had a pre-existing relationship with the Social Welfare Council in Shichigahama Town, also in Miyagi, and built on this by assisting the town's Disaster VC.[19]

Support to encourage volunteer participation

Also important was the provision of support to encourage those who had reservations about participating as volunteers. When Disaster VCs were being established, there was some uncertainty as to whether or not sufficient numbers of volunteers would step forward. Certain reports suggested that numbers would fall after Golden Week and there were concerns that the volunteer presence would eventually taper off altogether (*Asahi Shimbun*, May 10, 2011, and June 19, 2011). Transport access to the stricken areas was less than convenient and participating volunteers incurred a significant personal burden in terms of travel and accommodation costs. Many actors had to devise creative strategies to ensure the flow of volunteer numbers.

The first strategy designed to promote volunteer participation was the provision of special volunteer buses. Almost immediately the disaster struck, groups such as Social Welfare Councils and NGOs/NPOs from all over the country organized group bus travel in order to facilitate the participation of all individuals, including those from places distant from Tōhoku, looking to assist. For example, the Corporate Social Responsibility Committee 1% Club (hereafter, the 1% Club) of the Japan Business Federation (Keidanren)[20] funded volunteer bus travel in collaboration with the Joint Committee for Voluntary Relief. Between the end of April and the end of August 2011, this provided travel to the disaster zone for 2,101 people (including staff) (Japan National Council of Social Welfare 2012). Various travel companies also organized bus services, which, in order to attract volunteers, were sometimes packaged with accommodation or visits to tourist attractions in regions close to the disaster zones. Some private companies encouraged volunteer participation by expanding leave provisions for staff engaged in aid activities. In a survey conducted among member companies of Keidanren, 26 responded that they had expanded existing leave systems to accommodate volunteers, 27 had introduced a new leave system, and a further 21 were operating a system that applied specifically to the Great East Japan Earthquake.[21]

Universities, including two-year colleges, and higher education institutions throughout the country organized the dispatch of student volunteers. For example, in conjunction with 146 universities throughout Japan, Iwate Prefectural University established the Iwate GINGA-NET Project, which saw 1,086 university students from inside and outside the prefecture dispatched to Sumita Town, Kamaishi City, Ōfunato City and Rikuzentakata City in Iwate during the summer of 2011 (Iwate GINGA-NET Project Implementation Committee 2012).

In an online survey conducted by the author,[22] 54.4% of all respondents indicated that they had participated in volunteer activities through some form of group plan,[23] and 58.9% spent time in Iwate, 52.2% in Miyagi, and 51.0% in Fukushima. As noted above, many volunteers came from outside the Tōhoku region, and it appears that for those for whom travel to disaster-affected areas came at a significant personal time and monetary cost, group travel was the preferred mode of access.

A second key strategy was the establishment of special volunteer accommodation facilities. Day-trip volunteering in sites along the Iwate coastline was extremely difficult for people from Tokyo and even being accommodated at an inland center such as Morioka City meant a three-hour round trip to coastal disaster locations. Although NGOs/NPOs provided tent accommodation, tent numbers were limited. Eventually, local government administrations and Social Welfare Councils in areas close to the disaster zone proposed and took the necessary action to release public facilities for use as volunteer accommodation. Sumita Town Social Welfare Council, for example, offered a community hall situated in Ōmata District in the town's west. Here, facilities known as Sumita Base opened on April 25 and housed 22,485 volunteers in the one year and five months that followed.[24] In collaboration with the Morioka City Social Welfare Council, Morioka City transformed the site of a former prefectural high school situated between Morioka City and Miyako City into Kawai Camp, which housed a total of 23,597 volunteers during the one year and eight months after July 26.[25] On March 21, Tōno City in Iwate Prefecture also released facilities for use as volunteer accommodation (Tōno City 2013).

As for accommodation facilities when they first visited affected areas, the above survey results show that second only to the 25.0% who stayed in inns, boardinghouses or hotels was the high rate of 19.1% of volunteers who made use of specially provided facilities or tents.[26] Broken down by prefecture, the use rate of these forms of accommodation was 23.5% in Iwate, 16.4% in Miyagi and 17.5% in Fukushima. The high percentage in Iwate Prefecture, where many facilities were available for volunteer use, suggests the level of commitment to assist in disaster relief by the municipalities and organizations concerned.

NGO/NPO trends

Following the Great Hanshin–Awaji Earthquake, NGO/NPO actors played an increasingly significant role in disaster volunteer activity. What specific contribution did these organizations make at the time of the Tōhoku disaster?

Volunteer actor diversity

The activity sphere of NGO/NPO actors was quite broad. Religious groups,[27] for example, engaged mainly in activities such as clearing away rubble, which might be termed regular or ordinary volunteer service. Professional NGOs/NPOs, on the other hand, offered more specialized services in fields such as education, health and the protection of disaster victim human rights. There were also specialist groups, including groups with specific occupational qualifications, that dispatched volunteers to provide expert skills and knowledge onsite.

Although there were significant numbers of volunteer actors registered through the Disaster VCs who engaged in routine day-to-day tasks, large organizations with their own volunteer registers, such as the Japanese Red Cross, also assisted with ordinary day-to-day volunteer work. Some set up operating bases in disaster areas and matched their contributions to the results of disaster victim needs surveys. The Japan YMCA, for example, initiated disaster support centers in Sendai City, Ishinomaki City and Miyako City, and provided day-to-day services such as emergency goods provision, evacuation center support and rubble clearance. By March 2014, a total of 37,234 people had provided support through these three YMCA centers.[28] The NGO Peace-Boat, which conducts boat tours promoting international exchange, established an office in Ishinomaki City. This group also created the Peace-Boat Disaster VC, which drew volunteers from inside and outside Japan. By March 2014 a total of 87,504 people had participated in this program (Peace-Boat Disaster Volunteer Center 2014). Many religious groups contributed independently, with reports that one group dispatched a total of 20,000 people to the stricken zone within three months of the disaster (Okamoto Masahiro 2013b).

NGOs/NPOs had a strong profile in the provision of specialized volunteer services, particularly those organizations with an international profile and experience providing human support overseas. These groups offered an extensive range of specialist services, including children's education and childcare, health services, public hygiene, support for foreigners in the disaster area and monitoring the human rights of those impacted by the nuclear disaster. Save the Children Japan, for example, provided psychological support for children in evacuation centers and support for the re-opening of schools in disaster-affected areas. With experience in regional healthcare in places such as Cambodia and Timor-Leste, the Citizens International Health Cooperative – known in English as SHARE (Services for the Health in the Asian & African Regions) – provided health consultancy for the homebound elderly, disabled and mothers with small children. This group also provided mobile medical support for those in temporary accommodation. Human Rights Now, which deals with human rights violations throughout the world, monitored radioactive contamination and related health problems, while also seeking improvements in matters such as evacuation center conditions for women and the protection of disaster victim human rights. This group also drew on its observations of disaster area conditions to make recommendations to government (Japan NGO Center for International Cooperation 2012). Reports indicated that this was the first time that many of these international NGO/NPO actors had been involved in domestic activity.[29]

Although occupational group activity was a feature of the Great Hanshin–Awaji Earthquake response, by the time of the Tōhoku disaster there was an extraordinarily diverse range of occupational groups offering volunteer services. These included lawyers, scriveners, labor and social security lawyers, tax accountants, architects, doctors, nurses, pharmacists, hairdressers, and acupuncture and moxa massage specialists.[30] Among these were actors such as beauticians and hairdressers who offered hands-on aid to victims, while others such as lawyers offered consultancy services. The National Welfare Beauty and Barber Training Association, which supports the training of hairdressers and beauticians to provide services for those in receipt of care services and the disabled, dispatched hairdressers and beauticians from the Chūbu, Kantō and Kansai regions. These workers offered services intended to provide a degree of stress relief for disaster victims at evacuation centers and care centers in each of the three affected prefectures.[31] In terms of consultancy services, the Japan Federation of Labor and Social Security Attorney's Association conducted free telephone consultations through a Labor and Social Security Attorney Hot-Line operating in each of the three prefectures. This group also operated the Labor and Social Security Attorney Hot Caravan mobile consultation service. In this way, advice was provided on many issues, such as employment insurance and superannuation and healthcare and care matters, which were a source of anxiety for disaster victims (Kaneda 2012).

There were also groups, such as the Bar Association, that initiated new volunteer activities. The Bar Associations of the three affected prefectures conducted free legal consultations at disaster sites, while the Japan Federation of Bar Associations, the Japan Legal Aid Center, and various regional Federated Bar Associations all dispatched staff and assisted with local Bar Association consultation activities. The result was the aggregation of data from approximately 37,000 cases of legal consultation that could be drawn upon as a source of on-the-ground information for input into the design of legislation regarding the various problems experienced by disaster victims. This included the problem of the double loans[32] incurred by those who needed to seek funds to rebuild while still obligated to meet existing housing debt (Nagai 2012).

A noticeable trend in the activities of groups such as NGOs/NPOs in the aftermath of the Great East Japan Earthquake was the degree of mutual cooperation between actors. Immediately after the disaster, several hundred volunteers poured into Ishinomaki City, initially creating something of a volunteer task demarcation dispute. This led to the formation of a new collaborative organization known as the Ishinomaki Disaster Reconstruction Support Committee (now, Future Support Ishinomaki), intended to ensure information sharing and activity coordination between the various volunteer groups.[33] In Kesennuma City, too, a collaborative committee with similar objectives and a focus on NGOs/NPOs was regularly convened.[34]

These new organizations did not merely coordinate the activities of other groups. Some, such as Tōno Magokoro Net (The Tōno City Disaster Relief Network), involved themselves also in their own volunteer work. The Tōno network was an NPO formed by volunteer actors from inside and outside the prefecture and included NGOs/NPOs with Great Hanshin–Awaji Earthquake volunteering experience, local Social Welfare Councils, and local volunteer actors such as the local chapter of the Junior Chamber International, Japan. Looking to provide back-up for affected coastal communities, the group worked in conjunction with Disaster VCs to coordinate volunteers in the provision of services such as clearing rubble, care work and nursing, and the provision of psychological support for those impacted upon by the disaster (Japan National Council of Social Welfare 2012; Office of the Cabinet 2012).

A further prefectural-level development promoting information exchange and volunteer task coordination, with a particular emphasis on both metropolitan and local NGOs/NPOs, was the creation in each of the three affected prefectures of Reconstruction Cooperation Centers to provide volunteer support (Japan National

Council of Social Welfare 2012). At the national level, the Great East Japan Earthquake National Support Network, also known as Japan Civil Net, was formed with a base membership of NGOs/NPOs and private sector businesses involved in Tōhoku aid activities. In addition to supporting exchange and collaboration between NGOs/NPOs and the private sector, this group sought information sharing with government bodies and ensured that volunteer actor information was quickly available to the public on its website. Japan Civil Net also lost no time in collaborating with organizations such as local Social Welfare Councils and the Reconstruction Cooperation Centers in the three affected prefectures, and also with the Joint Committee for Voluntary Relief, to initiate regular meetings with and between national and local government bodies (Japan National Council of Social Welfare 2012).

Problems experienced

The above discussion provides a comprehensive overview of volunteer activities following the Great East Japan Earthquake. The Great Hanshin–Awaji Earthquake revealed inadequacies in the coordination of volunteer activity, which led to the establishment of the Disaster VC structure as a means of systematically processing individual volunteers during times of disaster. What problems, then, were apparent during the Great East Japan Earthquake and what recommendations for the future can be made?

The creation of a Disaster VC or NGO/NPO with carriage of volunteer activity gives rise to a range of organizational and operational questions. A particular problem common to many group actors who contributed to the Tōhoku relief effort was the availability of personnel.

For example, staff from the Yamada Town Disaster VC noted that not only did the VC have a shortage of staff with the requisite skills and knowledge at the time that operations commenced, but it also had an actual shortage of staff.[35] Since many small-scale communities were struck by the disaster, the Yamada Town example would certainly not have been unique. These staff shortages were also given as justification by Disaster VCs for limiting the numbers of volunteers processed. In addition to staff shortages related to day-to-day operations, many centers had to deal with support staff shortages. Immediately after the earthquake struck, experienced Social Welfare Council staff members from around Japan were dispatched to disaster sites. From July, however, many inexperienced people were sent to assist, which created problems even in basic areas of center administration. Disaster VCs accordingly requested the dispatch of staff members who were familiar with center procedures (Japan National Council of Social Welfare 2013). In Miyagi, where prefectural office staff members were completely overwhelmed with work, it became impossible to dispatch the personnel who had previously been identified as suitable for dispatch to a Disaster VC (Miyagi Prefecture 2012).

NGOs/NPOs found themselves in a similar situation. There were few local NPOs that operated in times of emergency and it proved simply impossible to secure adequate numbers of staff who could make an immediate contribution when the disaster struck. A survey of internationally active NPOs participating in volunteer activity indicated that the greatest difficulties experienced by these groups were managing staff dispatched to disaster sites and recruiting new staff (Japan NGO Center for International Cooperation 2012).

A second problem that emerged was the coordination of activities between different types of actors. According to the NGO/NPO survey referred to above, the second greatest difficulty encountered while undertaking volunteer work was 'building relationships with related organizations'. Japan Platform, whose objective is to aid international NPOs, had a presence in Miyagi Prefecture Disaster VCs and undertook to help by trying, for example, to match Disaster VCs and NGOs/NPOs. It would appear that overall, however, there were few examples of successful cooperation between the various agents such as Social Welfare Councils, different levels of government and NGOs/NPOs (Miyagi Prefecture 2012). In Iwate Prefecture, in fact, Disaster VCs in some cases declined assistance from NGOs/NPOs (Iwate Prefecture 2013), and there was considerable conflict between some Disaster VCs and NGOs/NPOs regarding perceived overlaps in spheres of activity.[36] A survey of local government administrations indicated that, one month after the disaster, these groups felt a 6.3% sense of disagreement towards Social Welfare Councils. This figure, however, more than doubled to 14.7% towards volunteer groups,[37] suggesting difficult relations between local administrations and NGOs/NPOs.

Measures to address problems

The types of staffing problems that occur at times of disaster will need to be addressed through the actors involved devising mechanisms that operate on a national level and not merely in disaster regions. For Disaster VCs, this will include the introduction of an effective system for the dispatch of volunteer coordinators. NGOs/NPOs will need to create a personnel bank system that enables organizations to draw upon experienced NGO/NPO personnel, including those from a particular group's broader networks, with overseas emergency aid experience (Iwate Prefecture 2013; Japan NGO Center for International Cooperation 2012). Regarding the problem of cooperation between groups, NGOs/NPOs, local government administrations and Social Welfare Councils will each need to liaise with outside organizations and compile a schedule for regular inter-group meetings to be conducted in non-disaster times. Miyagi Prefecture, for example, has announced plans for a volunteer reception system that involves NGOs/NPOs with a track record in disaster relief and not just Disaster VCs, while Iwate Prefecture has proposed a more collaborative approach between the prefecture and the municipalities' disaster relief headquarters (Japan National Council of Social Welfare 2012; Iwate Prefecture 2013; Miyagi Prefecture 2012).

If NGOs/NPOs are to win community trust, they will need to publically articulate the areas in which they work and the duties they wish to perform. In order to meet the needs of the various disaster-affected communities, private sector actors, too, will need to enter into stronger collaborative relations with municipalities and the NGOs/NPOs that are embedded in the regions (Japan NGO Center for International Cooperation 2012; Japan Business Federation Corporate Social Responsibility Committee 1% Club 2012). Each of these problems has emerged in the past and partially effective solutions have been proposed. As noted above, the Joint Committee for Voluntary Relief, a network organization formed from NGOs/NPOs and the private sector, dispatched large numbers of administrative support personnel to Disaster VCs. New cooperative actors, such as the Japan Civil Network, were created, while each of the three affected prefectures formed Reconstruction Cooperation Centers. Nevertheless, it proved difficult to solve the problems that emerged in the aftermath of the Great East Japan Earthquake 'on-the-run'.

Ultimately, these are systems problems that must be addressed outside times of crisis if effective solutions are to be found.

Conclusion

This chapter has sought to provide a detailed account of the kinds of actors who volunteered in the wake of the Great East Japan Earthquake and the nature of the relief that these actors provided. It has been noted that, following the establishment of Disaster VCs across the disaster affected regions, a total of more than 1,350,000 volunteer gave support in the three years to the end of March 2014. This activity, however, was only possible with the provision of widespread back-up support for volunteers and the Disaster VCs that processed them. Such support came from a range of sources, including the private sector. The chapter has noted the nature of the contributions made by NGOs/NPOs, which often worked independently, and also by occupational groups. Examples have been provided of the wide spread of activities undertaken by NGOs/NPOs and also of inter-organizational collaboration. Finally, attention has been given to problems that emerged in the volunteer sphere following the Great East Japan Earthquake. These included personnel shortages and tensions between the different actors involved. Measures have been suggested to address these.

Schwartz and Pharr (2003) argue that although Japanese civil society has passed through the formative stage of administrative guidance, it continues to develop and has yet to reach the degree of maturity seen in Europe or North America. Richard Samuels (2013) argues, however, that the explosion that occurred in volunteer activity in Japan following the Great East Japan Earthquake counterbalances any negative impact from the country's reactive administration or dysfunctional government. Citing the efforts of volunteers and Disaster VCs, cooperation between NGOs, and the systematic contribution made by the private sector after the 3.11 disaster, Samuels concludes that there could be no more vibrant civil society than that which has operated in Japan since that time (Samuels 2013: 17–19).

There is no doubt that, right from the point at which the Great East Japan Earthquake struck, various actors, many of whom came from groups and fields that had never previously been involved in volunteer activity, supported disaster victims in ways that are too numerous to mention. Most, furthermore, were cognizant of and ensured that they heeded the lessons learned from events such as the Great Hanshin–Awaji Earthquake and Niigata's Chūetsu Earthquake. The discussion of volunteer disaster activities provided above therefore gives strong support to Samuels' claim. Thus we can say that through the experience of the Great East Japan Earthquake, Japanese civil society has erased any previous image of immaturity and opened a path to new horizons.

Part III
Picturing Fukushima from the Data

12 Effects of the Nuclear Disaster: Evidence in the Data

Itō Yasushi

This chapter presents the basic data required to understand the damage caused by the Fukushima Daiichi Nuclear Power Station disaster. Although considerable data are available on this topic, here I have chosen that which indicates the structure and features of the disaster. This information thus provides a basic index to facilitate analysis of the damage from an economic point of view.

The most serious harm that can result from a nuclear power station disaster is damage to people's health caused by exposure to radioactive materials, which are scattered by the hydrogen explosion or other similar events. In the following section I present a summary of the results of the Fukushima Health Management Survey, which has been conducted on the residents of Fukushima Prefecture, and also a summary of the radiation doses workers were exposed to when containing the disaster.

In order to avoid harm from exposure to radiation, the national and local governments ordered people who resided near the Fukushima Daiichi Nuclear Power Station to evacuate. In addition to this, many evacuated voluntarily from areas where it was not compulsory to do so. In the third section below I show how many people moved out of the disaster area, including the number of people who evacuated voluntarily.

The evaluation of the large-scale population migration presented in the third section reveals that people who were forced to evacuate, as well as those who did not evacuate, suffered losses, such as loss of employment or a marked reduction in income, caused by the decrease in the population of the area. In the fourth section I show the amount of money that the Tokyo Electric Power Company (TEPCO) is paying for the actual damage caused by the disaster, as well as for expenses involved in taking action to avoid further harm.

In the fifth section I summarize the data on government spending as a result of the nuclear power station disaster. If a disaster occurs in a democratic country, each level of government, whether state or prefectural or the governing bodies of cities, towns and villages, is required to incur some expense in order to rescue people, ease suffering, and reconstruct damaged property or infrastructure. By investigating the scale and direction of the expenses, we shall come to understand what has been considered as damage by the national government and how it has carried out reconstruction work. We also note that the compensation for damages that TEPCO is responsible for is so large that the company is unable to pay in full, and therefore the national government has shared some of the expense. However, it is difficult to identify the structure of governmental assistance. I try to clarify the governmental share of expenses for the damage, which has not been provided in a transparent manner.

Finally, the data presented here are not necessarily the most current. Some data have been depicted to show why various subjects decided to take certain actions when they did, and I have left such data as is without updating it.

Health-related data

Those who were exposed to radiation caused by the nuclear power station disaster are divided into two groups: people who resided near the Fukushima Daiichi Nuclear Power Station and workers engaged in handling the disaster and decommissioning the nuclear reactor. There is a huge difference in the circumstances of these two groups.

The residents' health

The Fukushima prefectural government established an oversight committee to conduct the Fukushima Health Management Survey in May 2011, and began surveying the healthcare of prefectural residents with the aim of reducing anxiety about the health of prefectural residents and promoting healthcare in the future. The survey consists of two parts: the Basic Survey, conducted to estimate the radiation exposure for each individual, and the Detailed Survey, conducted to gauge overall health. An estimation of the exposed dose was arrived at in the Basic Survey by asking victims to answer detailed questions about their behavior patterns

Table 12.1: Basic Survey response rates of cities, towns and villages in Fukushima Prefecture as of March 31, 2015

	Number of subjects (a)	Number of responses (b)	Percentage response rate (c = b/a)
Northern area	504,045	150,628	29.9
Central area	557,259	134,016	24.0
Southern area	152,229	33,863	22.2
Aizu	267,205	55,953	20.9
Minamiaizu	30,788	6,180	20.1
Sōsō Districts			
Sōma City	37,372	13,040	34.9
Minamisōma City	70,013	29,844	42.6
Hirono Town	5,165	2,197	42.5
Naraha Town	7,963	4,137	52.0
Tomioka Town	15,751	8,566	54.4
Kawauchi Village	2,996	1,525	50.9
Ōkuma Town	11,474	6,016	52.4
Futaba Town	7,050	3,918	55.6
Namie Town	21,321	12,910	60.6
Katsurao Village	1,541	812	52.7
Shinchi Town	8,357	2,670	31.9
Iitate Village	6,587	3,428	52.0
Total	195,590	89,063	45.5
Iwaki	348,223	87,214	25.0
Total	2,055,339	556,917	27.1

Source: Prefectural Oversight Committee for the Fukushima Health Management Survey 2015a: 1.

during the four months immediately following the disaster, their answers being checked with the dose exposure map. The Detailed Survey consists of the following: a thyroid ultrasound examination, targeting people aged 18 years or younger at the time of the disaster; a medical examination, including differential leukocyte count, targeted at residents from the evacuated areas and those deemed high-risk as a result of the Basic Survey; a comprehensive health check, targeting residents of non-evacuated areas; a mental health and lifestyle survey, targeting residents of evacuated areas; and a pregnancy and birth survey, targeted at pregnant women and nursing mothers. By maintaining these surveys, the health of prefectural residents has been considered. The aim has been to understand not only the effects of radiation on health, but also the health-related effects of evacuation.[1]

The percentage of people who had responded to the Basic Survey as of March 31, 2015, was 27% of the population of Fukushima Prefecture. Although the response rate is relatively high in the cities, towns and villages in the Sōsō Districts, located near the Fukushima Daiichi Nuclear Power Station, it is very low in other areas (Table 12.1).[2]

Regarding the effective external exposure dose, as estimated from responses to detailed questions about behavior patterns during the four months immediately following the disaster, the highest figure was 66 millisieverts (mSv). When we exclude the workers who carried out radiation-related tasks, the highest figure was 25 mSv, and 94.8% of those surveyed returned figures less than 2 mSv. With reference to the report of the United Nations Scientific Committee on the Effects of Atomic Radiation (2008), the Prefectural Oversight Committee for the Fukushima Health Management Survey (2015a: 3) made the assessment that 'with the epidemiological investigation unable to confirm clear effects on health from exposure under 100 mSv, it is hard to find any effect on health caused by radiation, although we only have estimates for the effective external exposure dose over the first four months'.

Figures 12.1 and 12.2 show the cumulative distribution of the effective external exposure dose for residents in each district (excluding workers who carried out radiation-related tasks). While the figures for residents in the Sōsō Districts, located near the nuclear power station, are not as high as those of residents in the northern and central areas of Fukushima Prefecture, residents in some parts of the Sōsō Districts were still exposed to relatively high effective doses of radiation.

When we compare the figures for each city, town and village, we can see that of all locations, Iitate Village has the highest ratio of residents who show a high effective external exposure dose. We can also see that there are some residents in Namie Town who show a relatively high effective external exposure dose. The cumulative distributions of effective external exposure doses in the northern and central areas of the prefecture are higher than those of other districts. This is because there are many residents with relatively high effective external exposure doses in the cities of Fukushima and Kōriyama, which are highly populated. The high effective external exposure

Figure 12.1: Cumulative distribution of effective external exposure dose in each district of Fukushima Prefecture

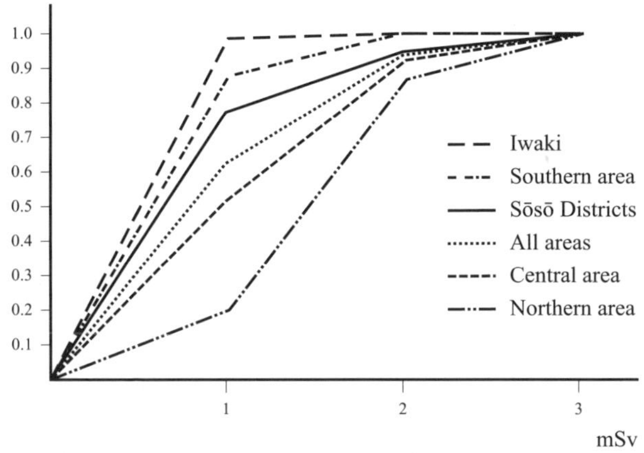

Source: Prefectural Oversight Committee for the Fukushima Health Management Survey 2015a: 9.

Figure 12.2: Cumulative distribution of effective external exposure dose in each district

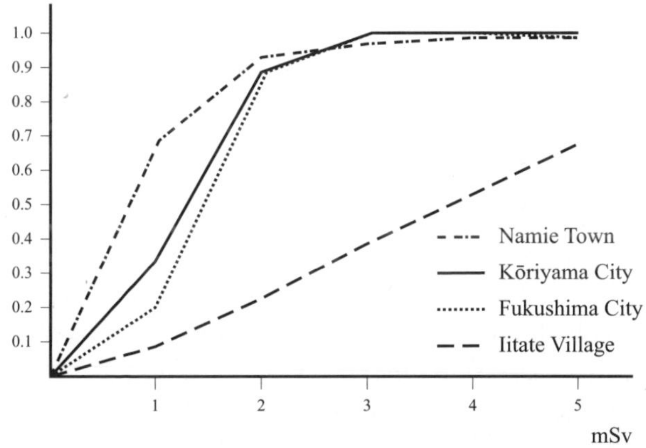

Source: Prefectural Oversight Committee for the Fukushima Health Management Survey 2015a: 9.

doses in these areas as a whole indicate that these areas lay in the path of radioactive fallout from the nuclear power station. Other instances of high doses are explained by the fact that in some parts of Iitate Village and Namie Town (and some parts of Minamisōma City, Katsurao Village and Kawamata Town) the evacuation order was delayed, and in the northern and central areas of the prefecture, no evacuation order was made at all.

In the Detailed Survey, on the other hand, the most notable result is from the thyroid ultrasound examination. Because the normal incidence of infant thyroid carcinoma is very low, it is easy to detect any mass occurrence. In the early stages of the Chernobyl nuclear disaster, an outbreak of infant thyroid carcinoma was confirmed in Ukraine and Belarus. It is considered that infant thyroid carcinoma will provide the first evidence that confirms effects on health in Fukushima, and the thyroid ultrasound examination was introduced for this reason. The tests have been conducted several times. In 2011 testing was conducted in 13 cities, towns and villages, concentrating on the Sōsō Districts, designated as priority areas by the national government (including Date City in the northern area of the prefecture). In 2012 testing was conducted in 12 cities, towns and villages located in the northern, central and southern areas, including Fukushima and Kōriyama cities. In 2013 testing was conducted in five cities and towns neighboring the districts where tests were conducted in the previous two years.[3]

The first tests, conducted from 2011 to 2013, were designated as primary tests to establish current conditions, as it was considered that even if residents were exposed to radiation, the effects on health would not yet have emerged, as only a few years had passed since the exposure.[4] The tests were conducted in two stages. If nodules or cysts bigger than a certain size were found by thyroid ultrasound examination in the first stage, a second stage test, which was more detailed, was ordered. Tables 12.2 and 12.3 show test results by year and area as of March 2015.

As of the end of March 2015, 112 people were diagnosed with malignant or possibly malignant thyroid carcinoma. Among them 99 people had an operation and 95 people had a confirmed diagnosis of papillary carcinoma. Three people were diagnosed with anaplastic carcinoma, and one person was diagnosed as having a benign nodule. The incidence of malignancy or the possibility of malignancy is about 33 to 40 per 100,000 people. Regarding these results, the Prefectural Assessment Section for Thyroid Ultrasound Examination presented the following summary at the 19th oversight committee meeting in May 2015:

> Regarding these test results, the number of people diagnosed with thyroid carcinoma is several dozen times more than the prevalence predicted by statistics on the incidence of thyroid carcinoma held by district cancer registration records. This could be explained as either an outbreak of thyroid carcinoma due to exposure to radiation, or as a case of over-diagnosis (diagnosis of thyroid carcinoma that was not life-threatening, or that had no obvious symptoms). The opinion was expressed that from the scientific evidence accumulated so far, while we cannot utterly deny the former, it is highly possible that the latter is the case.
>
> On the other hand, it was pointed out that even in the case of over-diagnosis, thyroid carcinoma that would still have threatened life in the short term or the long term was detected and cured early. (Prefectural Oversight Committee for the Fukushima Health Management Survey, Thyroid Ultrasound Examination Assessment Section 2015: 1)

The assessment that the number of people diagnosed with thyroid carcinoma is 'several dozen times more' than expected is based on an estimate made by Tsugane (2014) and the Cancer Information Services of the National Cancer Center (2014; from here on, the Center for Cancer Control and Information Services 2014). The National Cancer Center has published on the internet a file concerning 'data on the national incidence of cancer as estimated from district cancer registration' (Center for Cancer Control and Information Service 2016). Judging from data on the estimated numbers of patients with thyroid carcinoma in Japan from 2001 to 2010, population data and data on causes of death, the Center for Cancer Control and Information Services (2014) estimated that the number of people aged 18 years or younger who showed some symptoms of thyroid carcinoma in Fukushima should be 2.1 (male, 0.5; female, 1.6). This figure of 2.1 is an estimate of the number of people aged 18 years or younger who would be expected to show some symptoms of thyroid carcinoma without the effects of exposure to radiation in

Table 12.2: Thyroid ultrasound examination results (the primary test)

	Number of Stage 1 test subjects (a)	**Number of people directed to have a Stage 2 test**	**Number of people who actually had a Stage 2 test**	**Malignancy or possibility of malignancy (b)[4]**	**Ratio of malignancy and possibility of malignancy (b/a)[5]**
2011[1]	41,810	221	199	14	33.5
2012[2]	139,338	988	920	56	40.2
2013[3]	118,395	1,070	977	41	34.6
Total	299,543	2,279	2,096	111	37.1

Notes: 1: 13 cities, towns and villages where evacuation orders were made (Tamura City, Minamisōma City, Date City, Kawamata Town, Hirono Town, Naraha Town, Tomioka Town, Kawauchi Village, Ōkuma Town, Futaba Town, Namie Town, Katsurao Village and Iitate Village); 2: Fukushima City, Nihonmatsu City, Motomiya City, Ōtama Village, Kōriyama City, Kōri Town, Kunimi Town, Ten'ei Village, Shirakawa City, Nishigō Village, Izumizaki Village and Miharu Town; 3: other cities, towns and villages (Iwaki City, Sōma City, Aizu District etc.); 4: excluding one person who turned out to be benign after operation; 5: the ratio of malignancy or possibility of malignancy is the number per 100,000 people.
Source: Prefectural Oversight Committee for the Fukushima Health Management Survey 2015b: 8–9.

Fukushima.[5] As of March 31, 2015, 112 people had been diagnosed as having 'malignant or possibly malignant' thyroid carcinoma by the thyroid ultrasound examination.[6] As Tsugane (2014) pointed out, the number of people who actually had an examination was about 80% of all people who were supposed to have an examination. If those diagnosed as having malignant thyroid carcinoma only represent 80% of those who should have been examined, the real number of people with thyroid carcinoma in Fukushima would be 140. By comparing the estimate of 2.1 or the values adjusted by considering some factors and the projected figure of 140, an assessment was made that the number of people diagnosed with thyroid carcinoma in Fukushima was 'several dozen times more' than expected.

In addition, the estimate of 2.1 made by the Center for Cancer Control and Information Services (2014) was based on the number of thyroid carcinoma patients who presented for examination because they already showed some symptoms. The number of patients found by an exhaustive survey such as the health survey

Table 12.3: Comparison of thyroid ultrasound examinations (the primary test) in each district

	13 cities, towns and villages in evacuated area[1]	**Naka (Central) Street**[2]	**Hama (Beach) Street**[3]	**Aizu District**[4]	**Total**
Intended examinees	47,768	199,451	70,539	49,927	367,685
Stage 1 test subjects	41,810	169,116	55,516	32,791	299,233
Cystoscopy tests conducted	94	296	97	48	535
Diagnosed with malignancy or possibility of malignancy[5]	14	63	23	11	111
Ratio of people diagnosed with malignancy or possibility of malignancy (per 100,000 people)	33.5	37.3	41.4	33.5	37.1

Notes: 1: see note 1 for Table 12.2; 2: Fukushima City, Kōriyama City, Shirakawa City, Sukagawa City, Nihonmatsu City, Motomiya City, Kōri Town, Kunimi Town, Ōtama Village, Kagamiishi Town, Ten'ei Village, Nishigō Village, Izumizaki Village, Nakajima Village, Yabuki Town, Hanawa Town, Ayukawa Village, Ishikawa Town, Tamakawa Village, Hirata Village, Asakawa Town, Furudono Town, Miharu Town and Ono Town; 3: Iwaki City, Sōma City and Shinchi Town; 4: Aizuwakamatsu City, Kitakata City, Shimogō Town, Hinoemata Village, Tadami Town, Minamiaizu Town, Kitashiobara Village, Nishiaizu Town, Bandai Town, Inawashiro Town, Aizubange Town, Yugawa Village, Ynaizu Town, Mishima Town, Kaneyama Town, Shōwa Village and Aizumisato Town; 5: excluding one person who turned out to be benign after operation.
Source: Prefectural Oversight Committee for the Fukushima Health Management Survey 2015b: 10.

of prefectural residents would be expected to be greater than the numbers presenting voluntarily as a result of symptoms. It is called 'the effect of the screening'. This effect has been given much attention in the interim report, and the results were commented on as follows:

> At the present time, we cannot draw any conclusion as to whether or not thyroid carcinomas found by the test were caused by exposure to radiation. Considering...that according to the preceding surveys, the exposure doses in Fukushima were low compared with those in Chernobyl and no thyroid carcinoma was detected among children aged five or younger at that disaster, it is unlikely that the cases

Table 12.4: Results of the thyroid ultrasound examination (full-scale test) as of March 31, 2016

	Number of people who had a Stage 1 test (a)	Number of people who were supposed to have a Stage 2 test	Number of people who actually had a Stage 2 test	Malignancy or possibility of malignancy (b)	Ratio of malignancy and possibility of malignancy (b/a)[3]
2014[1]	158,698	1,277	1,025	48	30.2
2015[2]	109,071	784	320	9	8.3
Total	267,769	2,061	1,345	57	21.3

Notes: 1: 13 cities, towns and villages where evacuation orders were made (Tamura City, Minamisōma City, Date City, Kawamata Town, Hirono Town, Naraha Town, Tomioka Town, Kawauchi Village, Ōkuma Town, Futaba Town, Namie Town, Katsurao Village and Iitate Village) and others; 2: all other cities, towns and villages; 3: the ratio of malignancy or possibility of malignancy is the number per 100,000 people.
Source: Prefectural Oversight Committee for the Fukushima Health Management Survey 2016b: 9–10.

> of thyroid carcinoma were caused by radiation exposure. For the assessment of the effects of exposure to radiation a continuous long-term investigation is indispensable. (Prefectural Oversight Committee for the Fukushima Health Management Survey, Thyroid Ultrasound Examination Assessment Section 2015: 2)[7]

From April 2014, three years after the nuclear power station disaster, a second round of testing was begun as a follow up to the first round (the primary test). The second round of testing, called a full-scale examination, began with the intention of continuing for two years. The reason it is called a 'full-scale' examination is that if effects of low dose exposure to radiation do emerge, this should occur some time after exposure to radiation – that is, after the year 2014.[8] The methods of the test are basically the same as the preceding test.

The results of the full-scale test are shown in Table 12.4. The number of people diagnosed with malignant thyroid carcinoma per 100,000 people during the years 2014 and 2015 is about 20. Although this figure is lower than that found in the preceding test, it is higher than the number estimated without the effects of exposure to radiation. If the high number of people diagnosed with malignant thyroid carcinoma in the preceding test is the result of over-diagnosis (or the effect of the screening), as explained above, almost all patients with thyroid carcinoma should have been found in the primary test. Therefore, it was estimated that the rate of malignant thyroid carcinoma in the full-scale test should have become sharply lower than that in the primary test (Makino 2015: 938–9), but the number did not decrease sharply. The intended examinees for the thyroid ultrasound examination were people aged 18 years or younger in March 2011. When the full-scale test was conducted, they were older and more prone to contracting thyroid carcinoma. Hence it is rather difficult to interpret these numbers. Nonetheless, it is still indispensable to conduct a continuous and prudent investigation.

Radiation exposure for workers containing the accident

The exposure to radiation for workers engaged in containing the accident is much higher than for ordinary residents. Table 12.5 shows the distribution of cumulative exposure for workers at the Fukushima Daiichi Nuclear Power Station since March 2011, as revealed by TEPCO. According to the table, of 46,956 workers engaged in the clean-up from March 2011 to the end of March 2016, the highest exposure dose was 678.80 mSv and the average exposure dose was 12.83 mSv. Although the estimates of exposure for ordinary residents of Fukushima given above only showed the effective external exposure dose during the four months after the disaster, 99.8% of the residents were exposed to radiation of less than 5 mSv. By contrast, 47.8% of the workers engaged in radiation-related tasks were exposed to radiation greater than 5 mSv.

In the process of containing the disaster and decommissioning the nuclear reactor, employees will experience further exposure to radiation. Reducing workers' exposure to radiation is not an easy task, compared to that for residents near the nuclear power station who could be evacuated. In order to avoid the risk to health, it is indispensable to have proper radiation exposure management for workers. There are still some problems regarding a regulatory maximum for radiation exposure when managing labor, since laws such as the *Act on the Regulation of Nuclear Source Material, Nuclear Fuel Material and Reactors* (also known as the Reactor Regulation Act) for TEPCO employees and the *Industrial Safety and Health Act* (the ordinance on prevention of ionizing radiation dangers) for the workers of subcontracting companies are not unified (Network for Radiation-Exposed Work 2012: 19–25).

Table 12.5: Distribution of cumulative exposure for workers engaged in radiation-related tasks at the Fukushima Daiichi Nuclear Power Station

Distribution (mSv)	TEPCO employees	Workers of cooperating companies	Total
Over 250	6	0	6
Over 200~below 250	1	2	3
Over 150~below 200	26	2	28
Over 100~below 150	117	20	137
Over 75~below 100	321	312	633
Over 50~below 75	327	1,797	2,124
Over 20~below 50	633	6,513	7,146
Over 10~below 20	620	5,793	6,413
Over 5~below 10	507	5,442	5,949
Over 1~below 5	907	9,616	10,523
Below 1	1,247	12,747	13,994
Total	4,712	42,244	46,956
Maximum (mSv)	678.80	238.42	678.80
Average (mSv)	22.43	11.75	12.83

Note: Cumulative exposure dose from March 2011 to March 2016.
Source: TEPCO 2016a.

It is not necessarily the case that exposure to radiation through labor is always considered when the social cost of a nuclear power station is discussed. Moreover, harm from radiation exposure is not included among the nuclear damages subject to compensation as determined by the *Act on Compensation for Nuclear Damage*. In order to see the whole picture of the damage and loss caused by the nuclear power station disaster, we need to clarify the actual situation for radiation-exposed workers and related expenditure.

Data related to evacuation and population migration

Either under the direction of the national and local governments or through independent choice, many people evacuated from areas near the Fukushima Daiichi Nuclear Power Station in order

Figure 12.3: Numbers of evacuees in Fukushima Prefecture

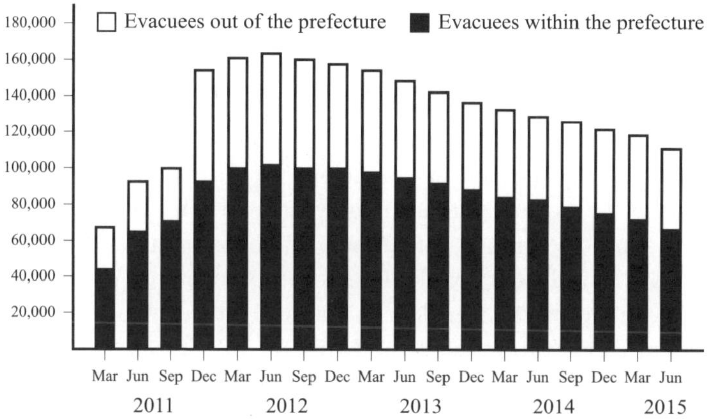

Source: Reconstruction Agency, 'The number of evacuees in Japan' (monthly reports).

to avoid exposure to radiation. The Fukushima Prefecture and the Reconstruction Agency released estimates of the number of evacuees based on reports from each town and village. Figure 12.3 shows changes in the number of evacuees for each three-month period following the earthquake disaster. The number of evacuees peaked in July 2012, when 99,521 people evacuated within Fukushima Prefecture and 62,084 people evacuated out of Fukushima Prefecture. More than 160,000 people evacuated in total. Since July 2012 the number of evacuees has gradually decreased. However, as of July 2016 there were still 49,140 evacuees within Fukushima Prefecture and 41,375 out of Fukushima Prefecture, with 90,000 people in total still living as evacuees.

These evacuee numbers are the sum of those who were evacuated compulsorily (evacuation by force) and those who evacuated voluntarily (voluntary evacuation). At first, Fukushima Prefecture reported evacuee numbers by dividing evacuees into two groups: 'compulsory evacuees' who evacuated from the towns and villages where an evacuation order was made, and so-called 'voluntary evacuees' who evacuated from the non-evacuation areas. The Dispute Reconciliation Committee for Nuclear Damage Compensation (2011b) recorded the number of voluntary evacuees progressively from immediately after the earthquake disaster until September 2011, as estimated by Fukushima Prefecture. In September 2011 there were 23,551 people who evacuated within the non-evacuation areas of the prefecture and 27,776 people who evacuated out of the prefecture – a total of approximately 50,000 people who voluntarily evacuated. Since then Fukushima Prefecture has stopped clearly reporting the number of voluntary evacuees. Evacuation orders were gradually lifted after 2012. If evacuees remain outside the evacuated areas after the lifting of evacuation orders, they are treated as voluntary evacuees.

The number of above-mentioned evacuees includes both those who transferred their resident cards and those who did not. The Report on Internal Migration in Japan Derived from the Basic Resident Registration from the Ministry of Internal Affairs and Communications, which is often used as an index of population movement, records the number of people who give notification when they move in and out of cities, towns and villages. Therefore, this report does not fully reflect the real number of people who evacuated (internal migration). Figure 12.4 shows the migration of people who were entering and exiting the four disaster-stricken prefectures (Iwate, Miyagi, Fukushima and Ibaraki) according to the Basic Resident Register. The excessive number of people who moved out of Fukushima Prefecture after 2011 is obvious, even when compared with the previous tendency for greater numbers to leave Fukushima than other prefectures, or when compared with the numbers leaving other disaster-stricken prefectures such as Iwate and Miyagi.[9]

Among all the pollution and environmental problems that Japanese people have experienced, such large-scale internal migration has never happened before. This massive migration is one of the features of this disaster.

Compensation from TEPCO

Regarding compensation for damage caused by a nuclear power station disaster, the nuclear operator that caused the accident (TEPCO in this instance) takes sole liability without fault in accordance with the *Act on Compensation for Nuclear Damage*. The scope of compensation is largely determined by the Dispute Reconciliation Committee for Nuclear Damage Compensation. When the actual amount exceeds that of financial security such as insurance (120 billion yen), the national government will give assistance as required.

Figure 12.4: Change in numbers of people migrating in and out of the four disaster-stricken prefectures

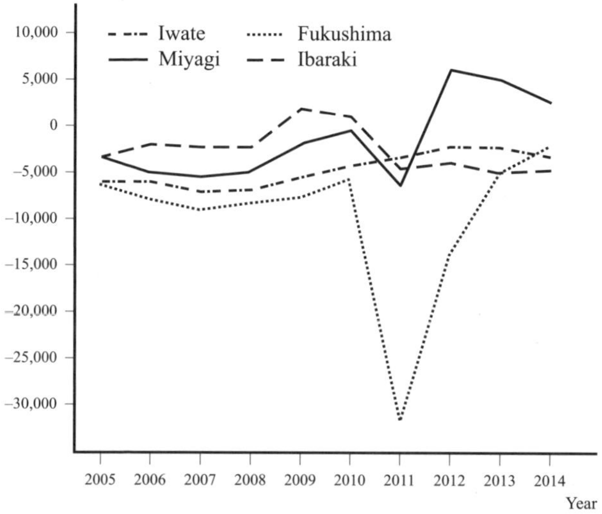

Note: The minus figures indicate reductions in population.
Source: Ministry of Internal Affairs and Communications, Report on Internal Migration in Japan Derived from the Basic Resident Registration (yearly report).

Table 12.6: Amount of compensation required to be paid by TEPCO (100 million yen) by the end of July 2016

To individuals	18,999
To corporations and sole proprietors	24,582
Common compensation	14,503
Decontamination	4,705
Total	62,791

Source: TEPCO 2016b.

In accordance with the guidelines of the Dispute Reconciliation Committee, the amount of compensation TEPCO is required to pay by the end of July 2016 is shown in Table 12.6.

Compensation to individuals is provided for evacuation costs from the areas where the evacuation order was made, for examination costs to determine exposure to radiation and the like, for loss of income caused by the difficulty in finding employment as a result of the evacuation, and for mental damage caused by the evacuation and the prolonged nature of the evacuation.[10] Compensation to corporations and sole proprietors is made for business losses, for losses caused by shipping restrictions on agricultural, forestry and fishery products, for damage caused by harmful rumors and for indirect damage.[11] Common compensation to both individuals and corporations is made for the loss or decrease in the value of properties and materials and for damage from insecure housing.[12] Regarding decontamination, compensation is made for expenses incurred in consequence of the *Act on Special Measures Concerning the Handling of Environmental Pollution by Radioactive Materials*.[13]

The amount of compensation determined to the end of July 2016 is about six trillion yen. This amount is too large for TEPCO to pay. As this situation was already anticipated in the immediate aftermath of the disaster, the government established an organization called the Nuclear Damage Compensation Facilitation Corporation (NDF) in July 2011 to assist with compensation. I discuss governmental assistance in support of compensation expenses through the NDF below.

Government spending

As part of its recovery and reconstruction support in response to the Great East Japan Earthquake, the government has made budgetary appropriations in the first, second and third supplementary budgets of the general account in 2011 and in the special account for the recovery from the Great East Japan Earthquake from 2012 onwards. Regarding all recovery and reconstruction works, the Board of Audit announces the actual budget and its detailed execution as part of its audit process. In this section I discuss government spending related to the nuclear power station disaster as detailed in Reconstruction Works in Response to the Great East Japan Earthquake from 2011 to 2014, as reported by the Board of Audit (2016). Although local governments also undertook expenditure in response to the nuclear power station disaster, the Board of Audit has reported only on national government spending (including subsidies and aid grants to local governments). However, the government designated the first five years after the

Table 12.7: Actual budget and amount of expenditure related to the nuclear power station disaster from 2011 to 2014

	Classified by account	**Actual budget**	**Amount of expenditure**	**Execution rate (%)**
2011	General account	980,888	783,207	79.8
2012	Special account for recovery	531,935	205,877	38.7
2013	Special account for recovery	1,162,911	553,113	47.6
2014	Special account for recovery	1,403,904	804,509	57.3
Total		4,079,638	2,346,706	57.5

Note: The unit for the actual account and amount of expenditure is one million yen.
Source: Estimated from Board of Audit (2013a, 2015a, 2015b, 2016) reports.

earthquake disaster as an 'intensive recovery period', and made financial provisions so that local governments would have 'no actual share' of the nuclear power station disaster-related works or of any recovery and reconstruction works resulting from the Great East Japan Earthquake. That is, during the past five years the greater part of local government spending on recovery-related works was made from subsidies or aid grants provided by the national government, and the actual expenditure incurred by local governments has been very small. Therefore, we can obtain a comprehensive picture of recovery expenses by investigating national government spending.

The actual budget and the amounts expended on nuclear power station disaster-related works during the four years from 2011 to 2014 are shown in Table 12.7, and Table 12.8 depicts the breakdown of expenditure by work.

According to Table 12.7, the total amount of the actual budget is about 4,080 billion yen, and the sum total of expenditure is 2,347 billion yen, an execution rate of 57.5%. When we examine the breakdown of tasks, for those such as the decontamination of polluted soil and the disposal of contaminated waste, 1.035 trillion yen was spent, which amounts to 57.4% of total expenditure. In addition, for Fukushima reconstruction work such as the return of evacuated residents and restoration of the evacuated areas, 133.2 billion yen was spent. Judging by the actual budget and the amount of expenditure, the government appears to attach great importance to decontamination and the return of residents.

Table 12.8: Breakdown of expenditure related to the nuclear power station disaster from 2011 to 2014 (unit is 100 million yen)

Item of works	Amount of expenditure	Classification 1	Ratio (%)	Classification 2	Ratio (%)
Urgent decontamination[1]	2,360	2,360	10.1		
Three works under law of special measures[2]					
Decontamination of polluted soil work	11,007			13,459	57.4
Construction of interim storage facility work	92				
Disposal of contaminated waste work	743	11,842	50.5	743	3.2
Fukushima reconstruction work					
Return and renewal work	94			94	
Infrastructure work	469			469	
Settlement in Fukushima work	59	1,331	5.7	59	5.7
Fukushima subsidy	709			709	
Other	7,929	7,929	33.8	7,929	33.8
Total	23,462	23,462	100.0	23,462	100.0

Notes: 1: decontamination work before the establishment of the *Act on Special Measures Concerning the Handling of Environmental Pollution by Radioactive Materials*; 2: three works under the *Act on Special Measures Concerning the Handling of Environmental Pollution by Radioactive Materials*; 3: this table follows the original table but I have made minor corrections to the totals.
Source: Board of Audit 2016: 144–5.

The figures in the above tables are taken from the supplementary general budget for 2011 and the Special Account for the Great East Japan Earthquake from 2012 onwards. However, the government has also used expenditure from the general account and other special accounts to deal with the nuclear power station disaster from 2012 onwards. Moreover, as mentioned above, the government has been providing assistance for compensation that TEPCO ought to pay through the NDF. Figure 12.5 shows the structure of governmental aid.[14]

In order to assist with compensation, the government issues 'delivery bonds' to the NDF. When TEPCO needs compensation money, TEPCO requests financial assistance from the NDF and the NDF requests delivery bonds to be redeemed. The funds for redeeming bonds are provided mostly by borrowing from the

Figure 12.5: Structure of the governmental aid through the Nuclear Damage Compensation Facilitation Corporation

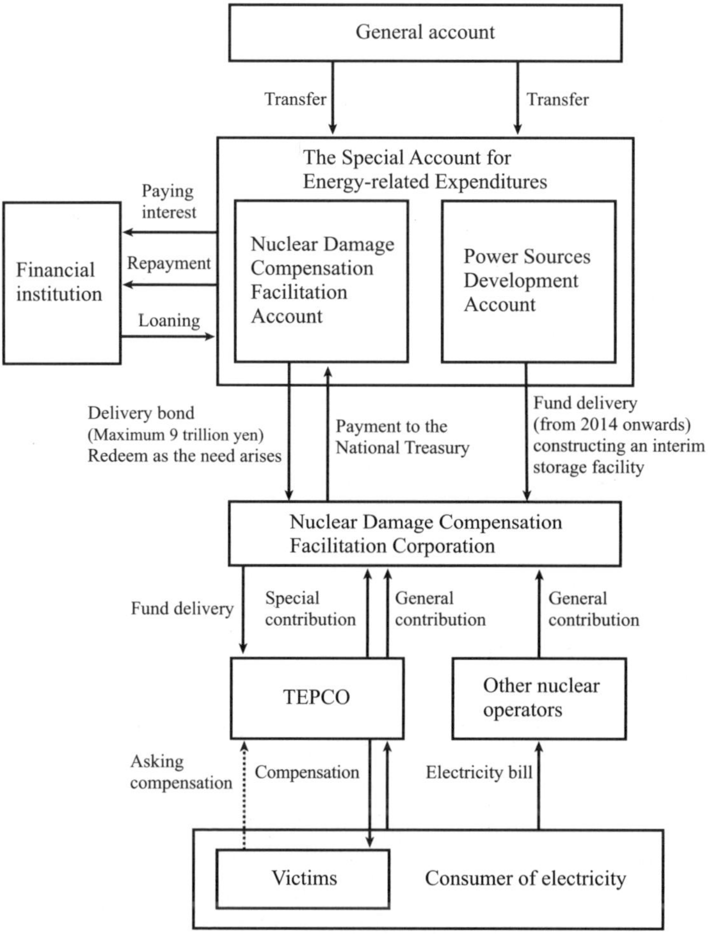

Source: Board of Audit 2015b: 21; Saitō Makoto 2015: 27.

private sector or by issuing short-term securities at the Nuclear Damage Compensation Facilitation Account of the Special Account for Energy-related Expenditures (SAEE).

When the NDF was established, the maximum amount to be raised by delivery bonds was set at five trillion yen, and its use was limited to compensation only. However, in December 2013 the government increased the maximum limit to nine trillion yen and decided that the

Table 12.9: Change in share of general and special contributions

	General contribution					
	Other than TEPCO (a)	TEPCO (b)	b/(a+b) (%)	Special contribution	Total sum of contributions by nuclear operators	TEPCO's share (%)
2011	53,130	28,370	34.81	0	81,500	34.8
2012	61,985	38,820	38.51	0	100,805	38.5
2013	106,260	56,740	34.81	27,100	190,100	44.1
2014	106,260	56,740	34.81	60,000	223,000	52.3
2015	106,260	56,740	34.81	70,000	233,000	54.4

Note: Unit of contribution is 1 million yen.
Source: Nuclear Damage Compensation and Decommissioning Facilitation Corporation website.

fund should also be allotted to other usages such as decontamination and so on (Nuclear Emergency Response Headquarters 2013: 12). Two-and-a-half trillion yen was scheduled to be allotted to decontamination expenses, and 1.1 trillion yen was allotted to the construction of interim storage facilities. The NDF will pay 2.5 trillion yen for decontamination expenses with the income gained from the sale of TEPCO shares, which the NDF obtained by investing 1 trillion yen. The 1.1 trillion yen for construction of interim storage facilities is to be paid from the general account through the Power Sources Development Account of the SAEE.[15]

TEPCO, which caused the nuclear power station disaster, is to pay a 'general contribution', together with other nuclear operators, and repay the compensation redeemed by delivery bonds in the form of 'mutual aid'. Furthermore, TEPCO pays a 'special contribution' on top of the general contribution to the NDF. Table 12.9 shows the amount of general and special contributions that TEPCO and other nuclear operators have paid so far. The amount paid by TEPCO from 2014 onwards, when it had to pay a special contribution, is about half of what all the nuclear operators have paid.

Although it is acknowledged that the general contribution can be transferred to electricity bills, the special contribution cannot be transferred. That is, ultimately the general contribution is paid by electricity consumers and only the special contribution is paid by TEPCO shareholders.

As the Nuclear Damage Compensation Facilitation Account of the SAEE manages the raising of funds to redeem the delivery bonds via borrowings, some interest liabilities are generated. By the end of 2014, 6.2 billion yen worth of interest liabilities were generated (Board of Audit 2015: 21) and this was paid from the general account. Thus, what is going to be paid by the nuclear operators, including TEPCO, is only the capital of the SAEE debt.

As discussed above, in spite of the official position that TEPCO is required to pay compensation for the nuclear power station disaster, the rather ambiguous governmental share (from the general account) and the share paid by electricity consumers (through their electricity bills) are, in fact, increasing.

13 The Cost Effectiveness of Radioactive Decontamination

Itō Yasushi

The accident at the Tokyo Electric Power Company (TEPCO) Fukushima Daiichi Nuclear Power Station caused by the Great East Japan Earthquake led to the release of a large amount of radioactive substances over a wide area. When the government designated a number of areas surrounding Fukushima Daiichi as 'evacuation instruction zones', many residents were forced to evacuate. High radiation levels were also identified in areas outside these zones, leading many other people to evacuate independently. Although the evacuation directive has been cancelled in some designated zones, radiation levels remain high in many areas. In January 2015 approximately 120,000 people (79,000 from areas currently designated as evacuation instruction zones, 20,000 from areas formerly designated as such and 21,000 from other areas) (MEXT 2015: 13) are still living as evacuees, with all of the attendant instability and uncertainty this entails.

Evacuees have three choices. They can return to their former homes, relocate somewhere else, or continue to live as evacuees (without, for example, purchasing a new dwelling) and wait to see what may happen in the future. As the third option is at best a temporary solution, in the medium to long term, evacuees can choose the first or the second option.[1] The return of evacuees to their former homes is predicated on a reduction in radiation levels. In order to achieve this, the affected areas will have to be thoroughly decontaminated. The government has placed decontamination and residents returning to their former homes on its central policy objective for addressing the evacuation issue. The phrase 'decontamination followed by return' appears in a variety of government documents.[2]

In carrying out this policy, the government has already spent a vast sum of money on decontamination. However, decontamination has not always proved to be effective in reducing radiation levels. This problem, along with the vast cost that will be involved in reducing radiation to the target levels, has led to widespread criticism of the government's current approach to decontamination. Three major questions are still being debated with regard to decontamination. First, there is the question of the level of radiation that is acceptable; that is, what are the target levels? Of course, the target levels have to be determined such that they are low enough to no longer have an effect on human health. However, a hypothesis is accepted that even low levels of radiation may have some health effects and so there is no way to determine an absolutely safe level. The enormous cost of decontamination in a situation of budgetary constraint means that the issue of cost cannot be ignored. Once a suitable target has been established, the second question is how to implement a decontamination program in order to achieve it. This question also brings up the related issue of who will bear the cost of decontamination. The third question relates to the purpose of decontamination. If the aim is to enable evacuees to get back to their former lives, is decontamination the best approach? If evacuees need to go home in order to get back to their former lives, then of course decontamination is necessary. However, not all evacuees believe that going home is the only way they can achieve their former standard of living. If there was another way to help the evacuees, then this alternative must be weighed up against the decontamination strategy.

Existing research on decontamination, other than explanation on the law for decontamination (Takahashi Yasufumi 2012; Nakajima 2013) and fact-finding surveys (Sato and Abe 2013), has generally taken the first or the second approach. The first approach looks at the costs and the benefits of decontamination. The Board of Audit (2013b) surveyed the expenditure for decontamination by the government in detail, as an original work. Yasutaka and Naitō (2013) came up with the detailed cost estimates that form the basis for this line of inquiry. Nakanishi (2014) used existing cost estimates to conduct a cost-benefit analysis of decontamination, arguing that decontamination is not cost-effective. Among the second approach, Suzuki Yoshinori (2013) and IGES (2013) highlighted the lack of flexibility in the government subsidies provided to local municipalities for decontamination. This makes it difficult for municipalities to match their decontamination efforts to local conditions. Isono (2015a) widely discusses the issues that actually occurred through implementing the decontamination policy. Furthermore, some studies discuss the issue of who will bear the cost. The government's original policy was for the state to defray the cost of decontamination initially and then later seek to recover these costs from TEPCO, the company responsible for the accident. However, in December 2013 the government assumed responsibility for these costs. Ōshima and Yokemoto (2014) have criticized this move, which they argue amounts to a policy of saving TEPCO in contravention of the 'polluter-pays principle'.

A third approach to research on decontamination would be to compare decontamination with other possible alternatives for restoring evacuees' quality of life. Although this approach has been touched on in the research (Hata and Mukai 2014; Isono 2015b), few studies have put this approach at the center of the analysis. This chapter aims to address this gap in the existing literature by comparing the government's decontamination policy with an alternative. The government's basic approach starts from the principle of seeking to return evacuees to their former homes. Many of the issues that have arisen in the decontamination process are derived from this basic premise. Therefore, I begin by looking at how the decontamination program was first set up and how 'decontamination and return' came to dominate the government's policy towards evacuees. Having done so, I analyze the cost effectiveness of the decontamination efforts that have been pursued thus far in terms of restoring the living standards of evacuees. I focus in particular on comparing alternative approaches to decontamination for evacuees from the evacuation instruction zones.

As an approach to reconstruction policy, decontamination emphasizes land and region. Reconstruction efforts have tended to remain trapped in this idea of the disaster-affected 'region'. Of course, the region has an important role as a kind of 'container' for social life but 'regional reconstruction' does not necessarily lead directly to 'individual reconstruction' (Yamashita et al. 2013). The alternative approach I examine in this chapter is one that places the individual at the center.

Decontamination policy

The evacuation instruction zones[3]

On March 11, the day of the Great East Japan Earthquake, cooling systems at the Fukushima Daiichi Nuclear Power Station ceased to function. In the midst of the crisis, the government issued an emergency evacuation order for residents living within two kilometers of the plant. The crisis escalated rapidly. The next day, there was a hydrogen explosion in reactor No. 1 at the plant. This was followed by hydrogen explosions in reactor No. 3 on March 14 and reactor No. 4 on March 15. The radioactive materials released in these explosions were distributed not only across Fukushima Prefecture but over a wide area of eastern Japan. On March 15 the government extended its evacuation order to cover a 20-kilometer radius surrounding the plant. People living within 30 kilometers were also instructed to remain indoors. In the past, evacuation orders had only ever been issued on a temporary basis. However, it was revealed that long-term evacuation would be needed, and some areas more than 30 kilometers from the plant were heavily contaminated with radioactivity due to atmospheric conditions, such as the wind direction. This led the government to declare on April 22, 2011, that everywhere within a 20-kilometer radius of the plant was a 'restricted area' and everybody living in this zone was forced to evacuate. Entry into the restricted area was forbidden.

A number of other regions, although more than 30 kilometers from the nuclear power station, have expected annual cumulative doses of radiation greater than 20 millisieverts (mSv). These areas, which include Iitate Village, the Yamakiya District of Kawamata Town, Katsurao Village, Namie Town and Minamisōma City, were designated as 'deliberate evacuation areas'. Residents of the deliberate evacuation areas were asked to evacuate within one month. On the same day, a further order was issued designating everywhere between 20 and 30 kilometers from Fukushima Daiichi that had not already been designated as 'deliberate evacuation areas' as 'evacuation-prepared areas in case of emergency'. This designation was supposed to facilitate rapid evacuations indoors or total evacuation from the area if required (Figure 13.1). A number of 'hot spots' with high radiation levels were also discovered outside the restricted areas and the deliberate evacuation areas. Individual households in these areas, where the annual cumulative dose was expected to exceed 20 mSv, were designated as 'specific spots recommended for evacuation'. Residents in these zones were provided with alerts and given support to evacuate.

The issue of decontamination first arose not in the restricted areas or the deliberate evacuation areas, where radiation levels were exceptionally high, but in areas outside the evacuation instruction

Figure 13.1: Evacuation instruction zones (as at April 22, 2011)

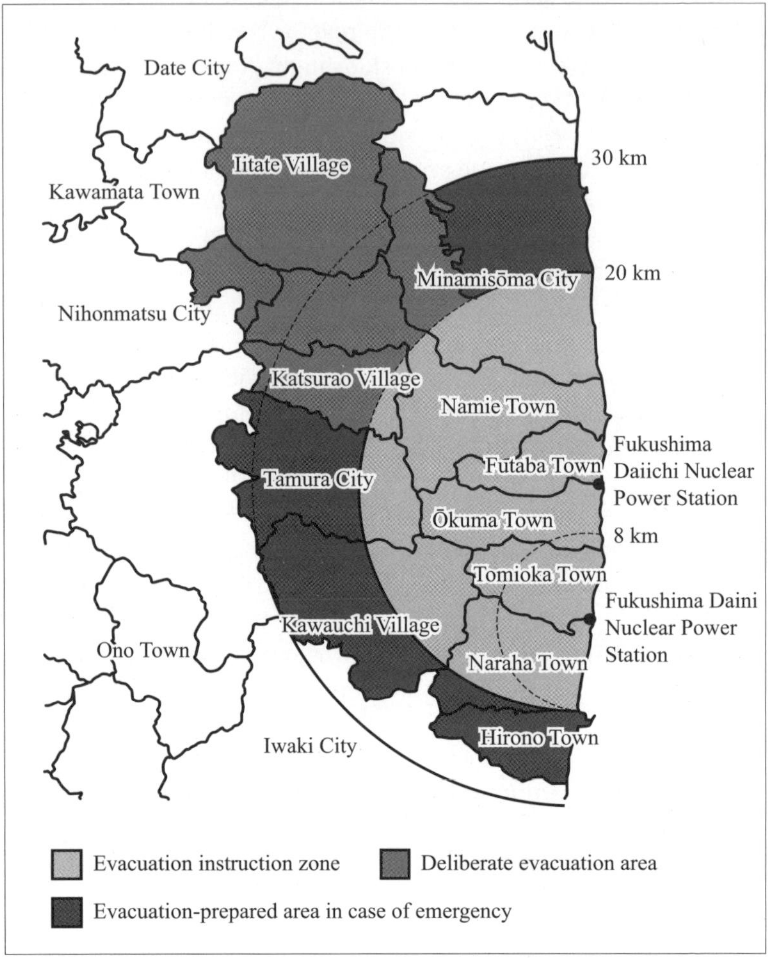

Source: Fukushima Prefecture n.d.

zones in Fukushima and neighboring prefectures where people were still living their lives as usual. For example, in Date City, Fukushima Prefecture, a significant amount of temporary accommodation was built for people who had been ordered to evacuate. However, high radiation levels were discovered in some parts of the city. As early as April 7, the city began carrying out decontamination measures such as removing topsoil from some school playgrounds. In Kōriyama City, too, between April and May 2011, topsoil was removed from playgrounds at a number of preschools and elementary and junior high schools where high radiation levels had been detected. The decontamination works undertaken by these municipalities helped to produce a storehouse of knowledge about decontamination techniques and led to a growing awareness of the need for decontamination.

The national government's approach to decontamination was outlined on August 26, 2011, in a policy statement on urgent decontamination procedures released by the Nuclear Emergency Response Headquarters, an agency that was established under the provisions of the *Act on Special Measures Concerning Nuclear Emergency Preparedness*. This statement acknowledged the government's responsibility for pursuing a nuclear energy policy and promised that it would work with prefectural and municipal governments and local residents to carry out decontamination. The policy included the following provisional goals:

1. Gradually reduce the number of areas where the annual cumulative dose of radiation exceeds 20 mSv.
2. In the long term, reduce the annual cumulative dose in areas where it is less than 20 mSv to less than 1 mSv.
3. Within two years, reduce the general public's estimated exposure to radiation by 50%.
4. Reduce children's exposure to radiation by 60% by thoroughly decontaminating schools and parks where children play.

The radiation levels adopted in the statement were based on the recommendations made in 2007 by the International Commission on Radiological Protection, which recommends 1 mSv as the maximum annual cumulative dose for individuals in 'planned exposure' situations (situations involving the planned introduction and operation of sources of radiation). It recommends that the maximum dose for 'existing exposure situations' (situations involving exposures from natural background radiation or residues from accidents) is 20 mSv.

The goals for reducing radiation dose rates outlined in the statement include reductions through 'natural decay'.

This statement, which was taken over by the *Act on Special Measures Concerning the Handling of Environmental Pollution by Radioactive Materials*, set out a framework for managing decontamination. The national government would assume direct responsibility for decontamination in designated 'special decontamination areas', while local municipalities would direct the decontamination works in designated 'intensive contamination survey areas' with support from the national government. The policy also stated that the government would defray the cost of decontamination in both the special decontamination areas and the intensive contamination survey areas and then seek to claim these costs from TEPCO in the form of damages. In other words, the cost of decontamination would be considered as nuclear damages under the provisions of the *Act on Compensation for Nuclear Damages (1961).* This Act specifies who is to be held responsible for nuclear damages and to what extent. Under the provisions of the Act (Article 3, paragraph 1), the nuclear energy company that is responsible for the damage, TEPCO in this case, is responsible for the costs incurred.

On September 30, 2011, the 'evacuation-prepared area in case of emergency' directive was lifted, and the government began to reorganize the evacuation instruction zone system. On December 26 the Nuclear Emergency Response Headquarters decided to revise the 'restricted areas' and 'evacuation instruction zones' and established a new system (Figure 13.2).

After ongoing radiation monitoring, on December 28, 2011, 11 municipalities that were designated as restricted areas or deliberate evacuation areas (Tamura City, Kawauchi Village, Minamisōma City, Iitate Village, Naraha Town, Ōkuma Town, Katsurao Village, Tomioka Town, Namie Town, Futaba Town and Kawamata Town) were designated as special decontamination areas. A further 104 municipalities in Iwate, Miyagi, Fukushima, Ibaraki, Tochigi, Gunma, Saitama and Chiba, where airborne radiation dose rates exceeded 0.23 microsieverts per hour (μSv/h) (equivalent to an annual cumulative dose greater than 1 mSv), were designated as intensive contamination survey areas.

The realignment further subdivided the special decontamination areas based on the estimated annual cumulative dose rate into 'evacuation order cancellation preparation zones', 'restricted residence zones' and 'difficult-to-return zones', with a different approach to decontamination in each zone.

- Evacuation order cancellation preparation zones: former evacuation instruction zones where, on December 26, 2011, annual cumulative radiation dose rates were confirmed to have fallen below 20 mSv. Entry into these zones is permitted but staying overnight is prohibited

Figure 13.2: Revised evacuation zones

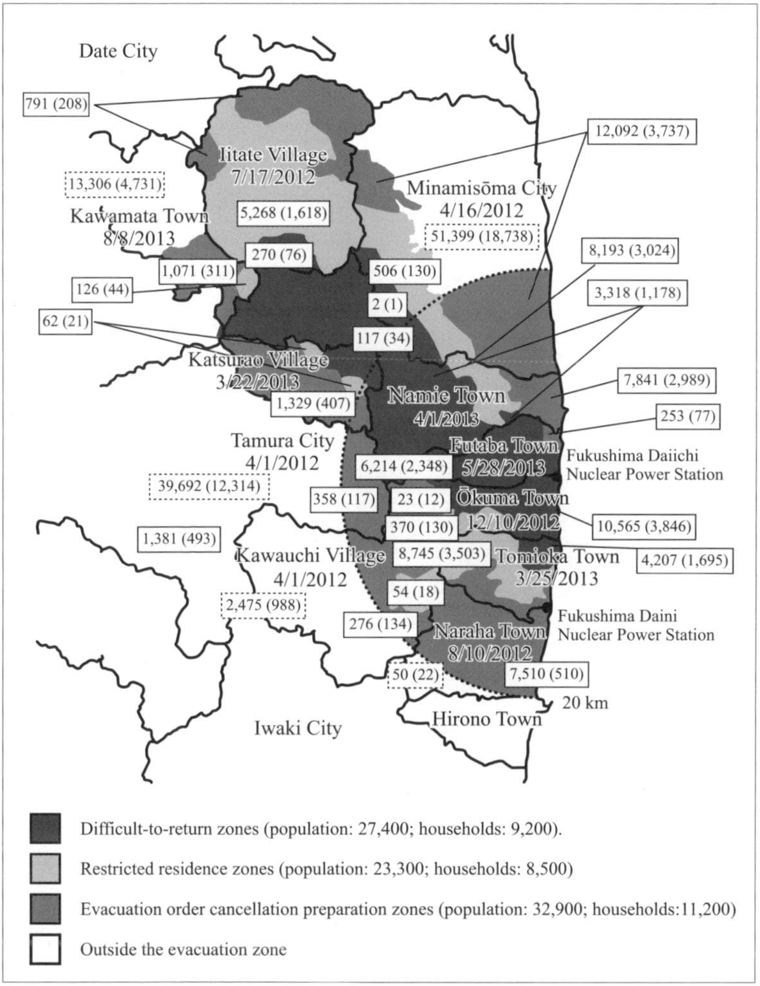

Note: The date in brackets beneath each geographical name is the date on which it was reclassified. The figures in rectangles give the population and the number of households (bracketed numbers). The figures in dotted rectangles give the population and the number of households outside the evacuation zone.
Source: METI 2014.

Table 13.1: Comparison of evacuation instruction zones

	Difficult-to-return zone	Restricted residence zone	Evacuation order cancellation preparation zone
Entry	×	○	○
Return to own home	×	×	×
Lodging permitted in certain cases	×	○	○
Business promotion	×	△[a]	○[b]
Reopening of businesses	×	△[a]	○[b]
Farming and forestry	×	×[c]	○

a Activities and residence essential to reconstruction and renewal are permitted.
b Activities targeting residents are prohibited but some preparatory activities are possible.
c Conditional on existing planting restrictions and the progress of decontamination.
Source: MOE 2015a: 28.

- Restricted residence zone: former evacuation instruction zones where, on December 26, 2011, annual cumulative dose rates were expected to exceed 20 mSv. In order to prevent further radiation exposure, residents are asked not to return to these zones. If the annual cumulative dose falls below 20 mSv, due to either decontamination or natural decay, then they can be reclassified as evacuation order cancellation preparation zones. Entry into these zones is permitted, but staying overnight is prohibited.
- Difficult-to-return zone: former evacuation instruction zones where annual cumulative dose rates are unlikely to fall below 20 mSv, even after five years, and where annual cumulative dose rates currently exceed 50 mSv. In principle, residence in these zones will remain restricted into the future. They cannot be reclassified for at least five years.

The different restrictions that apply in each zone are shown in Table 13.1.

Of the three zones, the priority for decontamination is the evacuation order cancellation preparation zones, where dose rates are comparatively low. The long-term goal is to reduce annual cumulative dose rates to less than 1 mSv. In the medium term, however, the goal was to reduce the public's exposure to radiation to 60% of August 2011 levels by August 2013. In the restricted residence zones, the goal was to reduce annual cumulative dose rates to less than 20 mSv by carrying out decontamination between 2012 and 2013.

In the difficult-to-return zones, where radiation levels are extremely high, the goal was to carry out experimental decontamination works in order to determine whether full-scale decontamination may be possible in the future. Further decontamination in these zones is dependent on the results of these experiments.

On January 26, 2012, the Ministry of the Environment (MOE) (2012a) released its decontamination policy statement for the special decontamination areas (the decontamination road map), as shown in Figure 13.3.

Even decontamination cannot eliminate radioactive substances. The best that can be achieved is to remove contaminated materials from residential environments and move them somewhere where they may be more easily managed. Decontamination produces soil scrapings and other radioactive wastes. The government plans to store this radioactive waste temporarily in 'provisional storage sites' located near the decontamination areas. The first decontamination works took place in locations where provisional storage sites were available. The waste is meant to be transferred to a large 'interim storage facility' that was to be constructed in Fukushima Prefecture by January 2015. After 30 years the material from the interim storage facility is to be transferred to a permanent facility located outside Fukushima Prefecture. However, developing the interim storage facilities and transferring the waste from provisional storage sites are far behind schedule. By the end of March 2017, the volume of contaminated soil transported to the interim storage facilities was approximately 230,000 cubic meters, which accounted for only a few per cent of the volume of soil generated from decontamination (MOE 2017: 25–6).

Appropriations and expenditure on decontamination

In 2011 decontamination costs were paid out of the general account in the supplementary budget. Since 2012 they have been paid out of the Special Account for Reconstruction from the Great East Japan Earthquake. Table 13.2 gives an overview of the decontamination budget for 2011–14.

Table 13.2 separates the appropriations and expenditure for decontamination carried out directly by the national government in the special decontamination areas and government assistance for decontamination carried out by local municipalities in the

Figure 13.3: The decontamination road map for new evacuation order areas

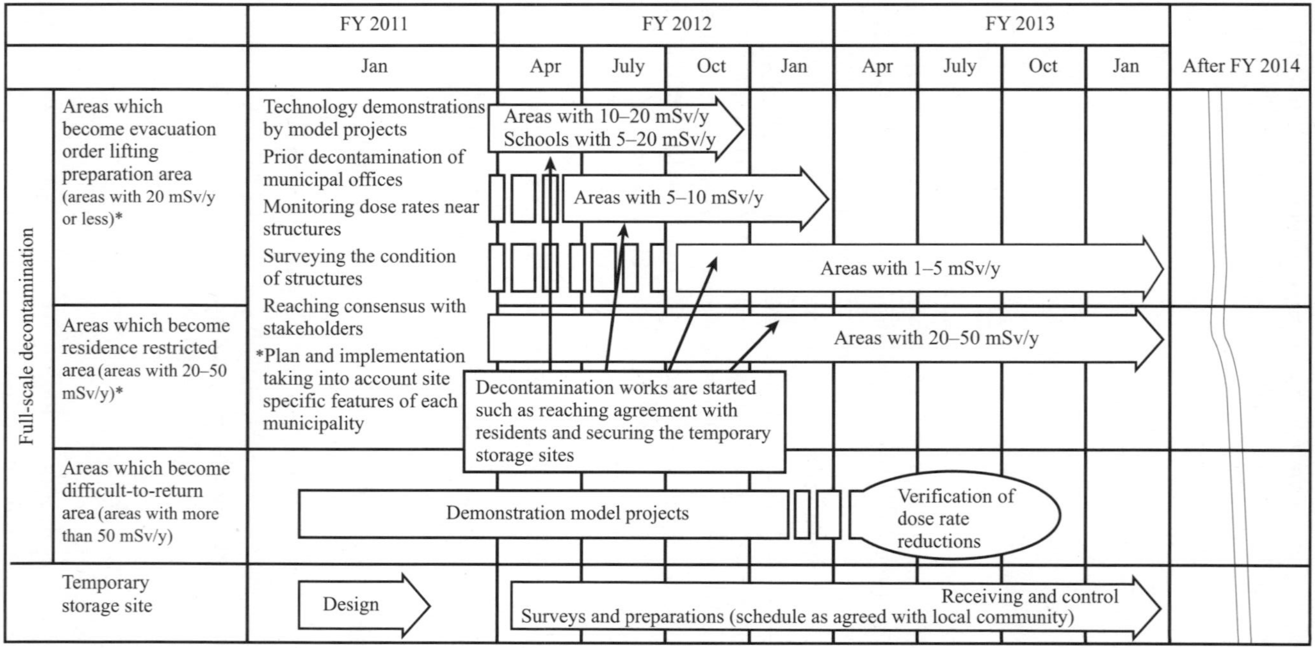

Note: FY = financial year; mSv/y = millisieverts per year.
Source: MOE 2012a.

Table 13.2: Appropriations and expenditure for decontamination (millions of yen)

	2011		2012		2013		2014		Total 2011–14	
	Appropriation	Expenditure	Appropriation	Expenditure	Appropriation	Expenditure	Appropriation	Expenditure	Appropriation	Expenditure
Direct government expenditure	297,106	202,311	321,411	74,284	427,516	174,320	417,336	240,722	1,463,369	691,637
Budget execution rate (%)		68.1		23.1		40.8		57.7		47.3
Support for municipal decontamination	104,723	72,602	137,009	95,662	293,227	280,499	148,770	140,229	683,730	588,992
Budget execution rate (%)		69.3		69.8		95.7		94.3		86.1
Total	401,829	274,914	458,421	169,946	720,743	454,819	566,106	380,951	2,147,099	1,280,629
Budget execution rate (%)		68.4		37.1		63.1		67.3		59.6

Note: Travel and some other expenses related to decontamination are not included.
Source: Compiled based on data in Board of Audit (2013b) and the Ministry of Finance's Budet and Accounts Database

Table 13.3: Progress of decontamination in the intensive contamination survey areas (as at January 2015).

	Commenced	Completed
Public facilities	Near completion	Approx. 80%
Residences	Approx. 90%	Approx. 60%
Roads	Approx. 70%	Approx. 40%
Agricultural and pastoral lands	Near completion	Approx. 70%
Forests (in residential areas)	Approx. 70%	Approx. 60%

Source: MOE 2015b: 7.

intensive contamination survey areas. A total of 2.147 trillion yen was appropriated for decontamination in the four years since 2011. However, over the same period the budget execution rate for decontamination was just under 60%. The execution rate for decontamination carried out directly by the national government was particularly low. Difficulties in negotiating with landowners and other local interests to carry out decontamination, leading to long delays, probably best explain the low execution rate.

On the other hand, the execution rate for the local municipality assistance budget was relatively high. However, this is because the money is counted as having been spent the moment it is handed over to the municipalities. For municipalities in Fukushima Prefecture, these funds are paid into the Fukushima Prefectural Health Management Fund's decontamination account and then distributed to the municipalities. Between 2011 and 2014 a total of 769.6 billion yen was accumulated in the decontamination fund, of which 702.2 billion yen was spent on decontamination (Fukushima Prefecture 2016: 5).

Effectiveness of decontamination

The first full-scale decontamination works were conducted in the intensive contamination survey areas (Table 13.3). Decontamination also gradually got under way in the special decontamination areas. It has already been completed in a large number of evacuation order cancellation preparation zones and restricted residence zones

Table 13.4: Progress of decontamination in the special decontamination areas (as at April 2014)

	Residential land	Agricultural land	Forests	Roads
Tamura City				
Completed	100	100	100	100
Commenced	100	100	100	100
Naraha Town				
Completed	100	100	100	100
Commenced	100	100	100	100
Kawauchi Village				
Completed	100	100	100	100
Commenced	100	100	100	100
Iitate Village				
Completed	13	5	11	3
Commenced	100	40	45	28
Kawamata Town				
Completed	66	12	31	0.3
Commenced	100	100	100	100
Katsurao Village				
Completed	100	1	99	1
Commenced	100	100	100	100
Ōkuma Town				
Completed	100	100	100	100
Commenced	100	100	100	100
Minamisōma City				
Completed	2	1	8	0.3
Commenced	99.9	65	79	65
Tomioka Town				
Completed	2	0.3	1	45
Commenced	100	100	100	100
Namie Town				
Completed	3	4	6	5
Commenced	11	15	14	23

Note: For both commencement and completed, figures refer to land area (%).
Source: Compiled from MOE (Environmental Management Bureau) 2014: 3.

(Table 13.4). However, as March 2016, decontamination works in the difficult-to-return zones, where radiation dose rates are still high, had hardly begun.

The MOE issued a report on the effects of decontamination in December 2013 (MOE 2013). An analysis of approximately 250,000 data items obtained during decontamination carried out between March 2012 and October 2013 by both national and

Table 13.5: Reduction in radiation exposure after decontamination

	Reduction (%)	Effectiveness of decontamination (%)
Target	50	10
Special decontamination areas	67	27
Intensive survey areas	62	22
Total	64	24

Source: MOE 2014: 3.

municipal governments showed that the airborne dose rate prior to decontamination was 0.36–0.93 μSv/h. After decontamination, this had dropped to 0.25–0.57 μSv/h.[4] In areas where the airborne dose rate was less than 1 μSv/h, 1–3.8 μSv/h and more than 3.81 μSv/h prior to decontamination, the dose rate was reduced by 32%, 43% and 51% respectively. In other words, decontamination was more effective in areas that had a higher airborne dose rate prior to decontamination. However, in areas where it was already relatively low, the rate of reduction was also low. Once the dose rate had reduced to some degree, it was very difficult to reduce it any further.

If we look at cumulative exposure rates in the two years from August 2011, excluding the natural rate of decay of 40%, a reduction of 27% was achieved in the special decontamination areas and 22% in the intensive contamination survey areas, exceeding the initial targets (Table 13.5).

Why the emphasis on decontamination and return?

Following the enactment of the *Act on Special Measures Concerning the Handling of Environmental Pollution by Radioactive Materials*, the government made decontamination the central platform of its reconstruction policy. The phrase 'no reconstruction without decontamination' appeared in government publications, such as a pamphlet put out by the MOE.[5] At the same time, evacuees who were seeking to relocate on a long-term basis received practically no support. However, at least in theory, it should have been possible to formulate a policy response that could support evacuees to relocate, even while carrying out decontamination. How did 'decontamination followed by return' become the government's default policy position?

Suzuki Hiroshi (2013) argues that the demand to return home as quickly as possible was an expression of Fukushima residents' anger towards TEPCO and the government, and the government was then forced to respond to this demand with promises of decontamination.

The National Diet of Japan Fukushima Nuclear Accident Independent Investigation Commission (NAIIC 2012) and Ōshika (2013) are full of examples of ineptitude on the part of both TEPCO and the government in their response to the disaster. In particular, NAIIC (2012: 464–81) found that the Nuclear and Industrial Safety Agency had failed in its duty as a regulator and fallen victim to 'regulatory capture'. Examples such as this make it easy to see why the victims of the disaster were so angry. In addition to the widely held perception of the government as inept in its response to the disaster, nuclear power had been a matter of national policy for decades. Therefore the government had little choice but to respond wholeheartedly to calls for assistance from the affected municipalities.

Fukuyama Tetsurō, who was responsible for the government's disaster response as the then Deputy Chief Cabinet Secretary, has touched on these issues in a book. In late March, when it became clear that radiation levels in Iitate and a number of other villages were exceptionally high, the government designated these areas as deliberate evacuation areas. When Fukuyama spoke with Iitate's mayor prior to issuing the order, the mayor told him that he would not be able to convince all of the villagers to evacuate and tried desperately to find an alternative, such as allowing business owners to continue to commute to the village (Fukuyama 2012: 156–62). While nuclear energy may have been a matter of national policy, the government did not cause the accident directly. Nevertheless, in issuing directives to the municipalities, the government also took on a number of their demands.

An episode that took place approximately one month after the disaster also sheds some light on how 'decontamination and return' came to form the basis of the government's policy for dealing with the evacuees. On April 13, 2011, Matsumoto Ken'ichi, who was a special adviser to Cabinet, reported that then Prime Minister Kan Naoto had said, 'it will not be possible to live in the area around the Fukushima Daiichi Nuclear Power Station for ten or twenty years'

(*Asahi Shimbun*, April 15, 2011, morning edn, p. 4). Kan later denied having made the statement and Matsumoto later retracted his claim. Nevertheless, it was roundly condemned in the media and was even the subject of questioning by opposition members in the Budget Committee of the House of Councillors on April 18. Not only the opposition but even the government's own Minister for Internal Affairs and Communications criticized Kan's statement (Miyazaki et al. 2013: 175). It is too late to determine exactly what Kan really said. Nevertheless, it was widely understood that it would be impossible to live in the region surrounding the Fukushima Daiichi Nuclear Power Station for a very long time. However, what the incident makes clear is that in April 2011, even though the situation at the nuclear power station was not yet under control, it was difficult to speak publicly about the fact that the area surrounding Fukushima Daiichi would be uninhabitable for an extended period of time and, therefore, that relocating some of the evacuees would have to be an option. Of course, one could still criticize Kan for not being more careful with his words. Had the prime minister made such a statement in a press conference or another public forum, such a criticism might have been applicable. However, the issue was with something the prime minister was reported indirectly to have said; it was not a question of how Kan said it.

Not long after, the idea of decontamination and return was raised in the Diet. On May 31, 2011, in a House of Representatives special committee on Reconstruction from the Great East Japan Earthquake, then Minister for Economy, Trade and Industry Kaieda Banri stated that 'our first priority is to stabilize the nuclear reactors. Once we have achieved that we need to decontaminate the surrounding region so that people can return to their homes as soon as possible. We will do our utmost to ensure that absolutely everybody gets to go home' (Minutes of the National Diet, [002/008], 177 House of Representative, No. 6). Around this time, the policy of decontamination and return became entrenched and other options virtually disappeared.

Differences over decontamination

One factor that complicates the issues of evacuation, decontamination and return is that evacuees vary in their sensitivity to radiation, both physically and psychologically. Furthermore, not all evacuees feel the

Figure 13.4: Evacuee households living separately

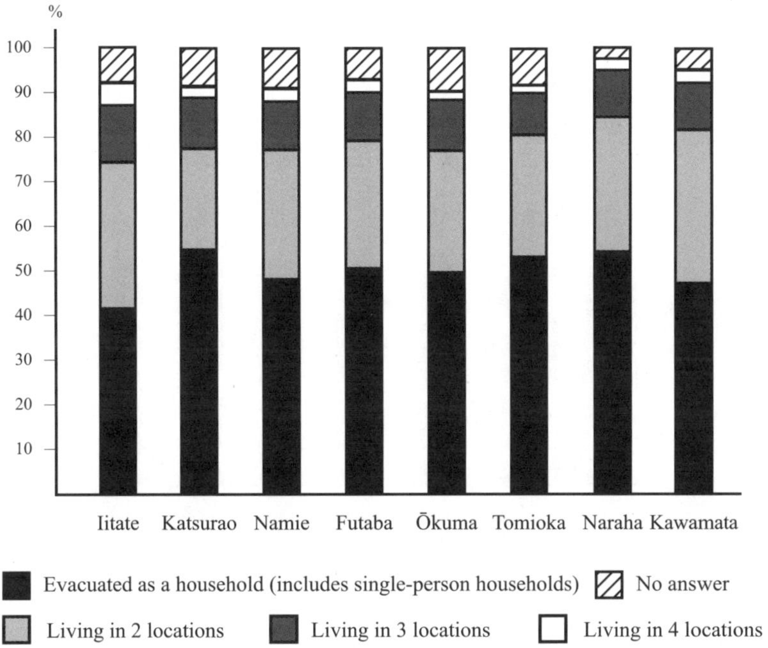

Source: Compiled based on Reconstruction Agency 2014: 6.

same way about the communities in which they once lived. These differences in turn impact on whether residents would prefer to return to their former homes or relocate.

It is generally accepted that the health risks associated with exposure to low-level radiation are much greater for younger people than for older people. The current situation, where many of the temporary housing facilities are only utilized by older people and grandparents, while children and their parents have already evacuated further away, is a reflection of this. The Reconstruction Agency and the municipalities that fell within the evacuation instruction zone conducted a joint survey of evacuees between August 2013 and January 2014. Figure 13.4 is based on data from this survey. Among evacuees from Katsurao, 32.6% of households had become separated, the lowest number in the surveyed municipalities. Iitate had the highest number of separated households at 50.5%, more than half (Reconstruction Agency 2014b).[6] In some families,

*Figure 13.5: Opinions on return in Iitate (*n *= 1,458, November 2013)*

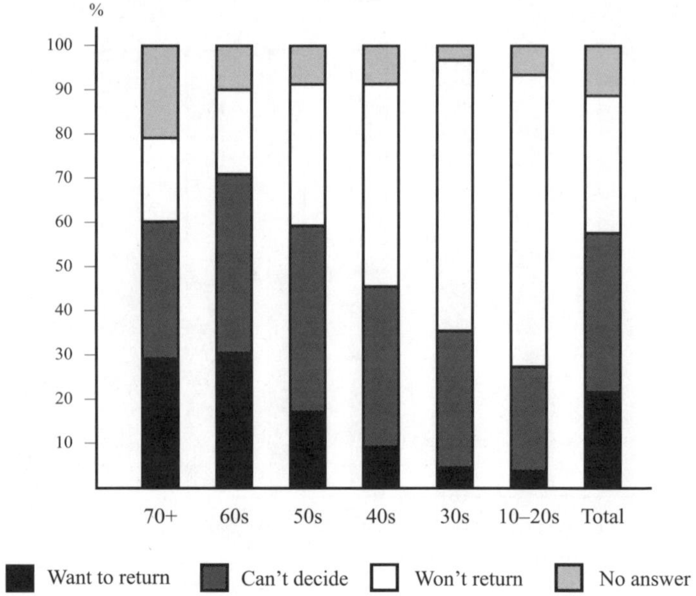

Source: Reconstruction Agency 2014: 15.

the primary wage earner lives alone near his or her workplace. The difference of the sensitivity to radiation among members of the household is connected with this separation of the family unit. It can also be seen in the generational divide on the question of return.

Figures 13.5 and 13.6 show the results of a survey of people by generation in Ōkuma and Iitate. These figures clearly show that, while older people have a strong desire to return, many younger people and people with children do not.[7]

These results are directly related to attitudes towards the decontamination that will need to be carried out if the evacuees are to return. Among the generations who have a strong desire to return, support for decontamination is strong. This aligns with the opinion of a majority of mayors and municipal councillors in the affected areas. If not enough people return to their former municipalities, the mayors and councillors will lose their political base. Therefore, they want to see as many people return as possible. People who want to return home are afraid that if not enough people share their

*Figure 13.6: Opinions on return in Ōkuma (*n *= 2,764, October 2013)*

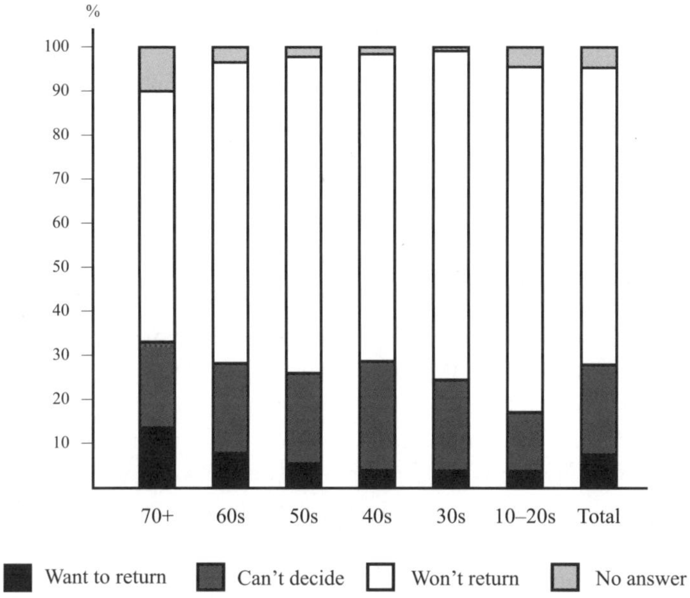

Source: Reconstruction Agency 2014: 14.

wish, then the government will not carry out a sufficient level of decontamination. They also worry that it will be difficult to maintain infrastructure and public services without the former population base. These fears give them an incentive to call for as many people as possible to return and to lobby local politicians. This issue can even be seen in calculations of the cost-effectiveness of decontamination. In order to determine the cost-effectiveness of decontamination, the overall cost divided by the number of returnees is often used as an indicator. However, if this method of calculation was emphasized too much, it would only add further pressure on evacuees to return.[8]

When the national government tries to gauge the evacuees' own opinions, their formal processes for doing so center on the opinions of municipal mayors and councillors. This has helped to put return at the center of the government's policy for evacuees. During a meeting with advocates from Fukushima, Suzuki Tatsujirō, the then deputy chairman of the Japan Atomic Energy Commission, explained that 'although some national politicians have been asking individual disaster victims their opinion, there is no institutional mechanism for guaranteeing that these opinions will be heard. That is why politicians tend to value the opinions of municipal mayors the most highly.'[9]

However, many younger people feel that unless radiation dose rates fall substantially, they cannot and will not return. They are skeptical about the amount of money being spent on decontamination, especially when its effectiveness is unclear. Nor are young people the only people who feel this way. In some cases, even after decontamination, there is no prospect of radiation levels falling sufficiently to allow evacuees to return. Figures 13.5 and 13.6 show that many young people would rather relocate. The longer the evacuees have to wait, the more likely they are to give up on going home. As a result, there are those who would like to see some of the vast funds that are being allocated for decontamination redirected to help people who want to relocate. However, if each individual household is allocated assistance separately, they can deal with the situation as individuals. They are also likely to move to different locations. People who want to relocate therefore have relatively little incentive to band together to demand support. For some, the idea of relocating is associated with a feeling of guilt for betraying or leaving behind their homeland.[10] The affected municipalities therefore have relatively little incentive to actively support those who want to leave permanently and a strong incentive to ask the national government to decontaminate the area so as to permit them to stay.

In reality, evacuees' feelings are extremely complicated. This is exemplified in the responses to a survey of high school age and older residents conducted by the municipality of Namie in November 2011. In this survey respondents were asked a number of questions, including their opinions about returning home. The 3,616 people who said that they 'would not return even if the evacuation order is lifted' were then asked, 'even if you plan to relocate, do you think it is still necessary to decontaminate Namie and try to rebuild?' Their answers are displayed in Figure 13.7. More than half said that they thought that both decontamination and rebuilding were important. It is difficult to know how to interpret these responses. Do evacuees want to restore their 'precious hometown' to its former state even if they will not return? Is it a manifestation of guilt over having 'left their hometown behind'? Or is it anger at TEPCO, the company responsible for the accident, and a desire to hold it to account by

Figure 13.7: Opinions of Namie residents who do not wish to return on decontamination and rebuilding

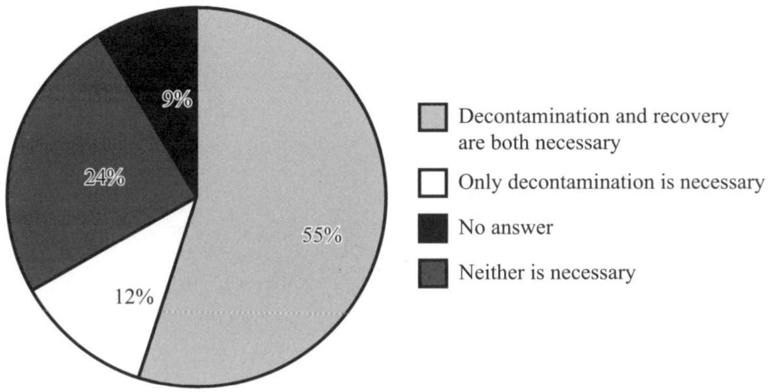

Source: Compiled from data in Namie Town 2011: 20.

doing everything possible to make it pay?[11] These complicated feelings indicate one reason why decontamination and return has become the key to the government's policy response.

The examples of Ōkuma and Iitate

The disaster-affected municipalities have put a number of demands for assistance to the national government. By looking at Ōkuma and Iitate, in particular, in this section I examine the affected municipalities' attitudes towards issues of return and relocation.

Almost the entire area of the municipality of Ōkuma has been designated as a difficult-to-return zone. Table 13.6 is a list of demands put by Ōkuma to the national government since the nuclear accident.[12] Fukushima Daiichi is located in Ōkuma. As such, it has been recognized as an area to which it will be very difficult to return. While it may have taken some time for residents to accept this fact, the majority now do so. However, most of the demands the municipality has made on the national government relate to issues of adequate compensation and decontamination. Of the 16 claims made by the municipality, ten touch on decontamination. In particular, specific demands for decontamination were made on October 24, 2012, and February 15, 2013. The municipality

Table 13.6: Demands put to the national government by Ōkuma

Date	Mentioned decontamination	Relocation assistance
April 28, 2011		
June 29, 2011		
July 13, 2011	○	
August 5, 2011		
August 30, 2011	○	
December 1, 2011		
January 18, 2012	○	
February 21, 2012		
October 24, 2012	●	
January 23, 2013	●	
February 16, 2013	●	○
February 5, 2014	●	
March 8, 2014	○	
May 22, 2014		
June 2, 2014	○	
May 28, 2015	●	

Note: ● indicates a specific request; ○ indicates the word 'decontamination' is used.
Source: Compiled by the author based on the website of Ōkuma municipality.

only clearly demanded support for relocation in its eleventh claim, made on February 15, 2013, in which it called on the government 'to consider the long-term evacuees who choose not to return and help them to build a new community where they can enjoy secure accommodation and employment opportunities'.[13]

Despite being more than 30 kilometers from the Fukushima Daiichi Nuclear Power Station, Iitate was heavily contaminated with radioactivity due to the prevailing atmospheric conditions at the time of the accident and was designated as a deliberate evacuation area. The municipality has been even more aggressive in its pursuit of a return policy. Immediately after evacuation, the mayor of Iitate began trying to find a way to get the villagers home. On September 7, 2011, he lodged a formal request with the national government to decontaminate the village. By September 18 he had already announced a decontamination plan worth 324.2 billion yen (Iitate Village 2011). Under this extremely ambitious plan, decontamination of residential areas was to take place over the following two years, with the goal of allowing some villagers to return, at least temporarily, after that. The mayor of Iitate consistently advocated for an early return. He proposed a number of plans detailing various specific measures in an attempt to achieve this. Some of them did mention providing support to people who do not wish to return home but they did not contain any specific policy for assisting them. Return was clearly the principal objective of these proposals.[14] However, even after two years, radiation levels had not gone down as far as was hoped and the evacuation directive for Iitate remained in place.

Support for relocation

Many of the affected municipalities have held on to their hopes for decontamination and return and so they have little incentive to provide support for those who wish to relocate.[15] Even if the municipalities want to offer support for relocation, they lack the human and financial resources and the necessary authority to enable them to carry out an adequate relocation policy. This highlights the important role of the national government in providing support for evacuees wanting to relocate. Or, to put it another way, if the government does not provide this support, it will be extremely difficult for them to rebuild their lives. However, the national policy is based on respecting the wishes of the municipalities. Therefore, decontamination and return has continued to be the government's main objective. It had provided little, if any, concrete support for people wishing to relocate permanently.

On December 20, 2013, the government's Nuclear Emergency Response Headquarters released a plan entitled 'For accelerating the reconstruction of Fukushima from the nuclear disaster'. This document outlined the government's intention to provide financial support to evacuees from the difficult-to-return zones – who will be unable to return home for an extended period of time – so as to enable them to restart their lives elsewhere and compensate them for the cost of securing accommodation. Compensation for property such as houses did exist as a form of compensation for relocation. However, under the previous plan, the amount of compensation was to be calculated based on the market value or the book value of the evacuated property. For evacuees who had older homes, in particular, the amount of compensation would be nowhere near enough to purchase a new home elsewhere. The new plan was really the first time the government had proposed a workable policy to assist people to relocate. In effect, this amounted to a reversal of the previous policy of returning all residents to their former homes.[16]

In response to this announcement, the then mayor of Tomioka, Miyamoto Kōichi (in Shindō and Tamura 2014: 23–4), reacted negatively, saying 'this is a change in policy from the previous objective of ensuring everybody could return home to what can only be termed a relocation policy. I am extremely worried that this will lead to a significant drop in the number of people seeking to return.' His comments reflect the serious depopulation problem faced by the affected municipalities. They also reflect the attitude of affected municipalities that they do not welcome assistance for relocation by the national government.

The costs and benefits of decontamination

With decontamination works already underway, it has become clear that reducing radiation levels to pre-accident levels is almost impossible, both technically and in terms of cost. This raises the question of how to establish a set of acceptable standards for decontamination. A number of possible approaches could be used to set these standards. In August 2011 the Dispute Reconciliation Committee for Nuclear Damage Compensation released its *Interim Guidelines on Determination of the Scope of Nuclear Damage Resulting from the Accident at the Tokyo Electric Power Company Fukushima Daiichi and Daini Nuclear Power Plants*,[17] which laid out the following basic standards for assessing the cost of decontamination:

> as a general rule, reasonable repair, decontamination and similar costs will be limited to the objective value of the property in question. However, with respect to cultural property, agricultural land, and other non-fungible forms of property, it will be possible to allow, as an exception, compensation in an amount that exceeds the objective value of the property in question, within reason.

According to a literal interpretation of these guidelines, there is no in-principle necessity to spend more than the market value of buildings or land on decontamination, unless the building is an irreplaceable historical artefact. The guidelines set the standard for decontamination on a per property basis by utilizing cost-benefit analysis, where the value of the asset is considered to be the benefit. However, if we consider decontamination not in terms of compensation for the value of the property but as a means of enabling evacuees to return home and rebuild their lives, this approach is unsuitable. Unless the same standard of decontamination is carried out across the entire area, then it is generally not possible to reduce radiation levels. Let us assume some highly contaminated buildings in the affected areas are extremely old and their market value and book value is close to zero. If we apply the principle of 'not carrying out decontamination works in excess of the market value' to these buildings, then money should not be spent on decontaminating them. However, the safety of people seeking to return to neighboring houses will be undermined unless decontamination is carried out. To put it another way, decontamination has significant positive externalities.

The *Act on Special Measures Concerning the Handling of Environmental Pollution by Radioactive Materials* does not require decontamination only to be carried out up to the value of the property concerned. While this shows that the decontamination program is not intended simply to recover the value of property, it leaves the decontamination program open to criticism on the basis of cost.

Costs of decontamination

In 'Appropriations and expenditure on decontamination' above, I discussed the amount of money appropriated for decontamination and how much has actually been spent. In December 2013 the Nuclear Emergency Response Headquarters released its preliminary calculations of the cost of decontamination, which included this data. These calculations, obtained from the MOE, are available in Ōshima and Yokemoto (2014) (Table 13.7). Based on these calculations, the total cost of decontamination and related expenses, including both the work that has already been carried out and planned future works, is estimated at 3,540 billion yen. This figure does not include the cost of permanently disposing of the contaminated soil generated by decontamination.

A number of other attempts have been made to figure out the total cost of decontamination. The most detailed of these was released by a research group at the National Institute of Advanced

Table 13.7: Cost breakdown for decontamination and medium-term storage (billions of yen)

Construction and management of medium-term storage facilities	1,060
Decontamination-related	1,740
Decontamination	1,070
Provisional storage facilities, volume reduction facilities	560
Technical development	110
Waste disposal	520
Decontamination etc. (Cabinet, 2011)	220
Total	3,540

Source: Ōshima and Yokemoto 2014: 4 (based on Ministry of the Environment data).

Table 13.8: Decontamination cost calculations (Yasutaka and Naitō 2013)

	Special decontamination areas	Intensive decontamination survey areas
Total surface area (km^2)	1,117	8,127
Surface area already completed (km^2)	295	1,516–2,290
Scenario 1 (billions of yen)	1,830	700–2,130
Scenario 2 (billions of yen)	2,030	1,030–3,100

Note 1: Scenario 1 is based on the standard cost of decontamination; Scenario 2 is based on the maximum cost.
Note 2: The variation in cost generally relates to the different decontamination methods utilized in regions where the annual cumulative dose is between 1–5 mSv (surface decontamination or localized decontamination).

Industrial Science and Technology in July 2013. The group looked at what decontamination might cost (= removal + provisional storage containers + provisional storage sites + volume reduction + medium-term storage facilities) in a number of different scenarios (Yasutaka and Naitō 2013). These figures are shown in Table 13.8. According to the group's calculations, the cost of decontamination in the special decontamination areas alone will amount to at least 1,830 billion yen. Together with the cost of decontamination being carried out by municipalities, their estimate of the total cost for a best case scenario is 2,530 billion yen and 5,130 billion yen for a worst case scenario. The cost of decontamination in the special decontamination areas does not vary much in either scenario, but for municipal decontamination efforts it varies by as much as three times.

These figures only include the costs of decontamination. If people are to return to these areas to live, a whole host of infrastructure will also have to be reconstructed. These are not direct decontamination costs but in order to fully realize decontamination and return, there is no doubt that the total cost will far exceed that of decontamination alone.

Benefits of decontamination

While uncertainties remain regarding the cost of decontamination, at least they are relatively easy to understand. The benefits, however, are not so easily defined.

In a dialogue with Nakanishi Junko (in Nakanishi 2014: 217), Iida Yasuyuki, an economist, stated that the benefits of decontamination are represented by the total value of land and the depreciable assets such as buildings that will be restored by decontamination. At present, it is not possible to live in the special decontamination areas. The market value of the land, buildings and other infrastructure in these areas is therefore zero. Other depreciable assets, such as rolling stock and immovable machinery, have already been out in the weather for several years and have not been maintained and will therefore have to be written off. Even decontamination cannot restore the value of these assets. According to Iida, the benefit of decontamination carried out in the special decontamination areas is therefore, at best, the value of the land, which is estimated at 1 trillion yen (Nakanishi 2014: 217–26). Even this figure is based on land values reverting to their previous levels after decontamination. However, this is an unrealistic expectation for land that has been subject to an evacuation order. Therefore, the actual value will be less.[18] According to the calculations made by Yasutaka and Naitō (2013) described above, it will cost at least 1.83 trillion yen to decontaminate the special decontamination areas alone. Iida and Nakanishi argue that, as the cost of decontamination will massively outweigh the benefits, the decontamination works currently being carried out (and those being planned) are not cost effective.[19]

However, as I discussed at the beginning of this section, while property values may be a valid means of calculating compensation payments, they are not necessarily appropriate for measuring the benefits of decontamination. We cannot say that spending money on decontamination is excessive simply because the cost will outweigh the recoverable value of the assets. After all, the purpose of decontamination is to enable evacuees to return home and rebuild their former lives. The benefits of decontamination are better thought of in these terms.[20] However, how can we place a monetary value on the evacuees' former standards of living that are to be resumed? This is a longstanding question in cost–benefit analysis but nevertheless a relevant one in this case.

In order to conduct a cost–benefit analysis of decontamination, we not only have to place a monetary value on the aggregated benefits of each individual's livelihood but also on the positive externalities of reviving regional communities. These externalities will actually have a major impact on returnees' quality of life. Awaji (2013) gives the following examples of the functions provided by a community: (1) a reduction in living expenses; (2) mutual aid, cooperation and welfare; (3) a substitute for and a complement to government administration; (4) personal development; and (5) environmental protection and conservation. If we take Awaji's fifth point as an example, restoring regional communities via decontamination and return could provide valuable public benefits, such as maintaining forests and waterways (including water storage functions, headwater conservation and erosion prevention). In order to put a monetary value on these public benefits, the usual method is to calculate the cost of carrying out these functions using the closest market substitutes, the so-called 'replacement cost approach' (Garrod and Willis 1999: 39–41). However, removing topsoil as part of decontamination works may actually destroy some of these public benefits (MOE 2012b: 11).

Clearly, evaluating the total value of the public benefits provided by decontamination is an extremely complex matter. Therefore, rather than proposing a number of hypothetical benefits of decontamination, which are themselves the subject of debate, and assigning a monetary value to them for the purposes of cost–benefit analysis, a more constructive approach would be to compare decontamination with an alternative method for restoring the evacuees' living standards (and the benefits this will provide) and see which approach is more cost effective. The alternative to decontamination that I compare here is relocation, the principal cost associated with which is that of obtaining a new residence.[21]

Comparing decontamination with relocation

In December 2013 the Dispute Reconciliation Committee for Nuclear Damage Compensation (2013a) released a fourth supplement to the interim guidelines, based on the framework for supporting relocation proposed by the Nuclear Emergency Response Headquarters (2013). In this document, the costs incurred in obtaining a new residence by evacuees from the evacuation instruction zones who wish to relocate are referred to as 'damages related to home securement'. The supplement specifies that the cost of compensation for relocation is to be borne by TEPCO under the provisions of the *Act on Compensation for Nuclear Damages*. Evacuees from the difficult-to-return zones are eligible for this compensation. However, for evacuees from the restricted residence zones and evacuation order cancellation preparation zones, this compensation for damages related to home securement is limited to 'persons for whom relocation is recognized as reasonable'.

The Dispute Reconciliation Committee proposed the following method for calculating the amount of compensation to be paid for damages related to home securement.

> Damages = {(area of new residential land × unit price) – (area of evacuated residential land × unit price)} × α
> + {(floor space of new residential building × unit price) – (floor space of residential building in the evacuation area × unit price)}× β
> Difficult-to-return zones: land $\alpha = 1$, building $\beta = 0.75$
> Restricted residence zones & evacuation order cancellation preparation zones: land $\alpha = 0.75$, building $\beta = 0.75$

In summary, evacuees who relocate will be compensated for the difference between the necessary cost of relocation and the value of their existing house and land, multiplied by adjustment factor α or β (1 or 0.75). The adjustment factor of 0.75 for land in restricted residence zones and evacuation order cancellation preparation zones is based on the idea that, once the evacuation order in these zones has been lifted, the land will recoup some of its original value. With regard to residential buildings, the Dispute Reconciliation Committee decided that the amount of compensation should be set at a level higher than the cost of obtaining land for public use and so the adjustment factor was set at 0.75 for all zones.[22]

The Dispute Reconciliation Committee for Nuclear Damage Compensation (2013b) has provided a sample calculation for the amount of 'damages related to home securement' to be paid to an evacuee from an average house in an evacuation instruction zone who relocated to an average detached house in an urban area in Fukushima Prefecture. Using the figures from this sample calculation, let us estimate the total cost of damages related to home securement (e.g. of providing support for relocation).

The average price of a parcel of land for a detached dwelling in an urban area in Fukushima Prefecture (X1) is 9.53 million yen and the average price of a parcel of land for a detached dwelling in the evacuation instruction zones (X2) is 5.23 million yen. The average value of a residential building in an urban area in Fukushima Prefecture (Y1) is 24.54 million yen and the average market value of a residential building in the evacuation instruction zone (Y2) is 9.40 million yen (for a breakdown of these values, see the appendix).

The average amount of damages related to home securement for a residence (including house and land) per home in the difficult-to-return zone can therefore be calculated as follows:

$(X1 - X2) \times 1 + (Y1 - Y2) \times 0.75$

Substituting the actual values gives:

$(9.53 - 5.23) \times 1 + (24.54 - 9.40) \times 0.75 = 15.65$ million yen.

The average amount of damages for home securement for a residence (including house and land) in a restricted residence zone or an evacuation order cancellation preparation zone can be calculated as follows:

$(X1 - X2) \times 0.75 + (Y1 - Y2) \times 0.75$

Substituting the actual values gives:

$(9.53 - 5.23) \times 0.75 + (24.54 - 9.40) \times 0.75 = 14.58$ million yen.

Records of Fukushima Prefecture show that there were 28,900 households in the evacuation instruction zones as at December 2013, when the guidelines for damages related to home securement were released by the Dispute Reconciliation Committee. Of these, 9,200 are located in difficult-to-return zones, 8,500 are in restricted residence zones and 11,200 are in the evacuation order cancellation preparation zones (METI 2014). If every evacuee household was to relocate to an urban area in Fukushima Prefecture, the total compensation payable under the damages related to home securement program using the figures above can be calculated as follows:

> Difficult-to-return zones
> 15.65 million yen × 9,200 households = 144.0 billion yen approx.
>
> Restricted residence zones and evacuation order cancellation preparation zones
> 14.58 million yen × 19,700 households = 287.1 billion yen approx.

These give a combined total of 431.1 billion yen.[23]

As discussed above, Yasutaka and Naitō (2013) estimate that decontaminating the special decontamination areas will cost at least 1.83 trillion yen. These calculations demonstrate that it would be far cheaper to provide relocation assistance to all of the households in the evacuation instruction zones and purchase a new detached house for each of them elsewhere than to decontaminate the evacuation instruction zones so as to enable people to return. In reality, not all households want to relocate and not everybody who relocates will want to purchase a detached dwelling. Some will choose to live in an apartment complex or to rent. If the cost of moving home owners who live in a detached dwelling is the most expensive cost, then the total cost of relocating all of the affected households (by paying damages related to home securement) would be less than 500 billion yen. If half of all evacuee households chose to relocate to a newly purchased detached dwelling, the total cost of damages related to home securement would be 215.5 billion yen, and the cost of decontaminating the special decontamination areas would be, by simple calculation, at least 915 billion yen (= half of 1.83 trillion yen). Then the total cost of providing relocation assistance and decontamination would be less than 1.2 trillion yen, a considerable saving.

Respecting evacuees' choices and reducing costs

As I discussed at the beginning of this section, decontamination must take place over a wide area in order for it to be effective. Even if half the number of evacuee households were to return, the cost of decontamination would not necessarily be halved. A similar point can be made with regard to the cost of rebuilding essential infrastructure. However, people who do wish to return home and who need to carry out renovations or rebuild their original homes may find that the cost of doing so exceeds the pre-accident value of their homes. In this case, returnees are also eligible for compensating damages related to home securement. Therefore, by providing assistance to relocate, at the very least it ought to be possible to significantly reduce the cost of dealing with the evacuees, as compared with a policy of promoting everybody to return. Furthermore, if fewer sites need to be decontaminated, it might also be possible to concentrate resources in the more important areas.[24] In the first place, providing adequate support for relocation means respecting each evacuee's choice.

In any case, although there are many things to be considered, respecting the decisions of those individuals who do not wish to return to highly radioactive areas and providing them with support to relocate may well be a less expensive method for helping them to rebuild their lives than carrying out decontamination on the assumption that everybody will return home. If the government continues to spend large sums of money on decontamination and fails to recognize the relatively less expensive relocation option, in spite of the large number of people who would like to relocate, it is open to accusations of overspending on decontamination.

There have been proposals to relocate entire communities to other regions (so-called 'home-away-from-home communities' or 'provisional communities'). It would also be possible to provide support for these options. However, by only providing support to 'the whole community', the government is failing to respect individual choice.

In this chapter, for aforementioned reasons, I have not attempted to assign a monetary value to the value of community. However, as Awaji (2013) has shown, communities do perform a number of valuable functions. For many people, the contaminated regions have an irreplaceable value, that of their 'hometown' (Yokemoto 2013b). That is why some people, maybe not many, have a strong desire to return and they are prepared to wait a long time for this to happen. Some evacuees, even if they do not wish to return, would like to see some kind of decontamination works carried out.[25] Therefore, if a monetary valuation of a community is conducted by methods such as the contingent valuation method (Garrod and Willis 1999), the value would be enormous. However, even if their community is regarded as priceless, some people are concerned about the health risks of returning to an area where radiation levels are higher than normal. These people want to relocate and start a new life elsewhere. Since individuals and families face different circumstances and have different preferences, both options, return and relocation, are equivalent.

Conclusion

The Fukushima nuclear accident led to the release of large quantities of radioactive contaminants over a wide area. The government quickly adopted a policy of decontaminating the affected areas so that evacuees could return home. Systems and frameworks for decontamination were created based on this position. In general, municipal governments have a strong incentive to see their former residents return in order to maintain the autonomy of their municipalities. The national government, having pursued nuclear energy as a matter of national policy, has had little choice but to follow the wishes of the affected municipal authorities. These factors led the government to adopt decontamination and return as its central policy response to the problem of evacuees. Adding to the pressure on the national government to act in accordance with the wishes of the municipalities was the strong criticism it received in the mass media and other sources over its response to the disaster, in particular to the nuclear accident.

However, it quickly became clear after the accident that it would be difficult for evacuees from the difficult-to-return zones and the restricted residence zones to go home. Surveys carried out by the Reconstruction Agency and other organizations have also shown that a significant number of people do not wish to return. In spite of this, the government had done little to provide concrete support to evacuees to enable them to relocate. Without such support, it is difficult for them to move away and rebuild their lives elsewhere. At the end of 2013 specific measures to assist people with relocation were proposed but even these are not unconditional. Leaving to one side the question of how far decontamination ought to be pursued and even allowing for a basic policy of helping evacuees return home, it should still be possible to respect not only the municipal governments and communities as a whole, but also the choices of individual evacuees by assisting them to relocate. As things stand, a lack of flexibility in the return policy and a lack of accessible support for relocation mean that some evacuees have been forced to put their lives on hold. The government's continued pursuit of a decontamination and return policy and its failure to respect individual choices, has led to overspending on decontamination. My criticism of the government's overspending on decontamination is not based on a primitive cost–benefit analysis, wherein the cost of decontamination exceeds the amount by which land values are expected to rise following decontamination. Nor do I consider the goal of reducing the cumulative radiation dose to 1 mSv to be unreasonable. My criticism of the decontamination policy in this chapter is based on a comparison with relocation, an alternative that ought to be considered more carefully.

Appendix: Calculating damages related to home securement

The Dispute Reconciliation Committee for Nuclear Damage Compensation (2013b) estimated the cost of damages related to home securement as follows. Based on data taken from the Land Price Public Notice System (now the Land Market Value Publication) published by the Ministry of Land, Infrastructure, Transport and Tourism's Land Appraisal Committee on January 1, 2013, and the *Land Price Research by Prefectural Governments* (Fukushima edition) published on July 1, 2013, the average land price for six cities in Fukushima Prefecture (Iwaki, Kōriyama, Fukushima, Aizuwakamatsu, Nihonmatsu and Minamisōma) is 37,813 yen/m^2. The average amount of land per house is 252 m^2. Therefore, an average sized house and land purchased for an average price in an urban area in Fukushima Prefecture costs:

37,813 × 252 = 9.53 million yen approx. (X1)[26]

Based on data taken from the *Monthly Marketing Report on Real Estate* published by the Ministry of Land, Infrastructure, Transport and Tourism, the average price per square meter of building a residential house in Fukushima Prefecture (for a owner-occupied detached dwelling) over the period from November 2012 to October 2013 was 165,800 yen. The average floor space in a detached owner-occupied home in Fukushima Prefecture is 148 m^2. Therefore, the average price of a newly built home (building only) in 2013 was:

165,800 × 148 = 24.54 million yen approx. (Y1)

The lot size used to calculate damages for dwellings in the 11 municipalities that fall within the evacuation instruction zone is 410 m^2, which is taken from the Housing and Land Survey 2008 by the Ministry of Internal Affairs Statistics Bureau as the average lot size of land for a detached owner-occupied dwelling in Fukushima Prefecture. Based on the property tax survey data taken from Municipal Finance Report 2008 by Fukushima Prefecture, land is valued at 8,922 yen/m^2. Therefore, the value of an average house block in the 11 affected municipalities for property tax purposes was 8,922 × 410 = 3.66 million yen approx. If we convert this to market price by applying the land coefficient of 1.43 we get 5.23 million yen approximately (X2).[27] In order to calculate the value of a residential building, the Dispute Reconciliation Committee takes the average price of a new building in the average year of construction (1975), which is 63,500 yen (the price determined by TEPCO), and multiplies this figure by the average floor space of 148 m^2 described above to get 9.40 million yen approximately (Y2).

14 Radioactive Contamination and Japan's Foreign Relations

Tsunekawa Keiichi

The Fukushima nuclear crisis had far-reaching consequences for Japan, both domestically and in its relations with other countries. In this chapter I explore the impact of the crisis on foreign relations in three areas: (1) the fall in the number of international visitors to Japan, (2) damage to Japan's external trade and (3) the diplomatic backlash against the discharge of contaminated water into the ocean.

On March 22, 2011, the growing crisis at the Fukushima Daiichi Nuclear Power Station saw the first sign of mitigation when a special truck known as a *kirin* (giraffe) – normally used to pump concrete at high-rise construction sites – started to pump water into the spent fuel pool of the No. 4 reactor. Hosono Gōshi, one of the main government figures responsible for dealing with the crisis, is on record as saying that 'we were over the worst of the nuclear accident by around the 20th' (Hosono and Torigoe 2012: 170). Fukuyama Tetsurō has also written that, by around March 25, 'we were finally able to establish some control over the situation inside the reactors' (Fukuyama 2012: 156). However, just as the crisis inside the plant appeared to be under control, other issues came to light. These included the radioactive contamination of food and the leakage of contaminated water from the plant. Fears of a large-scale release of radioactive contaminants from the reactors into the air may have abated, but the soil and water that have already been contaminated and the leakage of contaminated water accumulated within the reactor buildings pose a continuing challenge for Japan.

An effective means of dealing with radioactive contamination is necessary, not only to ensure the health and wellbeing of people living in Japan, but also to restore the confidence of the international community. After about March 15–17, people and governments outside Japan rapidly lost faith in the ability of the Japanese government and the plant's operator, Tokyo Electric Power Company (TEPCO), to maintain control of the situation at Fukushima Daiichi. As a result, many foreign residents and visitors left Japan, while others cancelled planned visits (Tsunekawa 2016b: 194–6). This lack of faith in the Japanese government also led many countries to impose unnecessarily strict and long-lasting restrictions on agricultural and marine product imports from Japan. Dealing with the huge volume of contaminated water onsite and recovering the melted fuel from the melted-down reactor cores at Fukushima Daiichi are technical feats that have not hitherto been undertaken anywhere in the world. If the clean-up is effective, then Japan will win the approval of the international community. If it fails and there are further leaks of radioactive materials into the air and the ocean, then Japan's reputation as a technological superpower will be dealt a further blow.

Falling visitor numbers

In the immediate aftermath of the earthquake, international arrivals to Japan halved and the effects of the disaster on international travel to Japan lingered for some time. Whether we look at the number of new 'short-term visitor' arrivals given in the immigration control statistics put out by the Immigration Bureau of Japan or at the 'foreign tourist' numbers calculated by the Japan Tourism Agency, annual visitor numbers did not return to their 2010 levels until 2013.

Figure 14.1 shows the total number of short-term visitors and includes a breakdown by country and/or region of origin. Looking at this graph, we can see that there was a dramatic fall in visitor numbers following the Global Financial Crisis (GFC) in 2008. By 2010 visitor numbers had bounced back and even surpassed pre-GFC levels. Then, in 2011, visitor numbers fell by almost 2.5 million. They increased by just over 2 million in 2012 but it was not until 2013 that they exceeded their 2010 level. The largest falls recorded in 2011 were for visitors from China and Korea.

As demonstrated in Table 14.1, statistics on international travelers, which include the number of foreign tourists, followed the same trend as for short-term visitors. The fluctuation in the number of short-term stayers and international travelers almost exactly mirrors the fluctuation in the number of tourists. This suggests that it was almost entirely foreign tourists who stayed away during this period.

Figure 14.1: New short-term visitor arrivals (1,000s)

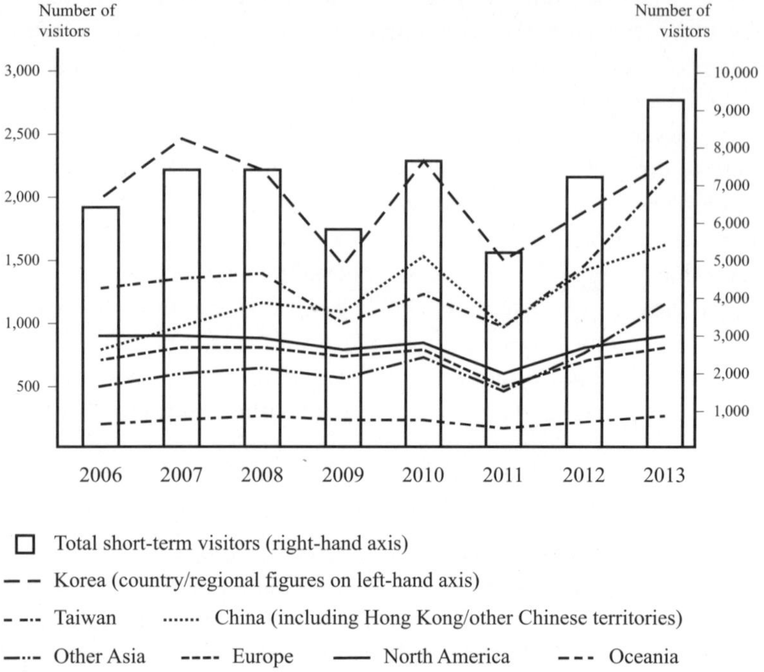

Source: Immigration Control Statistics (www.moj.go.jp/housei/toukei/toukei_ichiran_nyukan.html).

Table 14.1: Number of foreign tourists and tourism revenue

	International visitors[a]	**Tourists**	**International tourism revenue (billion yen)**
2007	8,346,969	5,954,180	1,461
2008	8,350,835	6,048,681	1,425
2009	6,789,658	4,759,833	1,170
2010	8,611,175	6,361,974	1,346
2011	6,218,752	4,057,235	998
2012	8,358,105	6,041,645	1,292
2013	10,363,904	7,962,517	1,631
2014	13,413,467	10,880,604	2,207
2015	19,737,409	16,969,126	3,305

Note: (a) The figures in this column indicate the number of non-Japanese people who entered Japan legally. Excludes foreign nationals such as permanent residents who reside in Japan long-term. Includes temporary landings by foreign travelers.
Source: Japan Tourism Agency 2014, 2016.

The decline in tourist numbers obviously had a significant negative impact on Japan's international tourism industry. In 2006 the tourism law was completely overhauled for the first time in 43 years and in 2007 Cabinet gave its approval to the Tourism Nation Promotion Basic Plan. The plan set a target of 10 million international visitors by 2010. The Japan Tourism Agency, an external agency of the Ministry of Land, Infrastructure, Transport and Tourism, was established in October 2008 to market Japan to draw foreign tourists. In 2010 the GFC contributed to international visitor numbers falling far short of the target of 10 million. Riding the wave of economic recovery, however, tourist numbers were trending upwards again when the Great East Japan Earthquake and the nuclear crisis occurred (Japan Tourism Agency 2009, 2011). As a result, Japan's tourism revenue (including passenger transportation) fell by 350 billion yen (Table 14.1).

The tourism industry incorporates a wide range of supporting industries, including transportation, accommodation, restaurant business, sport and entertainment, and has a major spillover effect on the rest of the economy. According to research carried out by the Japan Tourism Agency, in 2010 the industry directly added 11.5 trillion yen (2.4% of gross domestic product) to the economy and employed 2.29 million people (3.6% of employment). It also had a spillover effect of 25.2 trillion yen and 4.24 million jobs (Japan Tourism Agency 2012: 253). Of course, these figures account for domestic, as well as international, tourism. Nevertheless, international tourism made up 6.24% of all tourism revenue in 2010 (Japan Tourism Agency 2012: 190, 197), meaning that it added 1.57 trillion yen and 264,600 jobs to the economy. One-quarter of this was lost in 2011.

Nevertheless, as the radioactive fallout was confined to a limited geographical area and further pollution was not observed, except for contaminated water spillages, the radioactive contamination had a relatively small-scale impact on tourism. In 2013 the size of the tourism industry surpassed pre-crisis levels and the government realized its goal of 10 million international travelers. The relative weakness of the yen spurred a tourism boom and by 2015 international traveler numbers had nearly doubled to just under 20 million. At least for the moment, the issue of international travelers avoiding Japan due to radioactive pollution appears to have been resolved.

Contaminated food and import restrictions

In comparison with the decline in visitor numbers, the restrictions many countries have imposed on imports from Japan on the grounds of radioactive contamination have had a longer-term impact.

In 2010, prior to the disaster, Japan exported 500 billion yen worth of agricultural, forestry, marine and food products. Three-quarters of these exports went to Asia, in particular to Hong Kong, Taiwan and China. On March 12 Hong Kong began inspecting radiation levels in vegetables and fruits from Japan. Taiwan followed suit on the 14th, as did China on the 15th (*Asahi Shimbun*, March 17, 2011).

Compared with the rapid reactions of these countries to the food contamination issue, Japan was relatively slow to respond. Food that is contaminated with radiation can lodge in the body, causing internal radiation exposure. Internal exposure has a greater impact than radiation exposure from an external source. Considering the grave nature of the issue, a strategy ought to have been in place to respond to the potential a nuclear accident could have on Japanese food products. However, as the report compiled by the Government Investigation Committee on the Accident at the Fukushima Nuclear Power Stations of TEPCO (2012) makes clear, the *Food Sanitation Act* contained no standards limiting the level of radioactive contaminants in food. Furthermore, the confusion that prevailed in the aftermath of the disaster, combined with a lack of suitable equipment, meant that vegetables, raw milk and seafood products were not subject to radiation monitoring.

However, on March 15 the Fukushima prefectural authorities discovered grasses in an area more than 30 kilometers away from the Fukushima nuclear power station that contained radiation levels far in excess of the Guidelines for Limits on Ingestion of Radionuclides in Food and Drink that had been determined by a working group of the Nuclear Safety Commission as part of its emergency response framework developed in 1998.

Fearing the spread of harmful rumors about radioactive contamination, the Ministry of Agriculture, Forestry and Fisheries lobbied the Ministry of Health, Labour and Welfare (MHLW), which is responsible for food safety, to take action. Only then did the MHLW move to determine binding standards for radioactive contamination levels in food. The formal procedure for establishing such standards required approvals from the Food Safety Commission in the Cabinet

Office and the Pharmaceutical Affairs and Food Sanitation Council in the MHLW, as well as the Radiation Council in the Ministry of Education, Culture, Sports, Science and Technology. However, due to the urgency of the situation, the MHLW decided to bypass this process entirely and to adopt the Nuclear Safety Commission's standards almost without alteration.

The MHLW issued a directive to prefectural governments on March 17 establishing 'provisional safety standards' for food and raw milk (Table 14.2). Due to the particular sensitivity of the thyroid glands of nursing infants to radioactive iodine, the MHLW adopted the Food and Agriculture Organization (FAO)/World Health Organization (WHO) Codex Alimentarius standards of 100 becquerels per kilogram for milk and dairy products. However, even the Nuclear Safety Commission had not established guidelines for seafood products. Therefore, on April 5 the MHLW announced that it would apply the same provisional standards used for vegetables. On April 14 the Ministry of Agriculture, Forestry and Fisheries also established provisional standards for radioactive contaminants in animal feed (Government Investigation Committee on the Accident at the Fukushima Nuclear Power Stations of TEPCO 2012: 310–14, 318).

The first bans under these provisional standards were imposed on March 21. The prime minister, in his capacity as head of the Nuclear Emergency Response Headquarters, issued a ban on the shipment of spinach and *kakina* (a green leafy vegetable) from Fukushima, Ibaraki, Tochigi and Gunma prefectures and on raw milk from Fukushima. This was delivered to the governors of the respective prefectures under the terms of the *Act on Special Measures Concerning Nuclear Emergency Preparedness* (Government Investigation Committee on the Accident at the Fukushima Nuclear Power Stations of TEPCO 2012: 316). As monitoring of radiation levels in foodstuffs became more widespread, more and more items were discovered that exceeded the standards. From March 23 the bans were extended to shipments of other vegetables, as well as fodder and seafood products.

As Japan began placing shipment bans and imposing safety standards for radiation levels in food, other countries started to implement stronger controls on the import of Japanese food products. On March 19 the Korea Food and Drug Administration decided to carry out radiation inspections on Japanese agricultural and food products (Azuma 2015: 266). On March 24 Taiwan placed a ban

Table 14.2: Standards for radioactive contaminants in foodstuffs (becquerels per kilogram)

		Provisional		Final		
Nuclide	**Foodstuffs**	**March 17, 2011**	**April 5, 2011**	**April 1, 2012**	**Codex Alimentarius**	**European Union**
Radioactive iodine	Beverages	300		Unspecified		
	Milk and dairy products	300		Unspecified		
	Milk and dairy products (for infant)	100		Unspecified		
	Vegetables (excluding tubers and root vegetables)	2000		Unspecified		
	Seafood		2000	Unspecified		
Radioactive caesium	Beverages	200		10	1,000	1,000
	Milk and dairy products	200		50[a]	1,000	1,000
	Vegetables	500		100[b]	1,000	1,250
	Meat, eggs, fish, etc.	500		100[b]	1,000	1,250
	Infant formula			50	1,000	400
Permissible level of radiation from food in one year		5 millisieverts		1 millisievert	1 millisievert	1 millisievert

Notes: (a) Milk only. Limits for dairy products and vegetables are the same as for 'general foodstuffs'. (b) Classified as 'general foodstuffs'.
Sources: MHLW n.d.; Public Relations Office, Government of Japan 2014; Matsuo 2015: 260.

on the import of all foodstuffs from Fukushima, Ibaraki, Tochigi, Chiba and Gunma prefectures (Hayamaru 2014). On the same day, Russia's Federal Service for Surveillance on Consumer Rights Protection and Human Wellbeing also implemented a ban on the sale of any foodstuffs imported from Fukushima, Ibaraki, Tochigi and Gunma prefectures (Chuma 2013). By the end of March, all the major importers of Japanese food, including the United States, China, South Korea, Hong Kong and Taiwan, had imposed import bans on all or some foods from a number of prefectures. They also established testing regimes to measure radiation levels in food products. The United States, South Korea, Hong Kong and the European Union (EU) all required that food and fodder imports from Japan or from certain prefectures within Japan be accompanied by certificates confirming that radiation testing had been carried out. The EU went further, requiring region of origin certification for all food and fodder imports from prefectures where radiation monitoring certificates were not required.

Within a year, all the main importers of Japanese foodstuffs, apart from the EU, had expanded their import restrictions. Restrictions were imposed on more prefectures (including the Tokyo Metropolitan Area) and on more categories of foodstuffs. The number of countries requiring radiation testing certificates or region of origin certificates also grew. The EU did relax its restrictions slightly, removing one prefecture from the list of those requiring radiation monitoring certificates and exempting Japanese sake, shōchū and whiskey from the requirement. Nevertheless, these were rather small concessions.[1]

Continuing revelations about the scale of the contamination as a result of increasingly more effective radiation monitoring in Japan caused each of the major importers to expand their import restrictions by 2012. Japanese tardiness in implementing a certification regime was also a contributing factor.

The fact that Japan did not have established standards for safe levels of radiation in foodstuffs is symbolic of the broader failure to consider even the possibility that the radioactive contamination of foodstuffs might damage exports. That is why there were no procedures in place for certifying the level of radioactive contaminants in food prior to the accident. The Ministry of Agriculture, Forestry and Fisheries belatedly issued the directive to prefectural governors on the certification of agricultural, forestry, marine and food products for overseas export on April 21, 2011. This directive asked 'the relevant bureaus within each prefecture' to assume responsibility for providing certification. The ministry (actually, through the heads of the Regional Agricultural Administration Offices) only established a centralized certification system on April 1, 2013 (Ministry of Agriculture, Forestry and Fisheries 2014: 212).[2] The fact that each prefecture had a separate certification process and that importers each required different certification documents created further confusion. In addition, the sudden need to carry out radiation testing at a national level overwhelmed the existing infrastructural capacity. According to Laurent Kueny, a French Nuclear Security Authority official, this led to exceptional situations such as an offer (facilitated by the Japanese embassy and the French nuclear administration) to carry out testing of products (such as high-grade tea) for Japanese exporters at French radiation testing agencies (interview, Vienna, February 28, 2014).

The imposition of import restrictions not only on foodstuffs but on manufactured goods and even the ships and containers carrying Japanese merchandise led to further complications. Taiwan began testing samples of 658 categories of manufactured goods as early as March 22 (*Asahi Shimbun*, March 23, 2011). There were cases such as when towels made in Imabari, located in Ehime Prefecture far from the Fukushima plant, were stopped by customs at Rome. In another case, Chinese authorities refused permission for the unloading of containers carrying Japanese goods (*Nihon Keizai Shimbun*, April 10, 2011). In Europe, widespread fears that all goods coming from Japan were contaminated led the European Commission to advise member states on April 15 to adopt uniform testing standards for all maritime freight coming from Japan. Under these standards, further testing was to be carried out where radiation exceeding 0.2 microsieverts per hour was detected. The standards, unlike the measures for foodstuffs, were voluntary and member states did not have to implement them. England, Italy and Austria did not carry out inspections, while Germany, France and the Netherlands and some other countries did instruct their customs agencies to carry out radiation testing on manufactured goods (JETRO 2012). The French Nuclear Security Authority arranged for the emergency procurement of 2,500 examination devices and customs officials received training in their use. According to Kueny, while the scientific validity of such measures was unclear, they were effective in keeping people calm by giving them the impression that the situation was under control (interview with Kueny as above).

As the Japanese government was unable to keep abreast of the growing number of restrictions implemented by foreign countries, local Chambers of Commerce and Industry in different parts of Japan began issuing radiation level certificates in their own areas (*Asahi Shimbun*, April 14, 2011). By May 18 they had issued 1,520 such certificates, 65% of which were for manufactured goods, while 35% were for foodstuffs (*Asahi Shimbun*, May 19, 2011). The Japan Automobile Manufacturers Association adopted a unified testing regime in order to build trust in the reliability of the industry's own certification regime (*Nihon Keizai Shimbun*, April 16, 2011).

As mentioned above, from the middle of April the Japanese government worked to create a uniform certification regime in order to reduce the damage to Japan's international reputation. The government also launched a public relations campaign, asking importers to relax or eliminate their restrictions on Japanese goods. From March 20 the Public Relations Room of the Prime Minister's Office, in addition to organizing regular press conferences, made senior government officials available for one-on-one interviews. In his first interview with CNN, Chief Cabinet Secretary Edano Yukio argued that the import restrictions on Japanese foodstuffs lacked a scientific basis and called for an end to so-called 'harmful rumors'. The Japanese government made the same appeal during the G8 Summit meeting that took place in the French resort town of Deauville on May 26–27. The Japanese delegation was successful in having the following passage inserted into the Leaders' Declaration issued at the conclusion of the summit:

> The Prime Minister of Japan explained that his country would make every effort to minimize the uncertainty that the disaster might add to the global economy, including as a result of the nuclear accident. In particular, he committed to provide all relevant information regarding the nuclear emergency in a timely manner, and he ensured that products exported from Japan are safe. We stressed that measures on goods and travel should be based on scientific evidence. (Ministry of Foreign Affairs 2011)

Many international bodies also issued statements in line with the Japanese government's position. The International Civil Aviation Organization was the first to do so on March 18, when it announced that international air and sea transport to Japan could continue as usual and that there was no need to conduct radiation testing on people coming from Japan (*Asahi Shimbun*, March 20, 2011). The International Air Transport Association and the International Maritime Organization also put out a bulletin stating that 'there are no restrictions on voyages to Japan' (*Nihon Keizai Shimbun*, March 22, 2011).[3] On March 23 the FAO, WHO and the International Atomic Energy Agency (IAEA) issued a joint statement declaring that 'Japan has regulations in place relating to provisional regulatory limits of radioactivity in food. Food monitoring is being implemented, measurements of radioactivity in food are taking place, and the results are being communicated publicly' (FAO 2011).

The unspoken message was that the Japanese government's food safety policies ought to be trusted. At the beginning of May, the FAO and WHO issued a joint statement on the safety of Japanese seafood products. According to the statement, while caesium has a half-life of 2–30 years, it is excreted by marine organisms and so does not remain in the body – meaning that, in practice, it has a half-life of 5–100 days. Furthermore, because radioactive contaminants are dispersed by ocean currents, there would be no effect of the radiation on human health outside the immediate area of the nuclear reactor (WHO 2011).

In July 2011 the International Public Relations Committee in the Public Relations Room of the Prime Minister's Office was renamed the Committee for Countering Harmful Rumors and Coordinating Government and Agencies to Relaunch Japan Brand. At the Annual Meeting of the New Champions, a World Economic Forum event in Dalian, China (on September 14, 2011), and at the World Economic Forum's Annual Meeting in Davos, Switzerland (on January 26, 2012), the committee helped to organize 'Japan Night'. It also prepared questions and answers on Japanese food safety in English. Members of the Public Relations Room also participated in symposia and panel discussions at the Foreign Correspondents' Club of Japan and at the Chambers of Commerce and Industry of the United States, Australia and New Zealand in Japan, where they promoted the safety of Japanese food (Shikata 2014: 235–6; Shushō Kantei Kokusai Kōhō Shitsu 2012).

The Japanese government also tried to demonstrate that it had adequate food safety measures in place by creating stricter safety levels for radioactive contamination. The provisional measures that were adopted on March 17, 2011, later went through the MHLW's formal approval process. On April 1, 2012, the MHLW released the new standards (Table 14.2). The provisional standards were based on an annual dose of radiation of five millisieverts per year, higher than the one millisievert stipulated in the Codex Alimentarius. The new standards adopted the Codex level of one millisievert. In practice, the new standards were even more stringent than the Codex in presupposing that 50% (vis-à-vis 10% for the Codex) of the food concerned was contaminated (Matsuo 2015: 260).

Some countries responded to these efforts by reducing or relaxing their restrictions on Japanese food imports. However, the largest importers were reluctant to do so. By March/April 2013 only the United States and the EU had relaxed their import restrictions, and even then only slightly. The United States decided to restrict only those categories of imports that were already subject to shipment bans imposed by the Japanese government (although items were only removed from the United States' list some time after the Japanese government had done so). They also applied their own radiation standards for the testing of milk and dairy products and other agricultural products (excluding meat), standards that were less stringent than the Japanese ones. The EU dropped its import bans on goods from some prefectures and on some categories of goods and implemented a lower sampling rate for testing.

However, unlike the United States and the EU, the main importers in Asia maintained the same restrictions they had applied the year before, or even increased them. Taiwan and Hong Kong maintained the existing restrictions, while South Korea increased its restrictions on marine products while relaxing those affecting other food products. Like the United States, South Korea only imposed import restrictions on those products that the Japanese government had already subjected to shipment bans. Korea also removed the requirement for region of origin labelling for certain prefectures and product categories. But in the case of marine products, it increased the number of source prefectures that were subject to radiation testing from 13 to 20. China maintained extremely strict measures, banning all food and fodder imports from ten prefectures (including Tokyo) and requiring radiation monitoring certification and region of origin labelling on a wide range of food and fodder products. In addition, all Japanese marine products had to be accompanied by a quarantine inspection application form.

These import restrictions contributed to a decline in Japan's agricultural, forestry, food and marine exports in 2011–12. Figure 14.2 shows how, like Japan's tourism revenue, agricultural, forestry, marine and food exports fell in 2009 following the GFC. While they had begun to recover by 2010, they again fell in 2011. The slump continued in 2012. However, in 2013 the value of exports finally exceeded pre-GFC levels, as did the percentage of agricultural exports as a proportion of total exports in 2014. Many countries continue to maintain import restrictions on Japanese goods, but the effects of 'harmful rumors' of radioactive contamination appear to be gradually easing.

It remains to be seen whether import restrictions will continue to ease into the future. Some countries have actually increased them. South Korea responded to the discovery of highly radioactive water leaking from storage tanks into the ocean in August 2013 by extending import bans on marine products from Fukushima and eight other prefectures from 50 specific categories to all marine products, effective September 9, 2013 (*Asahi Shimbun*, September 6, 2013). In April 2015, after receiving the results of their own radiation testing, the Taiwanese authorities required radiation testing certificates to be affixed to all marine, tea, dairy, infant food and confectionary products from several prefectures (including Tokyo). These measures were added to the existing import ban on all foods (except liquor) originating from Fukushima and four other prefectures. Taiwan also decided to require all food products from Japan to carry region of origin labeling (*Asahi Shimbun*, April 17, 2015).

The contaminated water problem

As can be seen from South Korea's tightening of its import restrictions, importers are unlikely to relax their vigilance on Japanese food imports easily. One reason for this is the protracted problem of contaminated water at the Fukushima Daiichi Nuclear Power Station leaking into the ocean.

Ever since the accident, radioactive water that is escaping from the reactor and turbine buildings, from various contaminated water tanks, and from the dams that surround the tanks has been leaking into the ocean. At times it escapes directly into the sea, while at other times it seeps into the groundwater. The issue became the

Figure 14.2: Japanese agricultural, forestry, marine and food exports

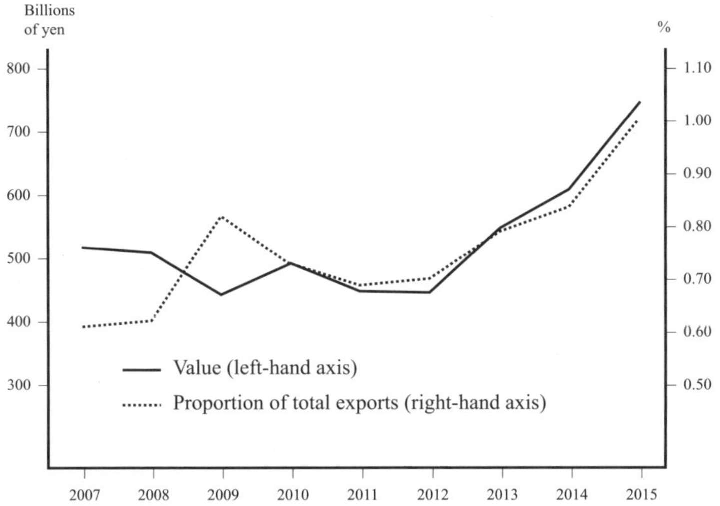

Source: Ministry of Agriculture, Forestry and Fisheries 2016: Appendix 14-i.

focus of intensive reporting in the newspapers in March–April 2011, July–August 2013 and May–August 2014.

The problem first came to light on March 24, 2011, when three workers laying cables in the basement of the No. 3 turbine building suffered radiation exposure after coming into contact with highly radioactive water. In the days following the incident, radioactive iodine was detected in seawater near the plant's wastewater outlet at levels 1,250 times the limit specified in the *Act on the Regulation of Nuclear Source Material, Nuclear Fuel Material and Reactors*. Further discoveries followed. Radiation levels greater than 1,000 millisieverts per hour were detected in the basement under the No. 2 turbine building and highly radioactive water was also discovered in a trench connecting the turbine building with the outside. Plant operator TEPCO planned to release low-level contaminated water that was being held in a storage facility at the plant into the sea to make way for the highly radioactive water inside the No. 2 and No. 3 buildings. The Democratic Party of Japan government initially refused to sanction the plan. However, when highly radioactive water was found to be leaking into the ocean from a crack in a pit near the

No. 2 reactor's seawater intake, the government had no choice but to give its approval. On April 4, 10,000 tons of water from storage facilities and 1,500 tons from subdrains under the No. 5 and No. 6 reactors were released into the sea (Government Investigation Committee on the Accident at the Fukushima Nuclear Power Stations of TEPCO 2012: 329–35).

The Chief Cabinet Secretary announced the decision to release low-level contaminated water into the ocean at a press conference held at 4.02 pm on April 4. Just prior to the press conference, TEPCO had made the same announcement. Having received the report, the Ministry of Foreign Affairs informed foreign diplomats of the planned release as part of the regular diplomatic briefing it carried out at 4.00–6.00 pm. However, neither South Korea nor Russia nor China, the countries that share the closest sea borders with Japan, were among the 51 countries that participated in the briefing. Furthermore, TEPCO's announcement that it would commence the discharge at 7.00 pm took place at 6.30 pm, half an hour after the conclusion of the diplomatic briefing. The Ministry of Foreign Affairs only notified the diplomatic corps that the release was to take place that day, via simultaneous faxes and emails, two minutes after the discharge had already commenced at 7.05 pm (*Asahi Shimbun*, May 11, 2011; Government Investigation Committee on the Accident at the Fukushima Nuclear Power Stations of TEPCO 2012: 358).

South Korea raised the loudest objections to the release of contaminated water into the ocean. A senior official expressed the government's displeasure that the disposal of contaminated water had been undertaken without prior warning (*Asahi Shimbun*, April 5, 2011, evening edn). Prime Minister Kim Hwang-sik, who was criticized for the lack of forewarning in the Korean parliament, refused to accept any responsibility of his government and condemned 'Japanese incompetence' (Azuma 2012a).[4] In China, too, the April 9 edition of the *People's Daily* included an international affairs column with the headline, 'The release of contaminated water should not be at Japan's discretion'. The column argued that, prior to making such a major decision, which had environmental implications that transcended national borders, serious consideration and prompt provision of information were necessary, and that there should have been adequate consultation with countries that might be affected (Azuma 2012b).

In response, the Ministry of Foreign Affairs conducted further detailed briefings to the Russian, Chinese and Korean embassies (*Asahi Shimbun*, May 11, 2011; Government Investigation Committee on the Accident at the Fukushima Nuclear Power Stations of TEPCO 2012: 358). However, the Japanese government maintained that the release of low-level radioactive water into the ocean did not violate international laws. First, the government asserted, contamination by land-based facilities is not covered by the London Convention on the Prevention of Marine Pollution by Dumping of Wastes and Other Matter (*Asahi Shimbun*, April 6, 2011). The government also maintained that it had informed the IAEA prior to the release, in line with its responsibilities under the Convention on Early Notification of a Nuclear Accident. This convention specifies that in the event of any accident involving facilities or activities of a member state 'from which a release of radioactive material occurs or is likely to occur and which has resulted or may result in an international transboundary release that could be of radiological safety significance for another State', the state should immediately notify the affected states 'directly or through the International Atomic Energy Agency' (IAEA 1986). The Japanese government had initially decided that, as the radioactive materials that were released into the ocean would be dispersed by ocean currents, there would be no safety consequences outside the immediate area where the release took place and, therefore, there was no obligation to notify under the convention. However, an official at the Nuclear and Industrial Safety Agency who saw the Chief Cabinet Secretary's press conference decided that it would be necessary to notify the IAEA, and had sent an email to that effect at 5.46 pm (Government Investigation Committee on the Accident at the Fukushima Nuclear Power Stations of TEPCO 2012: 357–8). There is also an obligation to notify under Article 198 of the United Nations Convention on the Law of the Sea. When questioned about this by a journalist during a press conference held in the afternoon of April 5, Chief Cabinet Secretary Edano replied that there was no immediate prospect of the release constituting a case 'in which the marine environment is in imminent danger of being damaged or has been damaged by pollution' as defined in the convention.

On June 17 TEPCO constructed a water decontamination system using technologies from four Japanese, American and French companies. Later, on August 18, the Simplified Active Water Retrieve and Recovery System (SARRY), a water purification system developed by Toshiba in conjunction with the United States company Shaw, commenced operation (Government Investigation Committee on the Accident at the Fukushima Nuclear Power Stations of TEPCO 2012: 340–1). These systems were designed to remove caesium from contaminated water. When used in combination, they had the potential to process 1,400 tons of contaminated water per day. If they were operating at full capacity, they should have been able to deal with the 450 tons per day of contaminated water that was accumulating in the basement of reactor buildings Nos 1–4 (*Asahi Shimbun*, October 20, 2011). However, immediately after the systems commenced operation, they suffered from leaks, power outages and malfunctions. As a result, there was no way to dispose of the highly radioactive water in the basement of the reactor and turbine buildings.

On March 30, 2013, another technology known as the Advanced Liquid Processing System (ALPS), a system capable of removing not only caesium but 62 other types of radioactive nuclides, commenced test operations. The three sets of ALPS were built with the capacity to deal with 750 tons of water, but they too were shut down due to malfunctions and leaks (*Asahi Shimbun*, March 19, May 20, June 22, 2014). Between September and October 2014 ALPS was expanded and its capacity was increased to 1,500 tons per day (*Asahi Shimbun*, September 18, 2014) but it remained unstable.

In the midst of all this, TEPCO acknowledged on July 22, 2013, that highly radioactive water that had leaked into a trench in the basement of the No. 2 reactor building had seeped into the groundwater and was flowing into the ocean. On August 10 it was revealed that the underground containment wall intended to prevent water from leaking into the sea was powerless to stop water flowing over the top of it. On August 20 and September 1, leaks of highly contaminated water from onsite containment tanks were also discovered. Some of this water was leaking into the sea through a drainage ditch. As discussed above, this is when South Korea clamped down on the import of Japanese marine products.

It also coincided with the final stage of the Japanese government's bid for the Olympic Games. On September 7, while attending the International Olympic Committee meeting in Buenos Aires, Prime Minister Abe announced that the situation at the plant was 'under control' and stated that 'the effect of radioactive substances in the nearby waters is blocked within 0.3 square kilometers of the plant harbor' (*Nihon Keizai Shimbun*, September 8, 2013). In reality, heavy rain from a typhoon was causing water to overflow from the dam surrounding the onsite containment tanks. Numerous other incidents occurred involving accidental leakages and overflowing from tanks. Behind this series of leaks was the fact that the amount of water kept on increasing and storage tanks were being constructed hastily to hold huge amounts of water.

Early on, TEPCO had planned to release decontaminated low-level radioactive water into the ocean and to divert uncontaminated groundwater before it came into contact with the reactor buildings and dump it into the sea (the groundwater bypass). However, these plans met with strong opposition from local fisherfolk and were never realized. As a result, tank after tank was built to contain the contaminated water.

When the IAEA monitoring team came to Japan to inspect the situation, it suggested on December 4, 2013, that discharging low-level contaminated water that was below safety limits into the ocean should be considered as an option (*Asahi Shimbun*, December 5, 2013). On March 25, 2014, the Fukushima Prefectural Federation of Fisheries Co-operative Associations consented to the 'groundwater bypass plan' and on May 21 TEPCO began discharging groundwater via the bypass. Nevertheless, the bypass did not make a significant dent in the 400 tons of groundwater flowing into the reactor building basements every day. On August 7, 2014, TEPCO and the government presented a new proposal to the fishing cooperatives that involved pumping the water that was accumulating below the reactors out through subdrains and, after decontamination, dumping it into the sea (*Asahi Shimbun*, August 7, 2014, evening edn). However, facing local opposition, work on the plan only began in September 2015 (*Asahi Shimbun*, September 15, 2015).

In the meantime, the Agency for Natural Resources and Energy entrusted a research and development consortium called International Research Institute for Nuclear Decommissioning with the task of finding a technical solution to the contaminated water problem. The institute called for proposals from both domestic and international companies. A total of 780 proposals were received, of which four were selected for further development: removal of radioactive materials from harbor seawater; capture of radioactive substances from soil; decontamination of flange-type water storage tanks; and unmanned boring technologies. The management of the project was then turned over to the Mitsubishi Research Institute, which issued, in March 2014, a second international call for proposals to implement these ideas. In June, 11 companies were selected to undertake the work. Three of the selected proposals were from consortia involving foreign companies: IBC Advanced Technologies, AREVA NC and SITA Remediation (Agency for Natural Resources and Energy 2014a, 2014b).

The contaminated water problem and reactor decommissioning are the most significant legacy of the Fukushima nuclear crisis. The Agency for Natural Resources and Energy's public tender for a solution to the contaminated water problem signifies that there is no known technological means of resolving it at present. As long as contaminated water continues to accumulate onsite, the discharge of water contaminated with radioactivity into the ocean will continue. Even if it does not directly impact the safety of other nations, it is having a negative impact on Japan's reputation in the international community. Every possible effort should be expended to find a solution.

Notes

Introduction

1 Kingston (2012: 7) summarizes these issues as follows: '1. Japan's political elite has consistently underwhelmed the public throughout this crisis, wasting too much time and energy on petty politics and failing to act efficiently or with sufficient urgency, leaving the public disenchanted with Tokyo and forcing the people of Tōhoku to rely on their own initiative…3. Local political leaders, community and neighbourhood associations, cooperatives, NPOs (non-profit organizations) and NGOs (non-governmental organizations), and the private sector have been the key actors in recovery, making up for central government inertia with timely, appropriate and generous interventions.'

2 One definition sees network governance as 'a relatively stable, horizontal articulation of interdependent, but operationally autonomous actors who interact through negotiations that take place within a relatively institutionalized community which is self-regulating within limits set by external agencies and contributes to the production of public purpose' (Torfing 2007: 5). In the field of politics, networks are typically an important constituent of the notion of governance. For example, Rhodes (1997: 53) defines governance as 'self-organizing networks that exist between organizations'. Governance theory also tends to emphasize the way society self-corrects in response to problems such as market failure and the hollowing out of the state. However, we believe that while self-correction by society is an important aspect of governance, we see an important role for governments (both central and local) in managing these networks and adapting them to social conditions. Furthermore, as we mentioned above, we emphasize the fact that Japan's integrated system of local administration, itself based on continental European systems, makes central–local relations particularly significant in the field of government. Our definition of network governance in this introduction includes the multi-layered networks that function through relationships between governments and quasi-governmental organizations, quasi-private organizations, neighborhood associations, employer groups and labor unions, civil society organizations such as NPOs and NGOs, private companies and ordinary residents.

3 In 2014 former Prime Minister Hosokawa Morihiro, with the support of former Prime Minister Koizumi Jun'ichiro, stood in the Tokyo gubernatorial elections on an antinuclear platform. Hosokawa announced his candidacy without much warning and did not become a major political force. Nevertheless, this intervention by two former prime ministers from

different political groupings (non-LDP and LDP respectively) gave the antinuclear movement a confidence boost.

4 According to the National Antinuclear Litigation Directory (www.datsugenpatsu.org/bengodan/list) put together by a number of antinuclear legal teams, as at December 16, 2016, 46 legal cases have been launched and 33 are currently before the courts.

5 The term 'system' is a political unit used in political systems theory (see Almond et al. 2010). However, the term was also used by the journalist Karel van Wolferen (1989: 43–9), who characterized a number of aspects of Japanese culture and politics (such as the ambiguous line separating the public from the private sphere, the regulation of capitalists, mutual restraint and its apparent paralysis) as 'the system'.

6 'Fukushima hairo, baishō hi 20 chō en' (20 trillion yen for decommissioning and decontamination at Fukushima), *Nihon Keizai Shimbun*, November 27, 2016, morning edn, p. 1. For further details, see Tokyo Denryoku Kaikau, 1F Mondai Iinkai, 2016.

7 'Monju hairo kettei' (Monju to be scrapped), *Asahi Shimbun*, December 22, 2016, morning edn, p. 1.

8 Law No. 156, 1999, Article 15, paragraph 4.

Chapter 1

1 The use of the term 'disaster-related bills' in this chapter follows that of the Cabinet Legislation Bureau.

2 Between 1947 and 2014 a conference committee has been initiated on 17 occasions to deal with specific bills, of which 15 were passed. However, 14 of these occurred between 1948 and 1952. Since then, only one bill has been passed by a conference committee (in 1994).

3 From the 2014 election, the House of Representatives has had 295 single-member constituencies, giving a total of 475 members. The number of single-member constituencies in the House of Councillors has varied due to boundary changes. In this chapter, I use the number of seats in the house in 2010.

4 For a more detailed discussion of the material presented here and in the subsequent section, see Hamamoto (2016).

5 The Japanese Communist Party (JCP) has significant policy differences from the other parties and hence was not a candidate for coalition negotiations. The JCP has never held power at the national level.

6 The same axes were extracted in surveys carried out in 2003 and 2005.

7 Interview with a Komeito Diet member, October 28, 2013.

8 In December Kan had approached the Sunrise Party of Japan and the New Renaissance Party to form a coalition. However, both refused due to significant policy differences and the administration's dwindling popularity.

9 Press conference held by LDP president Tanigaki Sadakazu, March 19, 2011.

10 Interview with Mori Yoshirō (*Sankei Shimbun*, July 7, 2012, morning edn).

11 The bill referred to here was for the *Nuclear Damage Compensation Facilitation Corporation Act.*

12 Speech by Tanigaki Sadakazu, August 11, 2011.

13 The JCP was the only party to vote against a bill on 40 occasions, as was YP on six and the SDP on one.
14 The definition of disaster-related bills used here is described in note 1.
15 The three nuclear energy-related bills rejected were a bill for special provisions for the disposal of waste generated by the Great East Japan Earthquake, a bill to partially amend the *Act for Establishment of the Ministry of the Environment* and other laws in order to reform the institutions and agencies responsible for ensuring nuclear safety, and a bill to establish a nuclear safety investigation committee.

Chapter 2

1 The ten electric utilities have combined sales of 15 trillion yen, of which TEPCO's sales make up 5 trillion, one-third of the total. TEPCO also exercises overwhelming influence within the FEPC. The most important task of the various utilities' staff seconded to the FEPC is to build strong relationships not with the Ministry of Economy, Trade and Industry (METI) but with TEPCO. When one of the other utilities wanted to put forward a proposal within the FEPC, 'the most important thing was to persuade TEPCO so that during meetings, TEPCO's representatives would signal their agreement with the proposal' (Kyodo News 2013a). The position of chairperson of the FEPC rotates between TEPCO, Kansai Electric Power Company (KEPCO) and Chūbu Electric Power (Chūbu). However, TEPCO staff members occupy all the important posts. In reality, the FEPC is simply an extension of TEPCO that lobbies politicians, public servants and the media in order to promote nuclear power (Arimori 2011: 150).
2 The 'nuclear village' is shorthand for the network of stakeholders that seeks to promote nuclear power. It includes both the utilities and the manufacturers in the nuclear and related industries, METI, the labor unions representing electricity industry workers, municipal governments in reactor host communities, university researchers and a section of the mass media.
3 The Central Research Institute of the Electric Power Industry (CRIEPI) is an 840-member research institute funded by the electric utilities and with an annual budget of 30 billion yen. One former member of CRIEPI has testified that research on responding to a complete loss of power to a nuclear power plant was 'taboo'. He explained that 'TEPCO and the other companies would not allow such research. There are four or five layers of safety measures in place within a nuclear reactor. There was no planning for what would happen in an event such as this one where all of these measures failed simultaneously. TEPCO acted as if "such an eventuality is impossible" and said that, "we will not provide money to fund such research"' (Shimura 2011: 201). Former members of the Japan Atomic Energy Research Institute and the former head of the Nuclear Safety Commission have also testified to these practices (NHK ETV Tokushū Shuzaihan 2013: 17, 270–1).
4 The TEPCO group includes numerous associated companies. There were 264 in July 2011 (including 166 subsidiaries and 98 associates) with a total

of 32,800 employees. Together with the parent company, the group has a total of 69,500 employees (Takeuchi 2013: 230–1).

5 For example, building a power station requires heavy electric machinery manufacturers to supply various pieces of equipment, information systems companies to build the control systems, manufacturers of electric power cables and steel manufacturers to supply the equipment for distribution systems, and general construction contractors to build the housing for the generators. There are also trading companies that procure fuel to generate electricity, and shipping and transportation companies to transport it (Takeuchi 2013: 235–6; Anzai 2012: 138–9; *Shūkan Daiyamondo*, May 21, 2011: 28–9).

6 Kigawada Kazutaka (former chairperson and president of TEPCO) was vice-chairperson of Keidanren and president of the Japan Association of Corporate Executives, and Hiraiwa Gaishi (former chairperson and president of TEPCO) was chairperson of Keidanren (*Shūkan Tōyō Keizai*, April 23, 2011: 41). The other utilities, too, are by far the largest companies in their regions and their executives have generally held the top or second-top posts in their local business federations (Takeuchi 2013: 236; *Shūkan Daiyamondo*, May 21, 2011: 33).

7 The comprehensive cost-accounting method takes into account the cost of construction, maintenance and operation of power stations and transmission systems; fuel costs; wage and employee insurance costs; corporation and property taxes; interest paid on loans; and dividends paid to shareholders. These costs are all designated as 'reasonable costs'. The price of electricity is set so as to cover all these reasonable costs plus a profit to the utilities. The profit margin is determined as a fixed percentage of the utilities' total capital investment in electricity generation, including fixed assets (including power stations, the grid etc.), fuel costs (including spent-fuel costs) and assets currently tied up in construction. To this total is added a fixed percentage for corporate profits. TEPCO's profit margin has been fixed at 3% since 2008 (Ōshika 2013: 261–2; *Shūkan Tōyō Keizai*, April 23, 2011: 48).

8 Honda (2005: 67) suggests that this is why the utilities have been so keen on investing in nuclear energy, which requires such enormous investment in infrastructure.

9 TEPCO has issued 5.2 trillion yen worth of corporate bonds, but its short- and long-term loans from the banks are only 2.3 trillion yen.

10 Although the *Dengen Sanpō* technically applies to all electric power generators, in practice communities that host nuclear power stations receive more than twice the amount received by communities that host similarly sized thermal and hydroelectric power stations. This account of the *Dengen Sanpō* is taken from Yoshioka 2011: 150–1; Honda 2005: 102–5.

11 Later, as the pace of new plant construction slackened and large-scale research and development operations were commercialized, the levy was reduced a number of times. Since 2007 the levy has been set at 375 yen.

12 In January 2011 Ishida Tōru (former commissioner of the Agency for Natural Resources and Energy) was employed by TEPCO as an adviser and was thought to be headed for vice-president. However, following the Fukushima nuclear accident, there was widespread criticism of *amakudari*

and Ishida was forced to resign in April. Former METI officials are not the only important figures who descend to posts in TEPCO. At least 50 former senior bureaucrats have descended to posts in the utility, while at least 121 have been employed at public service corporations it funds (Doi 2011: 82).

13 Ōshika (2013: 401) cites the testimony of a former director of the Electricity and Gas Industry Department in METI's Agency for Natural Resources and Energy who said that 'when TEPCO officials visited our offices, they would often say "if your children are having trouble finding a job, please don't hesitate to mention it to me and we will see what we can do"'. At least six former commissioners of METI's external agencies have children working for TEPCO.

14 JAIF's predecessor was established by the industry in 1956 to promote the use of nuclear power. It distributes information about nuclear power, conducts surveys and makes statements about nuclear energy.

15 It is possible to explain this by looking at the political – bureaucratic – corporate relationship on nuclear policy not as an 'iron triangle' but as a 'three-way deadlock': '"The nuclear industry" is in thrall to "the bureaucracy" which holds the "authority to grant a license and approval". "The bureaucracy", in turn, is in thrall to "the politicians" who have the "power to hire and fire".' Finally, '"the nuclear industry" gains access to "the politicians" by means of "elections and money"' (Asahi Shimbun Tokubetsu Hōdōbu 2014: 176).

16 The fate of nuclear energy under liberalization was also flagged as a problem. In Europe and the United States the construction of nuclear power stations ground to a halt following liberalization because, in addition to the difficulty in raising the enormous capital necessary to construct nuclear power stations, the risk of accidents is significant and private companies do not have the capacity to deal with radioactive waste (Komori 2013: 147).

17 The system known as 'pancaking', whereby electricity flowing across regional boundaries incurs a charge from each of the utilities in its path, was abolished and replaced with an integrated 'system-usage charge' (Takeuchi 2013: 201).

18 The electric utilities plotted to maintain their regional monopolies and avoid internal competition within the industry. In 2012 the only example of any of the nine utilities retailing electricity outside its region was Kyūshū Electric Power, which supplied electricity to a single supermarket in Hiroshima (Takeuchi 2013: 189–90). For further details on the three reforms initiatives see Arimori 2011: 152–4; Ōshika 2013: 266–81; Takeuchi 2013: 184–218; Saitō 2012: 260–75; Komori 2013: 147–53; Lee 2012: 192–3; *Asahi Shimbun Globe*, October 5, 2009.

19 Another bureaucrat also stated that 'if the utilities put a black mark against your name, you will not be promoted. The METI overseeing the electric utility industry is a fantasy. The industry controls the METI' (Kyodo News 2013a). One METI official has testified that 'the electric power industry whispers criticisms of individual bureaucrats in the Minister's ear. That's how they manage personnel inside the ministry.' One of his superiors stated that he had heard from the Minister, 'the electric utilities aren't happy with you' (Asahi Shimbun Tokubetsu Hōdōbu 2014: 179).

20 For further details see, *Asahi Shimbun*, October 23, 2014, morning edn.

21 At the time, the FEPC's public relations budget was 5 billion yen (Asahi Shimbun 'Genpatsu to Media' Shuzaihan 2013: 294).

22 In 2009 TEPCO's expenditure on entertainment reached 2.1 billion yen (Komatsu 2012: 124–5). This includes both purchasing tickets to fundraising events and entertaining in host communities (Shimura 2011: 160–1). Following the disaster, KEPCO took the lead in the industry's relations with government and politicians. After TEPCO was effectively nationalized, the company stopped purchasing tickets to fundraising events. However, Kandenkō, one of the major companies within the TEPCO group, continues to purchase these tickets and the directors now purchase more tickets individually so as not to draw attention to the company (Asahi Shimbun Tokubetsu Hōdōbu 2014: 174–5).

23 Arimori (2011: 152) also states that 'there are many children of leading politicians and high-level bureaucrats who have used their connections to secure employment with TEPCO'.

24 Kanō (2010: 97–9) has explained that he supported the Basic Law in order to counter what he saw as three imminent crises: market fundamentalism, unrealistic expectations about new forms of energy and the abandonment of the nuclear fuel cycle project.

25 This account of the circumstances that led to the enactment of the RPS Law and the *Basic Act on Energy Policy* is based on Lee 2012: 180–4, 189–92; Kyodo News 2013b; Saitō 2012: 271–4.

26 TEPCO sponsored three courses at the University of Tokyo with 395 million yen. In association with other companies (such as construction firms), TEPCO gave 306 million yen to a further three courses. These courses included lectures such as 'Social engineering the nuclear fuel cycle'. (These figures are based on the contributions made from October 2007 to September 2013. Other courses received 200 million yen in sponsorship from KEPCO, Hitachi, Mitsubishi Electric and Sumitomo Electric Industries.) TEPCO has also sponsored research and teaching at Tokyo Institute of Technology, Keiō University, Niigata University, University of Tsukuba and Yokohama National University, among others. The other utilities also sponsored courses at some of Japan's most prestigious universities and at universities that offer courses in nuclear engineering. Nuclear industry employees teach in these courses. TEPCO's head of corporate sales, Toshiba's former head of nuclear energy development and services, former technicians from Mitsubishi Heavy Industries and the head of METI's Nuclear Energy Division were all employed as guest lecturers at the University of Tokyo (Komatsu 2012: 89–90; Doi 2011: 94–9; *Shūkan Tōyō Keizai*, April 23, 2011: 41; *Shūkan Bunshun* special issue 'Tokyo Denryoku no daizai', July 27, 2011: 92). TEPCO also met frequently with seismologists and provided 'technical advice payments' of 50–80,000 yen for such meetings. These payments may have influenced the setting of safety standards (Soeda 2014: 76–7).

27 According to Anzai, 'notices were distributed in the research institute saying "don't speak to Anzai". When researchers from other universities visited me, they were driven away by my colleagues who told them "don't come in". A TEPCO employee would grab the seat next to mine and report on what I was doing to the company in great detail.' He was followed

when giving public lectures and these were recorded and passed on to TEPCO. On one occasion, a TEPCO employee told him, 'why don't you take a sabbatical to America for three years? We'll pay your expenses.' At a party held in 1975 to celebrate the 15-year anniversary of the founding of the Department of Nuclear Engineering, one professor who took to the stage to address the crowd stated, 'the fact that we graduated Anzai Ikurō is the only black mark against our name' (*Shūkan Tōyō Keizai*, June 11, 2011: 69; *Shūkan Daiyamondo*, May 21, 2011: 37; Kyodo News 2013c).

28 Other pronuclear unions within Rengō are the Japanese Electrical, Electronic and Information Union, Rengō's fourth-largest union with 572,516 members, and the Japan Federation of Basic Industry Workers' Unions, the seventh-largest federation with 250,211 members (Japanese Trade Union Confederation, 'Rengō kōsei soshiki ichiran').

29 Denryoku Sōren was highly valued by the DPJ because, come election time, the federation would loan union members with extensive electoral experience to the campaign to assist with everything from putting up posters to telephone canvassing (Asahi Shimbun Tokubetsu Hōdōbu 2014: 174).

30 One former utility executive explained that 'while the LDP have long supported nuclear power, it is important for us to have people who understand it on the other side of the chamber as well. By working from within, they can keep down the antinuclear currents that tend to crop up in the opposition parties' (*Asahi Shimbun Digital*, November 25, 2012).

31 For this reason, Yoshioka Hitoshi argues that the *Dengen Sanpō* have not been particularly effective in securing new sites for reactor construction. Instead, the payments have functioned as reparations for communities that already host nuclear facilities (Yoshioka 2011: 152; Honda 2005: 143–4).

32 *Shūkan Daiyamondo*, May 21, 2011: 36, 42. For this reason, host communities that were struggling to maintain public buildings constructed using the compensation provided under the *Dengen Sanpō* began calling for alternative ways to spend the money. In 2004 the government responded by creating a unified system of compensation funding for reactor host communities and reforming the complex rules for how the compensation funding could be spent. It modified the purposes for which the funds could be used and extended the time period during which they could be accessed. The reforms ensured that host communities would receive generous compensation, not only during the construction phase, but even after reactor operations commenced. Furthermore, the government expanded the list of projects that were eligible for funding. Initially, use of the funds was restricted to the construction of new public buildings, plans for luring industry and industrial modernization. The new list included softer 'community stimulus' projects such as the development of local specialty goods, event funding, welfare spending, support for nonprofit organizations and human resources expenditure. It was now possible not only to use the funds to maintain facilities that were initially constructed with compensation funding, but other facilities constructed with the support of other government agencies or by the municipalities themselves. They could also be used to pay for staff of the facilities (Asahi Shimbun Aomori Sōkyoku 2005: 13–14).

33 As at March 2012 the FEPC had provided a total of 18.46 billion yen (5 billion during construction and 5–6 hundred million per year thereafter) to the Mutsu-Ogawara Regional Industry Promotion Foundation. These funds were collected from the utilities by the FEPC and donated to the foundation to be distributed to every local municipality in Aomori Prefecture. The reason for these donations was that the then prefectural governor, who supported the nuclear fuel cycle project, was struggling in the 1991 gubernatorial elections. Investigations conducted after the election revealed that opposition to the nuclear fuel cycle project was high in municipalities that had no nuclear facilities and were not receiving any compensation from the government. The Aomori prefectural authorities asked the FEPC to begin distributing donations, which began in 1994 (NHK 2012).

34 The company also provided additional donations for certain large-scale construction projects, such as the national soccer training facility (J-Village) in Fukushima Prefecture (13 billion yen) and the construction of parks in Kashiwazaki City and Kariwa Village (10 billion yen). On TEPCO's donations, see Asahi Shimbun Tokubetsu Hōdōbu 2014: 142–3; *Asahi Shimbun*, September 15, 2011.

35 For specific examples of TEPCO paying commissions to construction contractors to purchase land and deal with opposition from local communities, see Asahi Shimbun Tokubetsu Hōdōbu 2014: 67–159.

36 It is said that TCIA maintains a database 'containing detailed data on local residents, including their subscriptions, educational background, health records, criminal records, political affiliations and ideologies'. In 1972 it was discovered that Hokuriku Electric Power had conducted background checks on 80 reporters from 14 newspapers and broadcasters in Toyama Prefecture. There are reports that executives at other utilities sneered at Hokuriku for its 'clumsy approach' to carrying out these checks (Asahi Shimbun 'Genpatsu to Media' Shuzaihan 2013: 304–7).

37 For example, in the late 1980s Hokkaidō governor Yokomichi Takahiro refused permission for the construction of a high-level nuclear waste facility in Horonobe Town, Teshio District.

38 Fukushima prefectural governor Satō Eisaku refused to assent to a nuclear power plant operating due to a lack of faith in the government and TEPCO, demonstrating the degree of influence prefectural governors can wield in disputes (Yoshioka 2011: 323–30; Satō 2011; *Shūkan Bunshun* special issue 'Tokyo Denryoku no daizai', July 27, 2011: 89).

39 Antinuclear lawsuits can be roughly divided into two categories. The first category of administrative suits has sought to reverse or nullify the development approval for power station construction. The second category involves civil suits seeking to halt construction or operations (Shindō 2012: 29, 38).

40 In a lawsuit seeking to revoke the planning permission for the No. 1 reactor at Shikoku Electric Power's Ikata Nuclear Power Station (Ikata Town, Nishiuwa District, Ehime Prefecture), it was revealed that the 13 members of the working group from the Reactor Safety Examination Committee had only visited the site on seven occasions and that their inspections had lasted no longer than one or two days. Most of the documents considered by the committee had been prepared by the utility or by the relevant

government agencies. Furthermore, many committee members were absent during the safety review meetings and no records were made of the proceedings (Asahi Shimbun 'Genpatsu to Media' Shuzaihan 2013: 95–7; Kaido 2011: 6–7).

41 In January 2003 the Kanazawa branch of the Nagoya High Court declared the planning permission for the Monju fast breeder reactor to be invalid and issued an injunction stopping the construction and operation of the plant. In March 2006 the Kanazawa District Court issued an injunction ordering the cessation of operations at the No. 2 reactor at Hokuriku's Shika Nuclear Power Station (Shika Town, Hakui District, Ishikawa Prefecture).

42 The Supreme Court's General Secretariat does not release information about personnel evaluations or transfers or salary details for judges. Shindō Muneyuki argues that, as a result of this lack of transparency regarding judicial appointments, individual judges follow the lead of the Supreme Court or of other more senior judges as they jockey for position and status within the judiciary (Shindō 2012: 148–9, 208–12; Isomura and Yamaguchi 2013: 170). In Japan judges are appointed to positions within the Ministry of Justice's Litigation Bureau and prosecutors are appointed to the bench as judges or assistant judges. Shindō discusses the negative effects of the mingling between judges and prosecutors that occurs as a consequence of this appointment system. When judges are seconded to the Litigation Bureau to represent the government as defendants, it strengthens their belief in the infallibility of government agencies. Shindō details cases where judges have been transferred to the Litigation Bureau to work as defense lawyers for the government in antinuclear lawsuits before returning once more to the bench (Shindō 2012: 149–58).

43 For specific examples, see Jōmaru 2012; Asahi Shimbun 'Genpatsu to Media' Shuzaihan 2013; Kanbayashi 2011; Sasaki 2011a, 2011c; Takahashi Atsushi 2011, 2012.

44 METI's Agency for Natural Resources and Energy spends enormous sums of money on public relations in order to persuade the public of the benefits of nuclear energy (Kanbayashi 2012).

45 This actually represents a budget cut when compared with the combined spending of 103.7 billion yen in March 2006, of which TEPCO contributed 29.3 billion (Asahi Shimbun 'Genpatsu to Media' Shuzaihan 2013: 283, 287).

46 Similarly, in Aichi Prefecture four of the five television stations had a Chūbu Electric Power executive on their boards and the remaining station had a Chūbu executive on its programming commission. Of the 202 companies listed in the Japanese Broadcasting Yearbook 2010, at least 61 listed an electric utility as a major shareholder or listed a utility executive on its programming commission (Asahi Shimbun 'Genpatsu to Media' Shuzaihan 2013: 320–1).

47 A particularly glaring recent example of the collusion between the utilities and the media is the fact that at the time of the Fukushima accident TEPCO chairperson Katsumata Tsunehisa was in China leading a group of journalists as part of the 'Chinese friendship committee', which started in 2001. Individual delegates paid just 50,000 yen to join the tour, which had an overall budget of 4–5 million yen. The remainder was divided

evenly between the Chinese and Japanese. On the Japanese side, half the money was provided by TEPCO, KEPCO and Chūbu Electric Power (Ōshika 2013: 17–22; Asahi Shimbun 'Genpatsu to Media' Shuzaihan 2013: 279–83; Takahashi Atsushi 2011: 70–2).

48 On the other hand, if reporters looked as if they might write something unfavorable about the utilities, they would sometimes receive a visit from the utilities seeking to persuade them against doing so. In August 2002 *Asahi Shimbun* reporter Okuyama Toshihiro began investigating a cover-up at TEPCO's nuclear reactors. He discovered TEPCO employees had produced a false report for MITI that stated that the company knew nothing about issues at the reactor. In December 2003 he wrote an article about the cover-up. He also began a follow-up investigation for a forthcoming book into how the cover-up had happened. Upon doing so, a TEPCO insider told him, 'we can buy up a lot of copies of your book' (Asahi Shimbun 'Genpatsu to Media' Shuzaihan 2013: 54).

49 The society made a number of other suggestions. It advised that primary school science and social science students be taught about nuclear energy and that middle school science and social science students be taught about how spent nuclear fuel was being 'recycled'. It recommended middle school science students be taught about the positive uses of radiation and that nuclear power is a safe means of producing energy, and that the curriculum should also cover the existence of natural background radiation, with science students conducting experiments to measure background radiation. Furthermore, the society suggested that middle school social science students be taught about the growing use of nuclear energy around the world.

50 Some specific examples of these recommendations include removing the phrase 'nuclear power carries the risk of major damage over a wide area if an accident does occur' and moving a discussion of the 1986 Chernobyl disaster from the main text to a shorter reference in a footnote. In another case, MEXT directed that a reference to the declining use of nuclear power in Europe be removed and that a note be added explaining that renewable energy still had problems.

51 See also Saitō 2011: 52–9; Kyodo News 2013c.

Chapter 3

1 'Baseload power' refers to an energy source that has a low generation cost (low operating cost) and can produce a stable supply of energy 24 hours a day.

2 The stress tests involved 'a comprehensive evaluation of the safety of an existing nuclear power station'. The tests used computer analysis to check whether a nuclear reactor could withstand a worse-case scenario such as the loss of external power or a natural disaster. There were two stages of the stress test evaluation. The first was conducted after a reactor had passed its routine maintenance inspection. In this stage, the reactors were assessed to see whether they could withstand a tsunami or earthquake that was larger than its design basis. The second stage was a comprehensive safety evaluation of a currently operating reactor to determine whether it should be allowed to continue to operate or ought to be shut down.

3 The demonstrations opposing reactor restarts were organized by the Metropolitan Coalition Against Nukes. They began in April 2012 with approximately 300 people but swelled to 200,000 by June 29 (according to organizers; police estimated that there were less than 20,000 present) (Anzai 2012: 31–2, 217).

4 The Monju fast breeder reactor commenced operations in 1995 but it was shut down a little over three months later due to a sodium leak. It has not been operated since. However, even while in shutdown, it has to be kept warm so that the sodium does not harden. This costs more than 50 million yen per day (Asahi Shimbun Keizaibu 2013: iii).

5 Edano has testified that 'our hands are tied unless Aomori agrees. If we cannot unload the high level nuclear waste in Aomori then we will have an international incident on our hands' (Ōta 2015: 104–6).

6 Former Tokyo University Chancellor and former Education Minister Arima Akito was president; former chair of Nippon Steel and chair of the Japan Atomic Industrial Forum Imai Takashi was the chairperson; former administrative vice-minister of METI and non-executive director of Hitachi Mochizuki Harafumi was deputy chair.

7 Controversy over the phrase 'assuring national security', which was said to 'officially recognize the potential to produce nuclear weapons', led to a supplementary resolution, which inserted the sentence, 'this does not overturn the three non-nuclear principles or non-proliferation' (*Asahi Shimbun*, July 14, 2012, morning edn).

8 The bill to amend the *Act on the Regulation of Nuclear Source Material, Nuclear Fuel Material and Reactors* included the introduction of a 'backfit system' that would suspend operations at reactors that did not meet the new safety standards. It also added the principle that 40 years was the operating life of a nuclear reactor. However, it also added an exception that, with the permission of the NRA, a 20-year extension could be granted in exceptional circumstances (Shiozaki 2012; *Asahi Shimbun*, February 24, 2012, morning edn; April 11, morning edn; April 12, morning edn; May 30, morning edn; June 14, morning edn; June 15, morning edn; July 14, morning edn).

9 There is a risk that if there is an active fault beneath a nuclear facility, then the facility or other equipment might be damaged not by tremors but by a crack opening up in the ground. In principle, the inspectors designated faults as active where the possibility that they had moved in the past 120–130,000 years could not be excluded.

10 See also *Nihon Keizai Shimbun*, April 11, 2014, digital edn; *Asahi Shimbun*, February 20, 2014, morning edn; February 26, morning edn; April 12, morning edn.

11 Tanaka explained on a number of occasions that 'even if a reactor meets the new safety standards, an accident might still occur' and that 'the authority does not determine whether or not a reactor should be restarted' (*Asahi Shimbun*, August 3, 2015, morning edn).

12 On March 28, 2017, the Osaka High Court ruled in favor of KEPCO's appeal pertaining to a temporary restraining order and overturned the Ōtsu District Court's injunction to suspend the operation of reactors No. 3 and No. 4 at the Takahama Nuclear Power Station. KEPCO restarted its Takahama No. 4 reactor on May 17 and No. 3 reactor on June 6.

13 In the election, Yoneyama Ryūichi, who promised to take a tough stance against restarting the nuclear plant, was elected governor.
14 However, Mitazono has changed his position and permitted Kyūden to restart them.

Chapter 4

1 See Kubo (2015, 2016) for previous research of this project. This chapter largely refers to those analyses.
2 In this chapter, each party is abbreviated as follows, unless noted: Democratic Party of Japan: DPJ; Japanese Communist Party: JCP; Japan Restoration Party: JRP; Liberal Democratic Party: LDP; New Komeito Party: Komeito; New Party Daichi: NPD; New Party Nippon: NPN; New Renaissance Party: NRP; People's New Party: PNP; Social Democratic Party: SDP; Sunrise Party: SP; Tomorrow Party of Japan: TPJ; Your Party: YP.
3 For Japanese politics under the DPJ government, see Kushida and Lipscy (eds) (2013).
4 The number of districts that fielded candidates for each party was as follows (in descending order): JCP, 299; LDP, 288; DPJ, 264; JRP, 151; TPJ, 111; YP, 65; SDP, 23; Komeito, 9; NPD, 7; PNP, 2 (Ministry of Internal Affairs and Communications, 'December 16, 2012 General election of the members of the House of Representatives – national review of the Supreme Court judge flash report of results'). In the districts with low numbers of candidates, it is reported that they voted on candidates' personality or policies rather than political parties (Natori 2014).
5 See Pekkanen et al. (2013) for information on the 2012 general election and Pekkanen et al. (2016) for the 2014 general election.
6 At the party leader debate held before the announcement of the election and hosted by the Japan National Press Club, leaders were asked, 'please list in order the three parties which you believe have similar policies and concepts (with your own party)'. Joining the parties that each leader ranked highest in response to this question makes these four blocs (Japan National Press Club website).
7 The prime minister's power in the parliamentary Cabinet system is influenced by the power base within the ruling party (Takayasu 2009). Not limited to the ruling party, in order to win interparty competition, political parties must be unified through competition within the party (Mori 1997, 2001).
8 The following description originally comes from Kubo (2016), which referenced joint press conferences and debates at the Japan National Press Club, in addition to party websites and newspaper reports. For party leadership elections in Japan, see Uekami (2008, 2010, 2011, 2015).
9 The DPJ Central Committee for Leadership Election Administration, 'Public notice of results of the leadership election candidate announcements', August 27, 2011, political views appended.
10 According to reports, 32 top representatives out of 47 prefectural chapters answered that the three-party agreement 'should continue' (*Asahi Shimbun*, September 14, 2012, morning edn).

11 The proportions of Japanese characters allocated to government and party management in their political views, excluding candidate backgrounds etc., were Noda, 3.6% out of 3,118 characters; Akamatsu, 14.9% of 1,856; Haraguchi, 20.1% of 2,305; and Kano, 24.4% of 2,246. Calculated based on the 2012 DPJ leadership election report (*Minshu*, September 12, 2012, extra issue).

12 However, it is thought to have been possible to criticize the DPJ for not including the consumption tax rate increase in either of the manifestos of the 2009 general election or the 2010 upper house election.

13 For example, the proportion of time spent addressing these issues at the campaign speech meeting was Abe, 42.7% out of 14 minutes, 30 seconds; Ishiba, 32.9% of 15 minutes, 9 seconds; Hayashi, 14.7% of 15 minutes, 54 seconds; and Machimura, 8.7% of 15 minutes, 31 seconds. Ishihara only referred to the issues of diplomacy and security as part of his reason for running. Measurements are based on the video of the 2012 LDP presidential candidate opinion speech meeting (September 14, 2012).

14 This is a part of 'the UTokyo-Asahi Survey (UTAS) conducted by Masaki Taniguchi of the Graduate Schools for Law and Politics, the University of Tokyo and the *Asahi Shimbun*'. For detailed data, see Kubo (2016).

15 The question was worded as follows. Regarding the ratio of nuclear power, two opinions were provided: A 'the percentage of nuclear power generation in overall electric power should be 0% by the 2030s', and B 'Nuclear power generation should be kept as a power source even after the 2030s', and the answers 'close to A' and 'closer to A' were counted as 'zero nuclear power' and the answers 'close to B' and 'closer to B' were counted as 'accepting nuclear power'. On restarting nuclear plants, the opinion provided was 'it is inevitable that we must resume operations in nuclear power plants that have been stopped for routine inspection', and the answers 'I agree' or 'somewhat agree' were counted as 'accepting the restart of nuclear plants' and 'disagree' and 'somewhat disagree' were counted as 'opposed to restarting nuclear plants'. Note that some of the data are based on small numbers, such as 'National' for the PNP.

16 Originally, the JRP summarized the common commitments with the YP on November 15, 2012, and stated, 'withdrawal from nuclear power'. However, two days later, the document, which was agreed between the JRP leader, Hashimoto Tōru, and the SP leader, Ishihara Shintarō, limited the expression to 'rule building' in relation to nuclear power plants. This is due to the reluctance of the SP towards transitioning away from nuclear energy, in contrast with the JRP's active stance. Although they held many talks on the issue of commitment wording, in the end the former JRP took the lead and settled on the expression 'to fade out' from nuclear energy (*Asahi Shimbun*, November 18, 2012, morning edn; December 2, 2012, morning edn).

17 'Each party questioned on public commitments and power to achieve', *Yomiuri Shimbun*; 'The responsibility of political parties: we do not need "fanatical politics"', *Asahi Shimbun*; 'Compete under the flag which will show the course this country should take', *Nihon Keizai Shimbun*; 'Look directly at the crisis and compete for direction', *Sankei Shimbun*; and 'Compete to rebuild this country: we cannot afford wrong choices', *Mainichi Shimbun* (all from November 17, 2012, morning editions).

18 The Committee on Fundamental National Policies, the House of Representatives (181th Diet, November 14, 2011).
19 The following analysis is based on Kubo (2015, 2016).
20 Campaign promises may be referred to in different ways, such as manifesto (DPJ), priority policy (LDP) etc. This chapter collectively refers to them as 'manifestos'. In cases where both the digest version and the detailed version were available, we referred to the digest version, as it is thought to be more accessible to voters (the LDP, YP, JCP and SDP). Data were obtained from each party's website and, if unavailable, from the Keiō University Graduate School of Media and Governance, Sone Yasunori Laboratory Manifesto Project. The following description is based on the headlines list created by Kubo (2015).
21 References used to analyze the content of the campaign promises in Japan were Tsutsumi and Uekami (2011) and Uekami and Tsutsumi (2015) for the DPJ manifestos, and Shinada (1998, 2006), Tsutsumi (1998, 2002) and Tsutsumi and Uekami (2007) for databases in relation to candidates' individual pledges.
22 See note 16.
23 The remarks were obtained from the Japan National Press Club and the Nico Nico Douga websites. For more details, see Kubo (2015, 2016).
24 Indeed, the newspapers on the day following the press club debate also compared the attitudes of the party leaders on these issues.
25 Based on whether issues are mentioned by multiple parties in a bloc or whether one party leader refers to it at both of the two debates.
26 On December 15, the final day of election campaigning, the TPJ leader, Kada Yukiko, appealed, 'nuclear power, consumption tax, the TPP (Trans-Pacific Partnership). Important issues are being hidden from discussion' (*Mainichi Shimbun*, December 16, 2012).

Chapter 5

1 This style was not shared by the antinuclear movement as a whole. Apart from the No Nukes Demonstrations that were organized primarily by young people, there were demonstrations called by antinuclear groups that had existed prior to the Fukushima accident and Energy Shift Parades held by environmental non-governmental organizations. These three types of movement occurred in parallel. Each tended to attract different participants in terms of gender, age and degree of social movement experience (Hirabayashi 2013).
2 While television is an important part of the mass media, Japan's major newsprint media and television stations are owned by the same large conglomerates. We can therefore assume that the stance each conglomerate adopted in its flagship newspaper is representative of the stance adopted by its affiliated television stations.
3 The *Asahi Shimbun* deliberately adopts this simple format for asking the question. By using the same wording in each poll, it is possible to capture the way people's ideas about nuclear power change over time.
4 For each statement, respondents were asked to select from one of five responses: 'Agree', 'Somewhat agree', 'Somewhat disagree', 'Disagree'

and 'Unsure'. These figures are aggregated from the favorable responses 'Agree' and 'Somewhat agree'.

5 In this survey respondents who said that they supported 'no particular party' were asked to name which party they would support if they were forced to choose. It should be noted that these results include these lukewarm supporters. Respondents were asked to choose from among the principal parties that contested the October 2012 and December 2014 elections.

In order to measure their opinions to nuclear power, respondents were asked to choose from the following five responses: 'Agree', 'Somewhat agree', 'Somewhat disagree', 'Disagree' and 'Unsure'. The figures for two statements, 'Restarting nuclear reactors needs a popular mandate' and 'I am sympathetic to the antinuclear power movement', include the favorable responses 'Agree' and 'Somewhat agree'. The figures for the statement 'It is inevitable to rely on nuclear power' include the negative responses 'Somewhat disagree' and 'Disagree'.

6 While these differences suggest the influence of political parties and newspapers on the respondents, they also reflect the fact that people choose to support political parties and read newspapers that are close to their own way of thinking. Therefore, caution is required when interpreting a causative relationship between the two.

Chapter 6

1 TEPCO, 'Genshiryoku songai baishō no goseikyū, oshiharai jōkyō' (Claims and payments for nuclear damage compensation) (as of January 23, 2015). Data on TEPCO's compensation claims are taken from the TEPCO website, <www.tepco.co.jp/>, viewed January 28, 2015. Interim payments that were later included in the final payout are not included.

2 For a comprehensive treatment of the nuclear damage compensation process, see Takahashi Shigeru (2014).

3 The Tōkaimura nuclear accident occurred in September 1999 at a uranium reprocessing facility located in Tōkaimura, Naka District, Ibaraki Prefecture. The facility was owned by the company JCO, formerly known as Japan Nuclear Fuel Conversion Co. The incident received a rating of four on the International Nuclear Event Scale (with seven being the highest possible rating).

4 The then Science and Technology Agency entrusted the establishment of this group to the Japan Atomic Industrial Forum Inc. (established in 1956 as an incorporated body, now a general incorporated association) (April 15, 2011, Dispute Reconciliation Committee for Nuclear Damage Compensation (First Session) appendix). Viewed January 28, 2015, <www.mext.go.jp/b_menu/shingi/chousa/kaihatu/016/>. All Dispute Reconciliation Committee documents cited in this chapter were downloaded from the Dispute Reconciliation Committee's website.

5 However, the Dispute Reconciliation Committee guidelines are not legally binding.

6 On the justice system and the history of the Justice System Reform, see Foote (2007). For further information on the debate on judicial

independence contained in Foote, see also Ramseyer and Rasmusen (2003) and Haley (2007).

7 The reports of the ADR Center and other ADR-related data quoted below were all downloaded from the Nuclear Damage Compensation Dispute Resolution Center website.

8 However, the number of applications for mediation through the ADR Center and the relatively small number of lawsuits that appear in the official statistics might be due to the number of joint claims. It is possible to apply for mediation through the ADR Center as a group. Therefore, the number of applications may be less than the number of people involved in claims. Similarly, because court actions may involve a large number of plaintiffs, the number of people affected by the court process may be much greater than the number of cases alone.

9 This section relies on press releases from TEPCO and websites such as that of the Dispute Reconciliation Committee. In the interests of clarity, I have not cited these individually.

10 TEPCO, 'Enkatsu na baishōkin no oshiharai ni muketa taiō ni tsuite' (Planning for a smooth compensation payment system), December 8, 2011.

11 Questions by Shina Takeshi (Democratic Party of Japan, First District, Iwate Prefecture) (Minutes of the Committee on Judicial Affairs, 177th House of Representatives session, March 30, 2011) and Mori Masako (Liberal Democratic Party, Fukushima Prefectural District) (Minutes of the Committee on Judicial Affairs, 180th House of Councillors session, August 28, 2012).

12 Reply to Mori Masako (Liberal Democratic Party) (Minutes of Committee on Judicial Affairs, 180th House of Councillors session, August 28, 2012).

13 Based on data received from the Tokyo District Court on December 16, 2014 (documents prepared by the general affairs chief). In data released since 2004, the largest number of judges was in 2012. The second-largest number was in 2013 and 2008. The smallest number of judges was in 2004, when there were 336. Until 2008 the Hachiōji branch court is included. Since then, the Tachikawa branch has been included.

14 This survey was carried out by the Social Scientific Research on the Great East Japan Earthquake 'Subcommittee on Politics and Policy'. The purpose of the survey was to understand nuclear damage compensation after the Fukushima nuclear accident. Surveys were sent out between October 2014 and January 2015. The subjects were 31 legal teams that provided support to victims of the nuclear disaster. Twenty-four groups replied (response rate: 77.4%). Of the respondents, the Fukushima Nuclear Accident Victims Legal Team was involved in two class actions, a suit by citizens of Iwaki and one by Fukushima evacuees. As these two actions were conducted separately, the lawyers returned one survey for each case. We treat these separately in this chapter, giving a total of 25 teams. For further details of the survey, see Ōkura (2015).

15 Minutes of the Committee on Economy, Trade and Industry, 187th House of Councillors session, October 16, 2014.

16 'Shihō seido kaikaku shingikai ikensho–21 seiki no Nihon o sasaeru shihō seido' (The Justice System Reform Council: a justice system for twenty-first century Japan). The documents on the Justice System Reform Council cited in this chapter were downloaded from the

Council's webpage, <www.kantei.go.jp/jp/sihouseido/index.html>, February 21, 2015.

17 A Japanese consumer group.

18 Kokka kōmuin no haichi tenkan, saiyō yokusei nado ni kansuru zentai keikaku' (Basic plan to reshuffle the public service and curb recruitment).

19 Minutes of the Committee on Judicial Affairs, 166th House of Councillors session, March 13, 2007.

20 ADR Kentōkai 'Minutes of the third ADR Kentōkai' (April 15, 2002), Appendix 4.1 (May 13, 2002). Viewed 2 May, 2015, <www.kantei.go.jp/jp/singi/sihou/kentoukai/03adr.html>. All ADR Kentōkai-related documents cited in this chapter are sourced from this website.

21 ADR Kentōkai 'Minutes of the third ADR Kentōkai' (April 15, 2002). Viewed 2 May, 2015.

22 Japan Federation of Bar Associations, 'Genshiryoku songai baishō funsō kaiketsu sentā no rippō ka o motomeru ikensho' (Submission calling for legislation to establish a Nuclear Damage Compensation Dispute Resolution Center).

23 Dispute Reconciliation Committee for Nuclear Damage Compensation, 'Minutes of the 12th meeting of the Dispute Committee', July 29, 2011.

Chapter 7

1 Personal communication during interview with a Reconstruction Agency staff member, April 26, 2014.

2 Personal communication during interview with a Reconstruction Agency staff member, April 26, 2014.

3 The Minister of State for Special Missions has carriage for the sorts of financial, economic and technical matters that fall under jurisdiction of the Cabinet Office and is equipped variously in order to do this.

4 Response by Prime Minister Noda Yoshihiko as recorded in the proceedings of the 179th Upper House Plenary Session, No. 11, p. 4, December 7, 2011.

5 At the time of the operation of the Reconstruction Design Council, three political scientists, centered around council chair Iokibe Makoto, held tripartite meetings and instituted council policies and decisions (Mikuriya 2014: 94).

6 Policy speech given by Prime Minister Abe Shinzō to the 183rd Diet, January 28, 2013, p. 8. Viewed April 21, 2017, <www.kantei.go.jp/jp/96_abe/statement2/20130128syosin.html>.

7 Record of press conference by Minister for Reconstruction Nemoto given between 2.20 and 2.32 pm on December 27, 2012, at the Reconstruction Agency press briefing room, p. 1. Viewed April 21, 2017, <www.reconstruction.go.jp/topics/241227.html>.

8 Reconstruction Promotion Committee, 'Towards the creation of a "New Tōhoku" (Recommendations)', April, 18, 2014, p. 2. Viewed April 21, 2017, <www.reconstruction.go.jp/topics/main-cat7/sub-cat7-2/20140425_01_teigen.pdf>.

9 Personal communication during interview with a Reconstruction Agency staff member, April 26, 2014.

10 During the tenure of an administrative vice-minister from MIC, there were two directors-general from MLIT and one from METI.

11 Reconstruction Agency, '(Reference) reconstruction response and related systems, (11 November, 2015)', p. 3. Viewed April 21, 2017, <www.reconstruction.go.jp/topics/main-cat1/sub-cat1-1/20151111_torikumi_seido.pdf>.

12 The 'ministry or origin' breakdown of staff attached to the director-general division was Cabinet Office, 1; MIC, 4; Foreign Affairs, 1; Finance, 2; Education, Culture, Sports, Science and Technology, 1; Health, Labour and Welfare, 2; Agriculture, Forestry and Fisheries, 3; METI, 4; MLIT, 11; and Environment, 1. These are the author's calculations based on information available in Yonemori (ed.) 2015 and on the Reconstruction Agency personnel page. Viewed April 21, 2017 <www.reconstruction.go.jp/topics/main-cat12/jinji/index.html>.

13 'Minutes of 5th Meeting of Reconstruction Council (10 January, 2013).' Viewed April 21, 2017, <www.reconstruction.go.jp/topics/20130314_gijiroku05.pdf>.

14 Iwate Prefecture website. Viewed April 21, 2017, <https://www.pref.iwate.jp/anzen/machizukuri/18197/001810.html>.

15 Miyagi Bureau of the Reconstruction Agency, 'Approaches for accelerating recovery and reconstruction', September 27, 2014. Viewed April 21, 2017, <www. thr.mlit.go.jp/Bumon/B00097/K00360/taiheiyouokijishinn/kasoku1/140927shiryou6>.

16 For the legislative processes before and after the Great East Japan Earthquake, see Hamamoto (2016).

17 Concerning the collaboration of multiple government agencies, see Bardach (1998) and Itō Masatsugu (2014, 2015). In addition, on the importance of government networks in the context of the official response to and support for reconstruction after earthquakes, see Soga (2016).

Chapter 8

1 In Japan local governments are independent entities that carry out local public administration within a specific region. Japan has two basic tiers of local government: prefectures and municipalities (including cities, towns and villages). Article 1-2 of the Local Autonomy Act defines the role of local government as the 'promotion of residents' welfare' and the development of administrative services. The chief executive (prefectural governor or municipal mayor) is charged with managing and executing local government business (Article 148) and is invested with personal responsibility to carry out these duties faithfully and has the executive authority to do so (Article 138, paragraph 2). The Local Autonomy Act does not provide a clear rationale for local governments giving support to other local governments that are located far away. Chapter 11, section 3, does regulate reciprocal cooperation between local governments, including provisions for the establishment of joint councils and shared facilities, administrative delegation and personnel dispatches. However, the law assumes that this kind of cooperation will take place between neighboring governments in response to regional issues.

2 Many of the dispatches from local governments outside the disaster zone after the Great East Japan Earthquake were considered legally either as short-term visits or as personnel dispatches under Article 252-17 of the

Local Autonomy Act. Although the latter stipulates that the chief executive of a local government can make a 'request for the dispatch of personnel' to another chief executive, nothing in the legislation compels the recipient of such a request to comply.

3 According to Organisation for Economic Co-operation and Development (OECD) data, public servants make up an average of 21% of the workforce in OECD countries. In Japan they make up just 8%: 'The size of public sector employment varies significantly among OECD countries. Nordic countries [such] as Denmark, Norway and Sweden report high public sector employment levels reaching near or over 30% of total employment. On the other hand, OECD countries from the Asian and Latin American regions rely less on public sector employees. Only around 8% of Japan's total employment is made up of public sector employment, while Chile and Mexico count just over 10%' (OECD 2015: 84).

4 By a 'system of interdependence' I refer to the dependence of central and local governments on one another in terms of the resources necessary for public administration, including personnel and finance.

5 Meiji University's Research Centre for Crisis Management conducted a survey of designated cities, core cities, special cities and special wards (it received 80 responses from the 123 municipalities contacted). It found that a total of more than 600 municipalities 'provided some form of assistance'. Of these, only 15 (2.5%) were signatories to sister-city or friendship agreements, while 90 (15%) had disaster management-related agreements. This means that only 17.5% of municipalities that provided some form of disaster relief did so based on an existing agreement. Ninety-four municipalities (15.7%) provided relief at the request of the JACM and 95 (15.8%) were responding to a request from a regional network. These figures show that the majority of dispatches occurred via arrangements made after the disaster, rather than depending on pre-existing agreements (Research Centre for Crisis and Contingency Management 2012). The majority of respondents to this survey indicated 'other' (260; 46%). More than half of these received a request from the MHLW (from nurses or other health professionals), their prefectural government or from a local mayors associations.

6 Although it sometimes took some time to match the requests for support with a local government body that was able to provide it, once established these relationships often formed the basis for ongoing one-to-one support. For example, the continuing deployment of personnel by Ibaraki, in Osaka, to Ōfunato in Iwate, started after Ibaraki was matched with Ōfunato's request. Public hearing on personnel, Ibaraki, October 8, 2014.

7 Hearing, Neyagawa general affairs division chief, October 8, 2014. In the morning of March 13, a request from the Japan Sewage Works Association also arrived at the municipal sewerage department.

8 Interview with the deputy mayor of Miyako, December 11, 2011 and December 8, 2012. Documents from Iwate's municipal affairs department dated December 3, 2012, and interview with Iwate prefectural officials, December 8, 2012.

9 Interview with the deputy mayor of Rikuzentakata, December 9, 2012.

10 Immediately after the disaster, Nagoya sent an advanced party to the Iwate coast on three occasions to verify the situation. On April 1 a survey

team led by Nagoya's deputy mayor was sent to Rikuzentakata. There they discovered that the urban area had been destroyed, more than one hundred local government employees (including volunteer firefighters) had been killed or were missing, and the local government was struggling to function. Upon receiving a request from Iwate Prefecture and Rikuzentakata, Nagoya decided on April 7 to focus its entire relief effort on supporting Rikuzentakata ('Ōen shimasu!! Tōhoku!! Rikuzentakata!' (We support you!! Tōhoku!! Rikuzentakata!!), leaflet, Nagoya, 2014). Nagoya called its administrative assistance program for Rikuzentakata 'all-of-government assistance'.

11 Interview with the deputy mayor of Kamaishi, December 9, 2012.
12 Interviews in Takatsuki and Ibaraki.
13 Interviews in Neyagawa and Takatsuki.
14 In Takatsuki dispatch workers who have returned from the disaster zone conduct training sessions for other municipal staff. Interview with head of welfare and health, Takatsuki, October 7, 2014.

Chapter 9

1 In addition to this fundamental provision, there are bodies established for prefectural governments and municipalities to jointly deal with specific issues.
2 There is also a national system of income tax separate from local income tax.
3 Wealthy local administrations do not receive the local allocation tax.
4 While evacuee support also comes from private organizations, this discussion deals with publicly funded support.
5 The comprehensive offering of public housing and private rental properties as temporary accommodation occurred for the first time after the Great East Japan Earthquake (see below).
6 In the case of the Great East Japan Earthquake, the remaining 10% to 50% of prefectural disaster relief finances also came from the central Treasury as disaster reconstruction special grants of local allocation tax.
7 For example, while elementary and middle schooling, which covers both childcare and compulsory schooling, is generally dependent on registration, there are occasions when school entry is possible, with some procedures, in authorities outside that in which one is registered.
8 For example, in the case of water reticulation, a user-pay service is possible even without resident registration.
9 Other services for which it is necessary to have evidence of being a disaster victim include the provision of funds to assist in returning to routine life and relief donation monies, and exemptions from tax, insurance and utility charges. On the other hand, the disaster victim aid necessary to open evacuation centers and to help place people in these centers is a matter of urgency. Furthermore, this is not a service that ordinary residents in normal circumstances require. The necessity to provide disaster victim evidence to gain entry was therefore waived.
10 At the peak of June 2012, evacuee numbers stood at 346,987 people (Reconstruction Agency 2012). However, because evacuee numbers published by the authorities were gathered and collated by evacuees

themselves reporting to government agencies, it has been pointed out that the real numbers were higher than the official figures published by government agencies (Nishikido and Harada 2013: 3–4). It was estimated by the government's Emergency Disaster Response Headquarters that the most number of evacuees was on the third day after the disaster when approximately 470,000 people were without a home.

11 Even at the time of the 1995 Great Hanshin–Awaji Earthquake, and during the 2000 Miyakejima volcanic explosions, there were victims who evacuated outside the local communities and prefectural jurisdictions (Tanami 2010, 2011).

12 Most voluntary evacuees came from Fukushima, although a number came from outside that prefecture. Some people, including Yamakawa (2013), refer to them as 'evacuees from outside the evacuation order zone', not 'voluntary evacuees' because they did not voluntarily evacuate, but were forced out by the impact of the nuclear accident.

13 The evacuee peak in Yamagata Prefecture was on January 26, 2012, when there were 13,797 evacuees (Yamagata Prefecture 2012).

14 Until June 2014, of all local administrations in Japan, Yamagata City received the highest number of evacuees, while Yonezawa City received the highest number on a per capita population basis (Wada 2015: 193).

15 Information relating to Yamagata Prefecture provided in the discussion that follows is based on documents from the July 18, 2013, Yamagata Prefecture Recovery and Evacuee Support Office Hearing. Information on Yamagata City is based on documents from the July 18, 2013, Yamagata City Evacuee Support Center Hearing and the March 25, 2014, Yamagata City Disaster Prevention Measures Section Hearing. Information on Yonezawa City is based on documents from the March 27, 2014, Yonezawa City Risk Management Office Hearing; the March 15, 2013, report entitled *Response to the Great East Japan Earthquake*, published by Yonezawa City Disaster Control Headquarters; and the Yonezawa City 2014 'On disaster relief for evacuees from the Great East Japan Earthquake'. Other sources referenced are noted.

16 The number of people subject to the evacuation order was approximately 78,200 (Yoshida and Harada 2012: 368).

17 In Yonezawa City on March 12, prior to the evacuation center operation, City Hall meeting rooms were made available to evacuees.

18 These were facilities at which evacuees were provided with a range of support, such as access to information and consultation services and the opportunity to have contact and engage with local residents. These facilities were established on July 1 in Yamagata and on June 6 in Yonezawa.

19 The voluntary evacuees given access to designated temporary accommodation came from Fukushima Prefecture. Voluntary evacuees from other prefectures were not eligible (Yamagata Prefecture 2011b).

20 This law was implemented to take account of the fact that the nuclear accident forced many people to evacuate from their homes and therefore from the local authority areas in which they had lived and to ensure that those who had been resident in the 13 designated Fukushima local authorities would be eligible to receive standard local government services from the municipalities that received them as evacuees.

21 The remaining 11 people (0.7%) came from hospitals and social welfare care facilities.
22 While the majority came from municipalities in Fukushima Prefecture, there were a small number also from municipalities in Miyagi and Iwate prefectures.
23 The disaster victim evacuee support unit of the Yamagata Prefecture-wide Support Measures Headquarters conducted these surveys from the 2011 administrative year. The survey format between the 2012 and 2014 administrative years was as follows: 2012, conducted between mid- and late October, distributed to 3,855 households with 1,275 responses (33.1%); 2013, conducted between late September and early October, distributed to 2,420 households with 850 response (35.1%); 2014, conducted between late September and early October, distributed to 1,706 households with 551 responses (32.3%).
24 Those without a Disaster Evidence Certificate needed one of a number of possible forms of evidence, such as a driver's license with a Fukushima address, to show that they had been residents in the disaster zone (Yamagata Prefecture 2011b, 2011c: 201).
25 The financial burden assumed by Fukushima Prefecture was to later be accommodated through disaster reconstruction special grants of local allocation tax by the central authorities.
26 This summary draws on the ideas of Takeuchi (2015).
27 According to Itō (2006: 30–5), 'cross-referencing' is a key concept that explains local government policy reform.
28 It might be noted, however, that the main reason for return to Fukushima was not diminished concerns about radiation contamination but economic burden (Wada 2015).
29 Many responses to the survey conducted between December 2011 and January 2012 by Yamanaka, Mori and Tanami concerning disaster victim evacuee aid provided by local authorities, with the exception of authorities in Iwate, Miyagi and Fukushima prefectures, indicated that the authorities surveyed gave aid to voluntary evacuees (Tanami 2012).
30 The point also needs to be made that this decision was not merely the result of central government finance concerns, but also due to the fact that Fukushima wanted the voluntary evacuees to return.

Chapter 10

1 The survey research team was organized by Professor Takaharu Kohara. The survey was addressed to earthquake disaster-affected municipalities and prefectural governments. Distribution and return of survey documentation was conducted by postal mail in accordance with postal survey legislation protocols. Interested respondents could also access an online version. Of the 1,325 subjects contacted to participate, 1,018 provided valid replies, giving a 76.8% valid response rate. The local areas surveyed were as follows: Hirono Town, Kuji City, Noda Town, Fudai Village, Tanobata Village, Iwaizumi Town, Miyako City, Yamada Town, Ōtsuchi Town, Kamaishi City, Ōfunato City and Rikuzentakata City in Iwate Prefecture; Kesennuma City, Minamisanriku Town, Onagawa Town, Ishinomaki City, Higashi Matsushima City, Rifu Town,

Matsushima Town, Shiogama City, Shichigahama Town, Tagajō City, Sendai City, Natori City, Iwanuma City, Watari Town and Yamamoto Town in Miyagi Prefecture; and Shinchi Town, Sōma City, Minami Sōma City, Namie Town, Futaba Town, Ōkuma Town, Tomioka Town, Naraha Town, Hirono Town and Iwaki City in Fukushima Prefecture.

2 In this survey, we asked whether the content of inquiries from residents was assumed, as compared with an ordinary inquiry. The ratio of each answer is as follows: 'All unexpected', 3.8%; 'Mostly unexpected', 20.0%; 'Partly unexpected', 59.7%; 'All expected', 15.8%; 'NA', 0.8%.

3 In this survey, we asked why they could not respond adequately to the inquiries (MA).

4 In this survey, we asked why they could not respond adequately to the inquiries (MA).

5 In this survey, we asked what kind of personnel is necessary in the event that work cannot be carried out sufficiently in the event of a disaster. The ratio of each answer is as follows. In the 37 municipalities: 'Public servants who were equipped with the necessary skills and knowledge', 56.2%; 'Private volunteers and non-profit organizations who were equipped with the necessary skills and knowledge', 4.1%; 'Any individual who had the skills and knowledge', 32.3%; 'Public servants who were not equipped with the necessary skills and knowledge', 4.5%; 'Any personnel', 1.7%; 'NA', 1.1%. In the three prefectures: 'Public servants who were equipped with the necessary skills and knowledge', 60.7%; 'Private volunteers and non-profit organizations who were equipped with the necessary skills and knowledge', 2.2%; 'Any individual who had the skills and knowledge', 29.7%; 'Public servants who were not equipped with the necessary skills and knowledge', 6.4%; 'Any personnel', 0.3%; 'NA', 0.6%.

Chapter 11

1 This spill occurred when the tanker sank in the sea off Shimane Prefecture on January 2, 1997. When the vessel broke up, the bow section beached on the coast near Mikuni Town (now part of Sakai City) in Fukui Prefecture. Large quantities of crude oil washed up all along the Sea of Japan coast, seriously impacting the natural environment and the lives of residents in the affected areas. See Fukui Prefecture website (www.pref.fukui.lg.jp/doc/kikitaisaku/kakonosaigai/totalpage.html), viewed June 27, 2016.

2 This was a Level 7 earthquake (measured on the Japanese scale of seismic intensity). In terms of direct impact, approximately 100,000 were people evacuated and around 12,000 homes damage. However, considerable additional economic costs were incurred, particularly to tourism as a result of issues such as misleading rumors and the suspension of the Tokyo to Niigata Jōetsu Shinkansen service. See Office of the Cabinet website (www.bousai.go.jp/kaigirep/houkokusho/hukkousesaku/saigaitaiou/output_html_1/case200404.html), viewed June 27, 2016.

3 This survey examined overall trends and issues in Great East Japan Earthquake disaster volunteering as at the end of April 2014. The survey was conducted between May 16 and 23, 2014, with sampling undertaken by online research survey company Net Monitor targeting women and men aged between 18 and 69 across Japan. The sample number was 31,210

from a parent set based on the 2010 national census results for women and men between 18 and 69 in compliance with regional breakdown and compositional ratio of age group levels. The total number of subjects was 118,694, giving a valid response rate of 39.5%.

4 Figures based on the survey referred to above. The percentage breakdown of residents from outside the Tōhoku region was Hokkaidō, 2.9%; Kantō (excluding Tokyo Metropolis and Saitama, Chiba and Kanagawa prefectures), 5.7%; Tokyo Metropolis and Saitama, Chiba and Kanagawa prefectures, 34.9%; Hokuriku, 4.3%; Tōkai, 9.9%; Keihanshin, 9.8%; Chūgoku, 3.9%; Shikoku, 1%; and Kyūshū and Okinawa, 5.8%.

5 The number of volunteers given here should be regarded as an estimate rather than a formally measured statistic (see, for example, Okamoto Masahiro 2013a).

6 The Fundamentals of Disaster Prevention Plan was based on the stipulations of Article 34, Clause 1, of the *Disaster Countermeasures Basic Act*, which clarifies the measures for disaster prevention structures of national and regional public authorities, and the responsibilities of those bodies, with a view to protecting people's lives and property in time of disaster. It is Japan's highest-level plan for disaster prevention and countermeasures.

7 This data is from a series of interview surveys conducted with the Japan National Council of Social Welfare, an organization established to facilitate contact between and provide support for each of the Social Welfare Councils around Japan.

8 Regional Disaster Prevention Plans were based on the stipulations of Article 42 of the *Disaster Countermeasures Basic Act*. It is the disaster prevention and countermeasures plan that operates in each region and is formulated by prefectural and/or metropolitan authorities in conjunction with municipalities.

9 Rather than the participation of the Social Welfare Councils of each municipality being requested by the national body or the central authorities, it was the local councils themselves that sought such involvement based on their experience both in supporting disaster prevention activity during non-crisis times and their experiences on the ground (interview survey of the Japan National Council of Social Welfare).

10 Unless otherwise noted, information on this topic is based on information provided in Japan National Council of Social Welfare (2012).

11 In telephone survey interviews conducted with the Tanohata Village and Fudai Village Social Welfare Councils, interviewees indicated that damage incurred was relatively slight and so no Disaster VC was formed.

12 This information came through internal documents and survey interviews conducted with Minamisōma City Social Welfare Council.

13 Golden Week is the name given to what is often, with an accumulation of weekends and designated public holidays, a week-long vacation period from late April to early May.

14 The criticisms that came from organizations such as NPOs are discussed in detail in Okamoto Masahiro (2013a). There was also considerable counter-criticism of those who made these kinds of critiques. For example, Arata (2011) pointed out that in circumstances in which the infrastructure was desperately inadequate, there was the danger that

an influx of volunteers would merely create chaos. Nihei (2012) argued that, given the high risks involved in some activities in which volunteers may have been involved, it was difficult to merely naively praise those who offered to help. From the Disaster VC perspective, reasons given for limits imposed included the extended time taken to search for missing persons, difficulty offering secure accommodation, fears relating to gasoline costs involved when volunteers used private cars to travel to the disaster zone, resident concerns relating to volunteers, limits to the capacity of Disaster VCs to needs-match volunteers, and Disaster VC onsite confusion caused by staff shortages (Japan National Council of Social Welfare 2013).

15 With the emphasis on a shift from recovery to reconstruction, there has been a spontaneous dissolution of Disaster VCs and these sorts of services are now offered through newly established organizational structures.

16 This information came from survey interviews with a number of Social Welfare Councils.

17 Japanese Social Welfare Councils had divided the country into eight regional divisions, referred to as 'regional blocks', and operated combined training workshops etc., which involved the Social Welfare Communities attached to the various blocks. When the Tōhoku disaster struck, these regional blocks worked as a single unit to offer relief activities.

18 Established under the Social Welfare Law, the Community Chest of Japan is a corporate body that raises funds to encourage private sector involvement in social welfare. Each prefectural and municipal jurisdiction has a local Community Chest of Japan fundraising branch coordinated by the national organization.

19 Information given during interview surveys with Yamamoto Town Social Welfare Council and Shichigahama Town Social Welfare Council.

20 Keidanren is the country's peak business body, with membership of 1,340 companies. It comprises 109 different industry groups, fundamentally divided into manufacturing and service, and 47 regional economy groups. It exercises strong influence in the political and economic spheres. The 1% Club is operated by Keidanren with the objective of independently committing at least 1% of member profits and disposable income to social service activities. See Keidanren webpage (www.keidanren.or.jp/profile/pro001.html or www.keidanren.or.jp/1p-club/report/).

21 By September 2011 the number of staff who had availed themselves of leave entitlements through this system had exceeded 4,000. This compares with the 2,761 staff who took leave in 2010 (Japan Business Federation Corporate Social Responsibility Committee 1% Club 2012).

22 This survey examined trends in day-to-day activities by Great East Japan Earthquake disaster volunteers as at the end of April 2014. The survey was conducted between May 23 and 27, 2014, with sampling undertaken by online research survey company Net Monitor targeting women and men between 18 and 69 across Japan. The sample number was 1,855, with participants grouped according to gender and the prefecture in which they had been active. Other categories included early term (March 12 to May 31, 2011), mid-term (June I, 2011, to February 28, 2012) and late term (March 1, 2012, to April 30, 2014). A 100-point allocation was used. As with the previous survey, the total number of subjects was 118,694.

23 In the breakdown of the 54.4% who volunteered as part of a group plan, 27% joined workplace groups (with 10.5% of these in occupational groups), 12.5% were part of a volunteer group, 5.7% came on Social Welfare Council plans and 5.0% on government administration plans, and 4.2% through travel companies.
24 Information given in telephone survey interview with the Sumita Town Social Welfare Council.
25 Information given in survey interviews with Morioka City and Morioka City Social Welfare Council.
26 Because a variety of accommodation was assumed for the first-time volunteers, a range of survey answer possibilities was provided. Excluding the 25.4% who made day trips, the accommodation breakdown was 25% for Japanese-style inns, boardinghouses and hotels, 19% for tents provided especially for volunteers, 16.9% for sleeping in a car or bring-your-own tent, 14.8% for staying with friends and relatives, 6.8% for evacuation centers, 5.0% for camp sites and 3.3% for other.
27 Among these were groups that undertook specialist training in order to be able to perform tasks not generally regarded as ordinary volunteer activity, including the operation of the heavy machinery necessary to remove large chunks of rubble.
28 Japan YMCA 2012 and telephone survey interview with Japan YMCA.
29 Out of 157 members or affiliates of the Japan NGO Center for International Cooperation, 59 conducted operations during the Great East Japan Earthquake. Of these, 60% or 34 organizations were first-time participants in a domestic disaster (Japan NGO Center for International Cooperation 2012).
30 Different schemes operated depending on whether groups came at government request or through the independent initiative of regional groups, and there was considerable difference in the degree of support that these groups received. Further research is required into the line that divided business and volunteers.
31 Japan Beautician and Barber Volunteer Project (RIBIBORA) homepage (www.fukuribi.jp/ribivora/).
32 'Double loan' refers to the fact that, given that debt had been incurred for a housing or business loan before the disaster, taking on an additional housing loan in the recovery and reconstruction process would have a significant financial impact on a victim. Without this second loan, however, victims would be trapped in their current immediate post-disaster circumstances and unable to return to their former ways of life.
33 In addition to survey interviews with multiple Social Welfare Councils, Japan National Council of Social Welfare 2012 and Office of the Cabinet 2012. As of May 2012, 344 groups had registered with the Ishinomaki Disaster Reconstruction Support Committee (Mirai Support Ishinomaki 2013).
34 Information gained in survey interviews with multiple Social Welfare Councils.
35 Information gained from survey interviews with staff of Yamada Town Social Welfare Council and from the internal council documents.
36 Information gained through survey interviews with a wide number of Social Welfare Councils.

37 From the same survey questionnaire on disaster and government systems from Chapter 10. However, it should be noted that since that survey focused on intergovernmental relations, it omits differences in levels of contact and opinion between other actors.

Chapter 12

1 There are some cases when illness induced by the stress of evacuation led to people's deaths, and these are included among deaths relating to the earthquake disaster.

2 Three districts (the Yamakiya area of Kawamata town, Namie town and Iitate village), where high exposure was estimated to have occurred, were targeted for conducting the survey prior to the remaining areas, and the response rate in these districts was high – 58.3% as of August 2013 (Prefectural Oversight Committee for the Fukushima Health Management Survey 2013).

3 In some cases targeted residents had a medical examination in later years.

4 Some researchers criticize this assumption. Tsuda et al. (2016) pointed out that in Chernobyl the minimum empirical induction time for thyroid cancer was 2.5 years for adults and one year for children.

5 While the estimation is based on the averaged data over 10 years from 2001 to 2010, the incidence of thyroid carcinoma has tended to increase recently. Therefore, the Center for Cancer Control and Information Service (2014) pointed out the figure of 2.1 may be an underestimation.

6 It has been pointed out that although the Center for Cancer Control and Information Services (2014) estimated the incidence of thyroid carcinoma using data from 2001 to 2010, the disease rate tends to increase over the long term, so the averaged incidence over ten years could be smaller than the current number.

7 In March 2016 the Prefectural Oversight Committee for the Fukushima Health Management Survey, which governs the thyroid screening section, announced an 'interim report' on the basis of test results gathered by December 2015, in which the number of firm diagnoses of thyroid carcinoma was slightly increased. The basic assessment remained much the same as that made by the thyroid screening section. See the Prefectural Oversight Committee for the Fukushima Health Management Survey (2016a). Also note that the 23rd meeting of the Oversight Committee held in July 2016 reported that the number of patients with 'malignant or possibly malignant thyroid carcinoma' had reached 116 in total. See the Prefectural Oversight Committee for the Fukushima Health Management Survey (2016c).

8 With regard to this point, see note 4.

9 The average yearly rate of population loss in the Fukushima Prefecture from 2005 to 2010 was 7,497 people, as estimated in the Report on Internal Migration in Japan Delivered from the Basic Resident Registration (yearly reports) by the Ministry of Internal Affairs and Communications.

10 In principle, as compensation for mental damage, 100,000 yen has been paid monthly per person. This includes compensation for the increase in living expenses caused by the evacuation.

11 Indirect damage was caused when there was no alternative to dealing with a person who had suffered direct damage from the nuclear power station disaster.
12 See Chapter 13 for damage related to home securement.
13 See Chapter 13 for details on decontamination.
14 The government share of funding for the NDF was 7 billion yen, TEPCO's was 2.3 billion yen, and 2.3 billion yen was provided by other nuclear operators.
15 From 2014 onwards, it is planned that about 35 billion yen will be granted every year by the Power Sources Development Account.

Chapter 13

1 For evacuees whose homes are very close to Fukushima Daiichi, in areas where they will have to wait ten years or more to return, the decision to prolong their evacuation is not simply a means of avoiding making a decision, but an important choice in and of itself. These evacuees need to be provided with dedicated public housing where the entire community can continue to live together or a kind of 'home away from home', rather than simply being left in temporary housing.
2 For example, the phrase appears in the Basic Guidelines for Fukushima Reconstruction and Revitalization adopted by Cabinet in July 2012.
3 The history and background of the framework and policy for decontamination in this section is mainly based on MOE (2015a) and Isono (2015a).
4 Airborne dose rates are measured at a height of 1 meter above ground. See MOE (2013). Because the effectiveness of decontamination depends on environmental conditions, the figures given in this report are only a guide.
5 On the cover of MOE (2011) is written the phrase, 'we are fully committed to decontamination under the slogan "no reconstruction without decontamination"'.
6 These figures were compiled based on a survey carried out in each of the affected municipalities in 2013.
7 The number of people who want to return to Iitate is higher than in other municipalities. One reason for this is probably the success of the village's unique community development strategy.
8 Nakanishi (2014) is one example where the effectiveness of decontamination is calculated on the basis of the size of the returnee population.
9 This is a statement in the Fukushima Fukkōkaigi Josen Bunkakai, November 11, 2012. See Satō (2012).
10 According to Interviewee Mr. 'I' of the non-profit organization Tomioka Children Future Network who evacuated from Tomioka to Tokyo, even many evacuees who are living temporarily outside Fukushima (rather than relocating) feel guilt for having left it behind. Interview with Mr. 'I', November 2013.
11 The interpretation of this kind of survey inevitably requires a great deal of care. Depending on what was happening at the time the survey was conducted, the responses may change. Even people who say that they 'plan to return home' may do so not out of a strong desire to return but because they see no other option.

12 From the homepage of the Ōkuma Town website, (www.town.okuma.fukushima.jp/fukkou/?page_id=665). For details of these demands, see Itō Yasushi (2016).
13 Ōkuma's 'First reconstruction plan', released in September 2012, contains detailed plans for supporting those who have decided not to return to Ōkuma. This indicates that the needs of those who wish to relocate have not been entirely neglected. However, the support proposed primarily calls on the national government to encourage compensation and provide information. The biggest problem facing evacuees who want to relocate is finding alternative housing. This issue is not addressed in the plan in any detail, probably due to budgetary constraints.
14 A significant number of villagers oppose returning if radiation levels do not go down sufficiently. See Hasegawa (2012).
15 While most of the affected municipalities have prioritized return, there are exceptions. Idogawa Katsutaka is the former mayor of Futaba Town, where the Fukushima Daiichi Nuclear Power Station is located. He established a temporary municipal office outside Fukushima Prefecture in Kazo, Saitama Prefecture. He did not advocate for an early return to the village and adopted a number of other positions that differed from the other municipalities. Idogawa explains that the hydrogen explosion in reactor No. 1, which irradiated the town's evacuees, had a major impact on him (in Kitamura 2013). However, Idogawa was criticized for establishing a temporary municipal office for Futaba outside Fukushima. Eventually a split developed between him and the rest of the council. In December 2012 the council proposed a motion of no confidence in him as mayor and he resigned in February the following year.
16 For a detailed discussion of this point, see the section 'Comparing decontamination with relocation'.
17 The Dispute Committee is responsible for deciding the specific scope of compensation to be provided under the provisions of the *Act on Compensation for Nuclear Damages*.
18 Iida estimated the market value of the land based on property tax valuations. He also notes that asset values in the Ōkuma and Futaba municipalities, which host the Fukushima Daiichi Nuclear Power Station, include TEPCO's enormous assets. Therefore, their real value would be somewhat lower.
19 The calculations were performed by Iida, but in their dialogue Nakanishi is in almost complete agreement with him.
20 An increase in asset values is certainly one of the benefits of decontamination. It would therefore be possible to think about these increases as a minimum value for calculating the benefits.
21 In order to conduct this comparison we have to assume that the utility created by returning to homes and that created by moving to a new area are the same. In other words, that there is no significant difference between the two locations.
22 In the case of the usual acquisition of land for public use, the current value of a 48-year-old wooden building is considered equivalent to about 50% of its value when it was constructed.
23 This is only a hypothesis. It is highly unlikely that all households would desire to relocate. If the nearly 30,000 households involved tried to find

housing in Fukushima's cities all at the same time, property values would increase dramatically. This would then increase the cost of securing accommodation.

24 However, in this case the cost of decontamination would increase somewhat.

25 For example, the survey of Namie residents who do not wish to return discussed in the section 'Differences over decontamination'.

26 The Dispute Reconciliation Committee assumes an upper limit of 250 m^2 floor area for the new residence. If the actual floor area is larger, the calculation is still carried out using the assumption of 250 m^2. Therefore, if a larger residence is desired then a greater proportion of the cost has to be borne by the purchaser. Here I use the average floor area of 252 m^2 for the calculation. In addition, a number of restrictions apply to cases such as where residences were situated on larger blocks of land in the affected area or the price of land in the relocation site is more expensive. For further information, see Dispute Reconciliation Committee for Nuclear Damage Compensation (2013b).

27 The price of land adjoining major roads is about 0.7 times the market price on average. For calculating the market price of land it is usual to multiply the price of land adjoining major roads by 1/0.7 = 1.43.

Chapter 14

1 For details of the restrictions imposed by the various countries on Japanese agricultural and marine products, see the Ministry of Agriculture, Forestry and Fisheries (2012–16) and the Ministry of Agriculture, Forestry and Fisheries (2017).

2 Certification for marine products was sometimes issued by the Fisheries Agency directly, while at other times the Fisheries Agency endorsed the preliminary certificates issued by prefectural fisheries bureaus. For further information, see the April 22 directive of the agency (Fisheries Agency 2011).

3 At the time, Delta Air Lines had suspended flights between Tokyo's Haneda Airport and Detroit and Los Angeles. Lufthansa had shifted its flights from Narita to Kansai and Chūbu airports, and Alitalia had shifted its flights to and from Narita to Kansai. Air France redirected its Narita–Paris route to Seoul (*Nihon Keizai Shimbun*, March 24, 2011).

4 South Korea's reaction to the decision was particularly strong due to its geographical location as Japan's closest neighbor. Historical grievances may also have been a factor. At the time, rumors were flying in Korea that radioactive rain would fall across the peninsula. On April 7 about 100 elementary schools in the Seoul area were closed temporarily (Azuma 2015: 262–3).

Bibliography

ADR Center (Nuclear Damage Compensation Dispute Resolution Center), 2012, 'Sōkatsu kijun ni kansuru kettei' (The determination regarding general standards), viewed April 4, 2017, <www.mext.go.jp/component/a_menu/science/detail/__icsFiles/afieldfile/2012/12/20/1329129_001.pdf>.

ADR Center, 2015, 'Genshiryoku songai baishō funsō kaiketsu sentā katsudō jōkyō hōkokusho (H26)' (ADR Center activity report), viewed May 14, 2015, <www.mext.go.jp/a_menu/genshi_baisho/jiko_baisho/detail/adr-center.htm>.

Advisory Committee on Nuclear Damage Compensation System (Genshiryoku Songai Baishō no Arikata ni Kansuru Kentōkai), 2008, *Genshiryoku songai baishō seido no arikata ni kansuru kentōkai, Dai 1 ji hōkokusho* (Investigating the nuclear damage compensation system), December 15, p. 21, viewed January 28, 2015, <www.mext.go.jp/b_menu/shingi/chousa/kaihatu/007/gaiyou/__icsFiles/afieldfile/2009/06/29/1279826_1_1.pdf>.

Agency for Natural Resources and Energy, 2014a, 'Summary results of the study on contaminated water management at TEPCO's Fukushima Daiichi Nuclear Power Station', viewed February 24, 2017, <http://dccc-program.jp/files/20140407l1J.pdf>.

Agency for Natural Resources and Energy, 2014b, 'News release 19 June 2014', viewed February 24, 2017, <www.meti.go.jp/press/2014/06/20140619005/20140619005.html>.

Akizuki Kengo, 2001, 'Partnership in controlled decentralization: local governments and the Ministry of Home Affairs', in Muramatsu Michio, Farrukh Iqbal and Ikuo Kume (eds), *Local Government Development in Post-war Japan*, New York: Oxford University Press, pp. 63–84.

Aldrich, Daniel P., 2008, *Site Fights: Divisive Facilities and Civil Society in Japan and the West*, Ithaca, NY: Cornell University Press.

Aldrich, Daniel P., 2014, 'The limits of flexible and adaptive institutions', in J. Kingston (ed.), *Critical Issues in Contemporary Japan*, New York: Routledge, pp. 79–91.

Almond, Gabriel A., G. Bingham Powell, Jr., Russel J. Dalton and Kaare Strom, 2010, *Comparative Politics Today, A World View*, updated 9th edition, New York: Pearson/Longman.

Amakawa Akira, 1983, 'Kōiki gyōsei to chihō bunken', Gyōsei no tenkanki (special issue of *Jyuristo*) (Wide area public administration and decentralization in Japan).

Amakawa Akira, 1986, 'Henkaku no kōsō: dōshūseiron no bunmyaku' (A framework for change: the context of regional system theory), in Ōmori Wataru and Satō Seisaburō (eds), *Nihon no chihō seifu* (Provincial government in Japan), Tokyo: Daigaku shuppankai, pp. 111–37.

Andeweg, Rudy, 2011, 'From puzzles to prospects for coalition theory', in Rudy Andeweg, Lieven De Winter and Patrick Dumont (eds), *Puzzles of Government Formation*, New York: Routledge, pp. 190–203.

Anzai Takumi, 2012, *Saraba kokusaku sangyō: 'denryoku kaikaku' 450 nichi no meisō to ushinawareta 60 nen* (Farewell to national industries: 450 days of confusion over electricity sector reform and the lost 60 years), Nihon Keizai Shimbun Shuppansha.

Arata Masafumi, 2011, 'Saigai borantia katsudō no "seijuku" to wa nanika' (What do we mean by disaster volunteering 'maturity'?), in Endō Kaoru (ed.), *Daishinsaigo no shakaigaku* (Sociology after the Great East Japan Earthquake), Kōdansha gendai shinsho, pp. 193–235.

Arimori Takashi, 2011, 'Keidanren oku no in "genpatsu shinjikēto" no yami: Tōden & Denjiren "zaikai", "seikai" shihai no ankoku shi' (Keidanren's inner sanctum, the shady-dealings of the "nuclear syndicate": a secret history of TEPCO & FEPC dominance in Japanese business and politics), *Bessatsu Takarajima* 1796: 150–6.

Asahi Shimbun Aomori Sōkyoku, 2005, *Kakunen manē: Aomori kara no hōkoku* (Nuclear fuel cycle money: report from Aomori), Iwanami Shoten.

Asahi Shimbun 'Genpatsu to Media' Shuzaihan, 2013, *Genpatsu to media 2: 3.11 sekinin no arika* (Nuclear power and the media 2: who is responsible for 3.11?), Asahi Shimbun Shuppan.

Asahi Shimbun Keizaibu, 2013, *Denki ryōkin wa naze agaru no ka* (Why do electricity prices go up?), Iwanami Shoten.

Asahi Shimbun Tokubetsu Hōdōbu, 2014, *Genpatsu riken o ou: denryoku o meguru kane to kenryoku no kōzō* (On the trail of special-interest of nuclear energy power: the structure of money and power in the electric power industry), Asahi Shimbun Shuppan.

Atsumi Tomohide, 2014, *Saigai borantia: Atarashii shakai e no guruupu dainamikkusu* (Disaster volunteering: group dynamics towards a new society), Kōbundō.

Awaji Takehisa, 2013, 'Fukushima genpatsu jiko no hōri o dō kangaeru ka' (Considering the legal principles behind compensation for the Fukushima nuclear accident), *Kankyō to Kōgai* (Research on Environmental Disruption), 43 (2): 2–8.

Awaji Takehisa, Yoshimura Ryōichi and Yokemoto Masafumi (eds), 2015, *Fukushima genpatsu jiko baishō no kenkyū* (Research on compensation for the Fukushima nuclear accident), Nippon Hyōronsha.

Axelrod, Robert, 1970, *Conflict of Interest*, Chicago, IL: Markham.

Azuma Kiyohiko, 2012a, 'Tōa shinpō ni okeru omona hōdō naiyō (2011nen 4gatsubun)' (Summary of newspaper reports in *The Dong-a Ilbo* for April 2011), background paper prepared for the Japan Society for the Promotion of Science research team on International Relations under the Triple Disaster.

Azuma Kiyohiko, 2012b, *Chūgoku 'Jinmin nippō' ni okeru higashi nihon dai shinsai, genpatsu kiki no hōdō* (Reports in the *People's Daily*'s on the Great East Japan Earthquake and the nuclear crisis), background paper prepared for the Japan Society for the Promotion of Science research team on International Relations under the Triple Disaster.

Azuma Kiyohiko, 2015, 'Itaku chōsa hōkokusho: Kankoku ni okeru higashi

Nihon dai shinsai, genpatsu kiki no shinbun hōdō' (Survey of newspaper reporting on the Great East Japan Earthquake and the nuclear crisis in South Korea), in K. Tsunekawa (ed.), *Dai shinsai, genpatsu kiki ka no kokusai kankei* (International relations of the earthquake and nuclear crisis), Tōyō Keizai Shinpōsha, pp. 262–8.

Bardach, Eugene, 1998, *Getting Agencies to Work Together: The Theory and Practice of Managerial Craftsmanship*, Washington, DC: Brookings Institute.

Baumgartner, Frank R. and Bryan D. Jones, 1991, 'Agenda dynamics and policy subsystems', *Journal of Politics*, 53 (4): 1044–74.

Baumgartner, Frank R. and Bryan D. Jones, 1993, *Agendas and Instability in American Politics*, Chicago, IL: University of Chicago Press.

Birkland, Thomas A., 1997, *After Disaster: Agenda Setting, Public Policy, and Focusing Events*, Washington, DC: Georgetown University Press.

Birkland, Thomas A., 2006, *Lessons of Disaster: Policy Change after Catastrophic Events*, Washington, DC: Georgetown University Press.

Board of Audit, 2013a, 'Higashi nihon daishinsai karano fukkō nado ni taisuru jigyō no jisshi jōkyō nado ni kansuru kaikei kensa no kekka ni tsuite' (Audit results relating to the implementation status of works for the recovery from the Great East Japan Earthquake), October.

Board of Audit, 2013b, 'Higashi Nihon daishinsai ni tomonau genshiryoku hatsudensho no jiko ni yori hōshutsu sareta hōshasei busshitsu ni yoru kankyō osen ni tai suru josen ni tsuite' (Decontaminating the environment after the release of radiation associated with the Fukushima nuclear accident), October.

Board of Audit, 2015a, 'Higashi nihon daishinsai karano fukkō nado ni taisuru jigyō no jisshi jokyō nado ni kansuru kaikei kensa no kekka ni tsuite' (Audit results relating to the implementation status of works for the recovery from the Great East Japan Earthquake), March.

Board of Audit, 2015b, 'Tokyo denryoku kabushiki kaisya ni kakawaru genshiryoku songai no baisyō ni kansuru kuni no shien nado no jisshi jokyō ni kansuru kaikei kensa no kekka ni tsuite' (Audit results relating to the implementation status of national government support for compensation for the nuclear disaster by TEPCO), March.

Board of Audit, 2016, 'Higashi Nihon daishinsai kara no fukkō nado ni taisuru jigyō no jisshi jigyō nado ni kansuru kaikei kensa no kekka ni tsuite' (Report on the audit of reconstruction works carried out in response to the Great East Japan Earthquake), April.

Center for Cancer Control and Information Service, 2016, 'Cancer registry and statistics', viewed March 13, 2017, <http://ganjoho.jp/reg_stat/statistics/stat/index.html>.

Center for Cancer Control and Information Services (Department of Cancer Registry and Statistics of the Cancer Information Services of the National Cancer Center, Japan), 2014, 'Fukushima ken ni okeru 2010 nen kōjōsen gan yūbyōsha sū no suii' (Estimation of the prevalence of thyroid carcinoma in Fukushima Prefecture in 2010), Prefectural Oversight Committee for the Fukushima Health Management Survey, Thyroid Ultrasound Examination Assessment Section (2014), Additional Material (The 4th meeting of Thyroid Assessment Section), November 11.

Chuma Mizuki, 2013, *Higashi Nihon dai shinsai ni kansuru Roshia no hōdō buri* (Newspaper reporting on the Great East Japan Earthquake in the Russian media), background paper prepared for the Japan Society for Promotion of Science research team on International Relations under the Triple Disaster.

CRIEPI (Central Research Institute of Electric Power Industry), 2003 (March), *JCO rinkai jiko no songai baishō (hoshō) shori no jissai ni miru jichitai no yakuwari to kadai (Chōsa Hōkoku): Y0212* (The role of local government in processing compensation for the Tōkaimura nuclear accident), viewed May 14, 2015, <http://criepi.denken.or.jp/jp/kenkikaku/report/detail/Y02012.htm>.

De Swaan, Abram, 1973, *Coalition Theories and Cabinet Formations*, Amsterdam and New York: Elsevier.

Dispute Reconciliation Committee for Nuclear Damage Compensation, 2011a, 'Tokyo Denryoku Kabushiki Gaisha Fukushima Daiichi, Daini genshiryoku hatsuden sho jiko ni yoru genshiryoku songai no hani no hantei nado ni kansuru chūkan shishin' (Interim guidelines on determination of the scope of nuclear damage resulting from the accident at the Tokyo Electric Power Company Fukushima Daiichi and Daini nuclear power plants), August.

Dispute Reconciliation Committee for Nuclear Damage Compensation, 2011b, 'Shiiyo 2-1: Jisyuteki hinan kanren data' (Materials 2-1. Data related to voluntary evacuation), 16th session, viewed August 10, 2016, <http://www.mext.go.jp/b_menu/shingi/chousa/kaihatu/016/shiryo/_icsFilcs/afieldfile/2011/11/11/1313180_2_2.pdf>.

Dispute Reconciliation Committee for Nuclear Damage Compensation, 2013a, 'Shiryō 1: Tōkyō Denryoku Kabushiki Gaisha Fukushima Daiichi, Daini genshiryoku hatsuden sho jiko ni yoru genshiryoku saigai no hani no hantei nado ni kansuru chūkan shishin dai 4 ji tsuiho' (Appendix 1: Fourth supplement to the interim guidelines on determination of the scope of nuclear damage resulting from the accident at the Tokyo Electric Power Company Fukushima Daiichi and Daini nuclear power plants), 39th session, viewed August 10, 2016, <www.mext.go.jp/b_menu/shingi/chousa/kaihatu/016/shiryo/__icsFiles/afieldfile/2013/12/26/1342848_1_1.pdf>.

Dispute Reconciliation Committee for Nuclear Damage Compensation, 2013b, 'Shiryō 2: Genshiryoku songai baishō no setai atari baishōgaku no shisan ni tsuite' (Appendix 2: Calculations for determining the amount of nuclear damage compensation payable per household), 39th session, viewed August 10, 2016, <www.mext.go.jp/b_menu/shingi/chousa/kaihatu/016/shiryo/__icsFiles/afieldfile/2013/12/26/1342848_3_1.pdf>.

Doi Yoshihira, 2011, *Genshiryoku mafia: genpatsu riken ni muragaru hitobito* (The nuclear mafia: the nuclear profiteers), Henshūkōbō Saku.

Döring, Herbert, 1995, 'Time as a scarce resource: government control of the agenda', in Herbert Döring (ed.), *Parliaments and Majority Rule in Western Europe*, New York: St Martin's Press, pp. 223–46.

Downs, Anthony, 1972, 'Up and down with ecology: the issue-attention cycle', *Public Interest*, 28: 38–50.

Endō Kaoru, 2012, *Media wa Daishinsai,Genpatsu Jiko wo Dō Katattaka: hōdō, netto, dokyumentarii wo kenshō suru* (How did the media report on the Great East Japan Earthquake disaster and the nuclear accident?: a study of reportage, the internet and documentary film), Tōkyō Denki University Press.

Endō Noriko, 2013, *Genshiryoku songai baishō seido no kenkyū: Tokyo Denryoku Fukushima genpatsu jiko kara no kōsatsu* (Research on the nuclear damage compensation system: reflections on the Fukushima nuclear accident), Iwanami Shoten.

Enerugii, Kankyō no Sentakushi ni Kansuru Tōrongata Yoron Chōsa Jikkō Iinkai (Executive Committee of a Deliberative Poll on Energy and Environmental Options), 2012, *Enerugii, Kankyō no Sentakushi ni Kansuru Tōrongata Yoron Chōsa: Chōsa Hōkokusho* (Report on a deliberative poll on energy and environmental options), viewed September 24, 2014, <www.cas.go.jp/jp/seisaku/npu/policy09/pdf/20120904/sanko_shiryo.pdf>.

FAO (Food and Agriculture Organization), 2011, 'Working together to support Japan and the global community', viewed December 21, 2013, <www.fao.org/news/story/en/item/53880/icode/>.

FDMA (Fire and Disaster Management Agency), 2013, *Higashi Nihon dai shinsai kirokushū* (Record of the Great East Japan Earthquake).

Feree, Myra Marx, 2003, 'Resonance and radicalism: feminist framing in the abortion debates of the United States and Germany', *American Journal of Sociology*, 109: 304–44.

Fisheries Agency, 2011, 'Yushutsu sareru suisanbutsu ni kansuru gensanchi no kakunin ni tsuite' (On the confirmation of region of origin for export marine products), viewed December 23, 2013, <www.jfa.maff.go.jp/j/kakou/export/pdf/suisanbutsushoumei_131225.pdf>.

Fishkin, James S., 2009, *When the People Speak: Deliberative Democracy and Public Consultation*, Oxford: Oxford University Press.

Foote, Daniel H., 2007, *Law in Japan: A Turning Point*, University of Washington Press.

Fukushima Prefectural Governor, 2011 (May 16), 'Request: concerning the use of private rental accommodation as emergency temporary accommodation outside Fukushima Prefecture following the Great East Japan Earthquake'.

Fukushima Prefecture, 2015 (June 15), 'Concerning extensions to the provision period of interim temporary accommodation related to the Great East Japan Earthquake', viewed August 1, 2016, <www.pref.fukushima.lg.jp/sec/16055b/260528-kasetukyouyoencyou.html>.

Fukushima Prefecture 2016 (May 18), 'Fukushima ken no zaisei jōkyō ni tsuite' (Fiscal conditions in Fukushima), viewed July 19, 2017 <https://www.pref.fukushima.lg.jp/uploaded/attachment/170049.pdf>.

Fukushima Prefecture, n.d., 'Hinan shiji no keii' (Development of evacuation instructions), viewed June 30, 2015, <www.pref.fukushima.lg.jp/download/1/01240331.pdf>.

Fukuyama Tetsurō, 2012, *Genpatsu kiki: kantei kara no shōgen* (Nuclear crisis: the view from inside the Prime Minister's Office), Chikuma Shinsho.

Garrod, Guy and G. Kenneth Willis, 1999, *Economic Valuation of the Environment: Methods and Case Studies*, Cheltenham: Edward Elgar.

General Secretariat of the Supreme Court of Japan (ed.), 2013, *Saiban dētabukku 2013* (Court data 2013), Hōsōkai.

Genshiryoku Shiryō Jōhō Shitsu (ed.), *Genshiryoku shimin nenkan 2015* (Citizen's nuclear energy annual 2015), Nanatsumori Shokan.

Gonoi Ikuo, 2012, *'Demo' to wa nani ka: henbō suru chokusetsu minshushugi* (What is a demo?: the metamorphosis of direct democracy), NHK Publishing INC.

Gordon, Andrew, 2013, *A Modern History of Japan: From Tokugawa Times to the Present*, updated 3rd edition, New York: Oxford University Press.

Government Investigation Committee on the Accident at the Fukushima Nuclear Power Stations of TEPCO, 2012, *Chūkan hōkokusho* (Interim report), Mediarando Kabushikigaisha.

Great Hanshin–Awaji Earthquake Reconstruction Follow-up Committee Editorial Board, 2009, *Tsutaeru Hanshin-Awaji daishinsai no kyōkun* (Lessons told by the Great Hanshin Awaji Earthquake), Gyōsei.

Green-Pedersen, Christoffer and Peter B. Mortensen, 2010, 'Who sets the agenda and who responds to it in the Danish parliament? A new model of issue competition and agenda-setting', *European Journal of Political Research*, 49 (2): 257–81.

Green-Pedersen, Christoffer and Peter Bjerre Mortensen, 2015, 'Avoidance and engagement: issue competition in multiparty systems', *Political Studies*, 63 (4): 747–64.

Gurūpu K21, 2011a, 'Genpatsu manē to seiji 1: tettei chōsa! jimintō no seiji shikin dantai ni denryoku 9 sha yakuin ga 1 oku en o damii kenkin!' (Nuclear money and politics 1: full investigation! dummy donations of 100 million yen to the LDP by executives at the nine utilities), *Bessatsu Takarajima* 1796: 87–91.

Gurūpu K21, 2011b, 'Genpatsu manē to seiji 2: hatsu kōkai! minshutō gi'in ni kenkin sareru zenkoku denryoku kei rōso no bakudai na kumiai hi!' (Nuclear money 2: revealed! how the electric power unions donated huge sums to DPJ politicians), *Bessatsu Takarajima* 1796: 92–5.

Gurūpu K21, 2011c, 'Denryoku gaisha ni yoru Kasumigaseki shihai no senpei, hatsu kōkai risuto! keisanshō, monkashō, naikakukanbō ni "amaagari" suru denryoku gaisha shain' (A list of the utility vanguard that controls Kasumigaseki: the utility employees who are seconded into METI, MEXT and the Cabinet Secretariat), *Bessatsu Takarajima* 1796: 96–7.

Gurūpu K21, 2011d, 'Kansai Denryoku "Ōsaka Mainichi Hōsō atsuryoku jiken" no shinsō' (The true story of how KEPCO pressured MBS Osaka), *Bessatsu Takarajima* 1821: 53.

Haley, John O., 2007, 'The Japanese judiciary: maintaining integrity, autonomy and the public trust', in Daniel H. Foote (ed.), *Law in Japan: A Turning Point*, University of Washington, pp. 99–135.

Hamamoto Shinsuke, 2007, 'Kojin tōhyō no teika' (The decline of personal vote), *Senkyo gakkai kiyō* (Review of Electoral Studies), 9: 47–66.

Hamamoto Shinsuke, 2016, 'Rippō: nejire kokkai ka no rippō katei' (Legislation: the legislative process in the divided Diet), in Tsujinaka Yutaka (ed.), *Daishinsai ni manabu shakai kagaku, dai 1 kan, seiji katei to seisaku* (Lessons for the Social Sciences from the 2011 Great

East Japan Earthquake, Volume 1 Political process and policy), Tōyō Keizai Shinpōsha, pp. 55–76.

Hamamoto Shinsuke and Nemoto Kuniaki, 2011, 'Kojin chūshin no saisen senryaku to sono yūkōsei: senkyoku katsudō wa tokuhyō ni musubi tsukunoka?' (Re-election strategies focusing on individuals and their effectiveness: does constituency campaigning lead to voting?), *Nenpō seiji gaku* (The Annuals of the Japanese Political Science Association), 2011 (2): 70–97.

Harada Hiroki, 2012, 'Shinsai fukkō no hōgijutsu toshite no fukkō tokku' (Special Reconstruction Zones as legal technique in earthquake disaster recovery), *Shakaikagaku kenkyū* (Journal of Social Science), 64 (1).

Harada Hiroki, 2013, 'Gyōsei hōgaku kara mita genshiryoku songai baishō' (Nuclear damage compensation from the perspective of administrative law), *Hōgaku Ronsō* (Legal Debates), 173 (1): 2–25.

Hasegawa Ken'ichi, 2012, *Genpatsu ni 'furusato' o ubawarete: Fukushima ken Iitate mura rakunōka no sakebi* (I lost my hometown to nuclear power: the cry of a dairy farmer from Iitate, Fukushima), Takarajima Sha.

Hasegawa Koichi, 2015, *Beyond Fukushima: Toward a Post-nuclear Society*, Melbourne: Trans Pacific Press.

Hata Akio and Mukai Yoshiyuki, 2014, *Itai-itai byō to Fukushima: kore made no 100 nen, kore kara no 100 nen* (Itai-itai disease and Fukushima: the last 100 years and the 100 years to come), Gotō Shoin.

Hayamaru Kazuma, 2014, *Higashi nihon dai shinsai o meguru Taiwan no hōdō* (Newspaper reporting on the Great East Japan Earthquake in the Taiwanese media), background paper prepared for the Japan Society for Promotion of Science Research Team on International Relations under the Triple Disaster.

Head of the Local Government Administration Bureau (Ministry of Internal Affairs and Communications), 2011 (November 15), 'Communique: concerning notices on the administration of special matters relating to disaster victim evacuee residents'.

Head of the Social Support Bureau (Ministry of Health, Labour and Welfare), 2011 (April 30), 'Concerning the provision of private rental accommodation as emergency temporary housing following the Great East Japan Earthquake'.

Hirabayashi Yūko, 2013, 'Nani ga "demo no aru shakai" o tsukuru no ka: Posuto 3.11 no akutivizumu to media' (What makes 'society with demonstrations': post 3.11 activism and the media), in Tanaka Shigeyoshi, Funabashi Harutoshi and Masamura Toshiyuki (eds), *Higashi Nihon Daishinsai to shakaigaku: daisaigai wo umidashita shakai* (The Great Eastern Japan Earthquake and Japanese sociology: how a country brought disaster upon itself), Minerva Shobō, pp. 163–95.

Hioki Masaharu, 2011, *Kakudai suru hōshanō osen to hōkisei: ana darake no seido no genjō* (The growing problem of radioactive contamination and the law: a system full of holes), Waseda Daigaku Shuppanbu.

Hirano Hiroshi, 2007, *Henyō suru Nihon no shakai to tōhyō kōdō* (Changes in Japanese society and voting behavior), Bokutakusha.

Hitomi Takeshi, 2011, 'Fukushima Daiichi genshiryoku hatsudensho jiko no songai baishō' (Nuclear damage compensation after the Fukushima accident), *Hōgaku Seminā* (Legal Seminar), 56 (12): 20–5.

Holt, Diane and Ralf Barkemeyer, 2012, 'Media coverage of sustainable development issues: attention cycles or punctuated equilibrium?', *Sustainable Development*, 20: 1–17.

Honda Hiroshi, 2005, *Datsu genshiryoku no undō to seiji: Nihon no enerugī seisaku no tenkan wa kanō ka* (The antinuclear power movement and politics: is it possible to change Japan's energy policy?), Hokkaidō Daigaku Tosho Kankōkai.

Honma Terumitsu, 2014, 'Genshiryoku songai baishō no sekinin shutai: kokusaku to shōgyō bēsu, taishō suru seizōbutsu, denryokusai' (The responsibility for nuclear damage compensation after Fukushima), *Keizai Kenkyū* (Economic Research), 6: 197–236.

Hori, M., Matsuda, T., Shibata, A., Katanoda, K., Sobue, T., Nishimoto, H., et al., 2015, 'Cancer incidence and incidence rates in Japan in 2009: a study of 32 population-based cancer registries for the Monitoring of Cancer Incidence in Japan (MCIJ) project', *Japanese Journal of Clinical Oncology*, 45 (a): 884–91.

Horie Takashi, 2014, 'Yoron', in Hiroshi Honda and Takashi Horie (eds), *Datsu genpatsu no hikaku seijigaku* (The comparative politics of opposition to nuclear power), Hosei University Press, pp. 90–108.

Horn, Murray J., 1995, *The Political Economy of Public Administration: Institutional Choice in the Public Sector*, Cambridge University Press.

Hosono Gōshi and Torigoe Shuntarō, 2012, *Shōgen: Hosono Gōshi* (Testimony: Hosono Gōshi), Kōdansha.

Howlett, Michael, 1997, 'Issue-attention and punctuated equilibria models reconsidered: an empirical examination of the dynamics of agenda-setting in Canada', *Canadian Journal of Political Science*, 30 (1): 3–29.

IAEA (International Atomic Energy Agency), 1986, 'Convention on Early Notification of a Nuclear Accident', viewed February 13, 2014, <www.iaea.org/sites/default/files/infcirc335.pdf>.

Idei Naoki, 2012a, 'Shinsai to ADR genshiryoku songai baishō ADR ni tsuite: Genshiryoku Songai Baishō Funsō Sentā no torikumi to tenbō' (Disasters and alternative dispute resolution: the Nuclear Damage Compensation Dispute Resolution Center initiative and its prospects), *Chūsai to ADR* (Journal of Japanese Arbitration and ADR), 7: 46–56.

Idei Naoki, 2012b, 'Songai baishō e no torikumi: genshiryoku songai baishō e no Nichibenren no torikumi' (Dealing with nuclear damage compensation: the Japan Federation of Bar Associations approach to nuclear damage compensation), *Hō to Minshushugi* (Japan Democratic Lawyers Association), 466: 54–6.

IGES (Institute for Global Environmental Strategies) FAIRDO, 2013, *Challenges of Decontamination, Community Regeneration and Livelihood Rehabilitation* (2nd discussion paper), July, viewed February 6, 2017, <http://pub.iges.or.jp/modules/envirolib/view.php?docid=4718>.

Iitate Village, 2011, 'Iitate-mura josen keikakusho' (Iitate decontamination plan).

Imai Akira, 2014, *Jichi saiken: genpatsu hinan to 'idō suru mura'* (Municipal reconstruction: nuclear evacuation and 'communities on the move'), Tokyo: Chikuma shobō.

Imai Ryosuke, 2011, 'Senkyo undō shishutsu no yūkōsei' (Does campaign spending matter?), *Nenpō seiji gaku* (The Annuals of the Japanese Political Science Association), 2011 (2): 11–32.

Inatsugu Hiroaki, 2000, *Jinji, kyūyo to chihō jichi* (Human resources, salaries and local autonomy), Tōyō Keizai Shinpōsha.

Inatsugu Hiroaki, 2001, 'Personnel systems and pay policies of local governments', in Muramatsu Michio, Farrukh Iqbal and Ikuo Kume (eds), *Local Government Development in Post-war Japan*, New York: Oxford University Press, pp. 154–84.

Inatsugu Hiroaki (ed.), 2012, *Dai kibo saigai ni tsuyoi jichitai kan renkei: genba kara no hōkoku to teigen* (Local government relationships that can withstand large-scale disasters: a report and proposal from the field), Waseda Daigaku Shuppanbu.

Inatsugu Hiroaki, 2013, *Jichitai gabanansu* (Local governance), Tokyo, Hōsō Daigaku Kyōiku Shinkōkai.

Institute for Renewable Energy Policies, 2012, 'Kan Naoto no shizen enerugī ron: kiraware sōri no oki miyage' (Kan Naoto's renewable energy policy: an unpopular prime minister's parting gift), Mainabi.

International Commission on Radiological Protection (ICRP), 2007, 'The 2007 recommendations of the International Commission on Radiological Protection', Publication 103, *Annals of the ICRP*, 37 (2–4).

Irie Hideaki, 2013, *Gendai chōteiron: Nichibei ADR no rinen to genjitsu* (Contemporary arbitration: the theory and practice of ADR in Japan and the United States), Tokyo: Daigaku Shuppankai.

Isomura Kentarō and Yamaguchi Eiji, 2013, *Genpatsu to saibankan: naze shihō wa 'merutodaun' o yurushita no ka* (Nuclear power and the courts: why did the judiciary allow a 'meltdown' to happen?), Asahi Shimbun Shuppan.

Isono Yayoi, 2011, 'Genshiryoku jiko to kuni no sekinin' (Nuclear power accidents and state responsibility), *Kankyō to Kōgai* (Research on Environmental Disruption), 41 (2): 36–41.

Isono Yayoi, 2015a, 'Josen: sono mondai to kadai' (Decontamination: issues and questions), in Awaji Takehisa, Yoshimura Ryōichi and Yokemoto Masafumi (eds), *Fukushima genpatsu jiko baisyo no kenkyu* (Research on compensation for the Fukushima nuclear accident), Nippon Hyosonsha, pp. 227–40.

Isono Yayoi, 2015b, 'Josen to "kenkō ni ikiru kenri"' (Decontamination and 'the right to health'), in Yokemoto Masafumi and Watanabe Toshihiko (eds), 2015, *Genpatsu saigai ha naze fukintona fukko wo motarasunoka* (Why is the nuclear disaster producing unequal reconstruction outcomes?), Mineruva Shobō, pp. 227–47.

Isozaki Hatsuhiko, 2012, 'Higashi Nihon daishinsai tokubetsu kuiki hō no igi to kadai (Part 2): Enkatsu, jinsoku na fukkō to chihō bunken' (Significance of and issues related to the Great East Japan Earthquake Special Zone Law (Part 2): smooth and swift recovery and the decentralization of authority), *Jichi sōken* (Local Government Research Monthly), 405: 26–56.

Ito Shigeru, 2005, 'Kenshō teema "Fukkō taisei: fukkō no suishin taisei"' (Inspection theme 'Reconstruction systems: systems to support reconstruction'), *Hyōgo ken – Fukkō 10 nen sōkatsu kenshō, teigen deetabeesu* (Inspection and suggestion database of ten years on reconstruction, Hyogo Prefecture), viewed April 21, 2017, <web.pref. hyogo.jp/wd33/documents/000038664.pdf>.

Itō Mamoru, Katsuhiro Matsui, Noboru Watanabe and Naoko Sugiura, 2005, *Demokurashii rifurekushon* (Reflections on democracy), Liberta-shuppan.

Itō Masaki, 2012, *Demo no media ron: shakai undō no yukue* (Demonstrations and the media: whither the social movement society), Chikuma Shobō.

Itō Masatsugu, 2003, *Nihongata gyōsei iinkai seido no keisei: soshiki to seido no gyōseishi* (The Independent Administrative Commissions in postwar Japan: their organization and institutionalization), Tokyo University Press.

Itō Masatsugu, 2014, 'Tajū bōgyo to takikan renkei no kanōsei' (The possibility of multiple defense and multiple agency collaboration), in Mikuriya Takashi and Iio Jun (eds), *Bessatsu Asuteion: 'saigo' no bunmei* (ΑΣΤΕΙΟΝ special edition: civilization 'after disaster'), Hankyū Communications, pp. 64–81.

Itō Masatsugu, 2015, 'Takikan renkei toshite no rookaru gabanansu: shūrō shien gyōsei ni okeru kanōsei' (Local governance as interagency collaboration: its possibility in job support services official support for work), in Uno Shigeki and Iokibe Kaoru (eds), *Rookaru kara no saishuppatsu – Nihon to Fukui no gabanansu* (Re-departing from the local: governance in Japan and Fukui), Yuhikaku.

Itō Shūichirō, 2006, *Jichitaihatsu no seisaku kakushin: keikan jōrei kara keikanhō e* (Promoting local government reform: from landscape ordinance to landscape law), Bokutakusha.

Itō Yasushi, 2016, 'Josen no hiyō tai kōka' (The cost effectiveness of radioactive decontamination), in Ueta Kazuhiro (ed.), *Hiyō no hōkatsu teki ha'aku* (A comprehensive overview of the damages and costs of the Great East Japan Earthquake), Tōyō Keizai Shinpōsha, pp. 165–207.

Iwasaki Tadashi, 2011, 'Higashi Nihon daishinsai fukkō kihonhō no seitei katei' (The legislation processes of the Great East Japan Earthquake Fundamental Reconstruction Law), *Jichi sōken* (Local Government Research Monthly), 394: 48–62.

Iwate GINGA-NET Project Implementation Committee, 2012, 'Iwate GINGA-NET purojekuto katsudō hōkokusho – 1198 nin ga tsunaida 2001 nen natsu no katsudō kiroku' (Iwata GINGA-NET project activity – an activity record of the summer of 2011 when 1198 people joined together as one).

Iwate Prefecture, 2013, 'Iwate ken Higashi Nihon daishinsai tsunami no kiroku' (Iwate Prefecture record of the Great East Japan Earthquake and tsunami).

Japan Business Federation Corporate Social Responsibility Committee 1% Club, 2012, 'Higashi Nihon daishinsai ni okeru hisaisha, hisaichi shien katsudō ni kan suru hōkokusho – keisaikai ni yoru kyōjo no torikumi' (Report of support activities for Great East Japan Earthquake disaster victims and sites – a mutual effort by the corporate world).

Japan Municipal Hospitals Association, 2011, 'Higashi Nihon daishinsai ni okeru kai'in byōin no shien jōkyō chōsa (Dai 3 kai) ni tsuite' (On the third survey of responses to the Great East Japan Earthquake by member hospitals), July 27.

Japan National Council of Social Welfare, 2012, 'Higashi Nihon daishinsai saigai borantia sentaa hōkokusho' (Report of the Great East Japan Earthquake disaster volunteer centers).

Japan National Council of Social Welfare, 2013, '2011.3.11 Higashi Nihon daishinsai e no shakai fukushi bunya no torikumi to kadai – shinsai kara ichi nen no katsudō o fumaete' (Contributions and issues in the social welfare field after the March 11, 2011, Great East Japan Earthquake – based on activities in the one year following the disaster).

Japan NGO Center for International Cooperation, 2012, 'Higashi Nihon daishinsai to kokusai kyōryoku NGO – kokunai de no arata na kanōsei to kadai, soshite teigen' (The Great East Japan Earthquake and Japan's international NGOs – new domestic possibilities and issues, with recommendations).

Japan YMCA, 2012, 'YMCA kyūen fukkō shien katsudō repooto' (Report of YMCA relief and reconstruction support activity).

Japan Tourism Agency, 2009, *Kankō hakusho Heisei 21nenban* (White papers on tourism 2009), Ministry of Land, Infrastructure, Transport and Tourism.

Japan Tourism Agency, 2011, *Kankō hakusho Heisei 23nenban* (White papers on tourism 2011), Ministry of Land, Infrastructure, Transport and Tourism.

Japan Tourism Agency, 2012, *Ryokō, kankō sangyō no keizai kōka ni kansuru chōsa kenkyū* (Research study of economic impacts of tourism in Japan), viewed February 24, 2017, <www.mlit.go.jp/common/000220421.pdf>.

Japan Tourism Agency, 2014, *Kankō hakusho Heisei 26nenban* (White papers on tourism 2014), Ministry of Land, Infrastructure, Transport and Tourism.

Japan Tourism Agency, 2016, *Kankō hakusho Heisei 28nenban* (White papers on tourism 2016), Ministry of Land, Infrastructure, Transport and Tourism.

Jasper, James M., 1988, 'The political life cycle of technological controversies', *Social Forces*, 67 (2): 357–77.

JETRO (Japan External Trade Organization), 2012, 'Genpatsu jiko ni tomonau Ōshū ni okeru Nihonhatsu kamotsu (kōgyōhin) heno hōshasen kensa ni tsuite' (On the radioactivity inspection of cargos of manufactured goods from Japan which is conducted in Europe due to the nuclear accident), viewed December 13, 2013, <www.jetro.go.jp/world/shinsai/manufacturing_inspection.html>.

Joint Committee for Coordinating and Supporting Voluntary Disaster Relief Activities, 2014, 'Higashi Nihon daishinsai katsudō hōkoku' (Report of Great East Japan Earthquake activities).

Jōmaru Yōichi, 2012, *Genpatsu to media: shinbun jānarizumu 2 dome no haiboku* (Nuclear power and the media: the second failure of newspaper journalism), Asahi Shimbun Shuppan.

Joppke, Christian, 1991, 'Social movements during cycles of issue attention: the decline of the antinuclear energy movements in West Germany', *British Journal of Sociology*, 42 (1): 43–60.

Kabashima Ikuo and Imai Ryosuke, 2002, 'Evaluation of party leaders and voting behavior: an analysis of the 2000 general election', *Social Science Japan Journal*, 5 (1): 85–96.

Kaido Yūichi, 2011, *Genpatsu soshō* (Antinuclear lawsuits), Iwanami Shoten.

Kaido Yūichi (ed.), 2014, *Han-genpatsu e no iyagarase zenkiroku: genshiryoku mura no hinsei o warau* (Harassment of the antinuclear movement: the sordid character of the nuclear village), Akashi Shoten.

Kainuma Hiroshi, 2011, *Fukushima ron: genshiryoku mura ha naze umareta no ka* (Fukushima: the origins of a nuclear village), Seido-sha.

Kan Naoto, 2012, *Tōden Fukushima genpatsu jiko: sōri daijin to shite kangaeta koto* (The Fukushima nuclear accident: the prime minister's view), Gentōsha.

Kan Naoto, 2014, *Kan Naoto 'genpatsu zero' no ketsui: moto sōri ga kataru Fukushima genpatsu jiko no shinjitsu* (Kan Naoto's decision to end nuclear power: the reality of the Fukushima nuclear accident as told by the former prime minister), Nanatsumori Shokan.

Kanbayashi Hiroe, 2011 'Dare mo kakenakatta terebi, shimbun, zasshi no fuhai: Tōden kōkoku & settai ni baishū sareta masukomi genpatsu hōdō no butai ura!' (The unwritten story of the corruption of the television, newspaper and magazine media: behind the scenes of the mass media's coverage of nuclear power, paid for by TEPCO through advertising and entertaining), *Bessatsu Takarajima* 1796: 50–7.

Kanbayashi Hiroe, 2012, 'Ketsuzei o tsukatta kokumin sennō: yarase kanchō "Keisanshō Shigen Enerugī chō" genpatsu suishin PR no daizai' (Brainwashing the public with their own taxes: the swindlers at METI's Agency for Natural Resources and Energy and the crimes of pronuclear PR), in Y. Ichinomiya, H. Koide, T. Suzuki, T. Hirose et al., *Genpatsu saikadō no fukai yami*, Takarajimasha, pp. 76–94.

Kaneda Osamu, 2012, 'Zenkoku shakai hoken rōmushi kai rengōkai no taiō' (The contribution of the Japan Federation of Labor and Social Security Attorney's Association), in Shinsai taio seminar jikko iinkai (Disaster Seminar Implementation Commission)(ed.), *3.11 Daishinsai no kiroku – chūō shōchō, hisai jichitai, kaku shigyō tou no taiō* (Records of the 3.11 disaster – the response of central government agencies, disaster affected municipalities and professional groups), Minjihō kenkyūkai (Civil Law Research Association), pp. 656–92.

Kanō Tokio, 2010, *Mittsu no hashi o kakeru: kokusei sankaku 12 nen no chōsen* (Building three bridges: the challenge of 12 years of participating in affairs of state), Nihon Denki Kyōkai Shimbunbu.

Katō Hisaharu, 2012, *Genpatsu terebi no kōya: seifu, denryoku gaisha no terebi contorōru* (The nuclear TV wasteland: how the government and the utilities control television), Ōtsuki Shoten.

Kawato Sadafumi, 2015, *Gi'in naikaku sei* (Parliamentary government), Tōkyō Daigaku Shuppankai.

Kikkawa Takeo, 2011, *Tokyo denryoku, shippai no honshitsu: 'kaitai to saisei' no shinario* (TEPCO, the truth about its failure: scenarios for its breakup and renewal), Tōyō Keizai Shinpōsha.

Kingston, Jeff, 2012, 'Introduction', in J. Kingston (ed.), *Natural Disaster and Nuclear Crisis in Japan: Response and Recovery after Japan's 3/11*, London and New York: Routledge, pp. 1–11.

Kingston, Jeff, 2014, 'Japan's nuclear village: power and resilience', in Jeff Kingston (ed.), *Critical Issues in Contemporary Japan*, New York: Routledge, pp. 107–19.

Kitamura Ryūkō, 2013, 'Intabyū, Idogawa Katsutaka, abunai basho o anzen to itsuwatteiru' (Pretending dangerous areas are safe: interview with Idogawa Katsutaka), *Toshi Mondai* (Urban Issues), 104 (10).

Kohno Masaru, 2009, 'Senkyo kekka kara mita Minshutō asshō, Jimintō taihai no kōzu' (The landslide victory of the DPJ as seen from the election results, composition of the LDP's major defeat), in Tanaka Aiji, Kohno Masaru, Hino Airo, Iida Takeshi and the Yomiuri Shimbun Public Opinion Survey Department, *2009 nen, naze seiken kōtai dattanoka: Yomiuri-Waseda no kyōdō tyōsa de yomitoku Nihon seiji no tenkan* (Why there was a change of government in 2009 – deciphering the changes in Japanese politics with a joint survey by *Yomiuri Shimbun* and Waseda University), Keiso Shobo.

Kojima Nobuo, 2011, 'Fukushima Daiichi genpatsu jiko ni yoru higai to sono hōritsu mondai' (Damage from the Fukushima Daiichi nuclear power plant accident and its legal problems), *Hōritsu Jihō* (Legal News), 83 (9–10): 55–65.

Kokumin kyōiku bunka sōgō kenkyūjō: Higashi Nihon Daishinsai to gakkō: shiryō shūshū purojekuto (National Institute for Education and Culture: The Great East Japan Earthquake and Schools: Document Collection Project), 2013, *Shiryōshū: Higashi Nihon daishinsai/genpatsu shinsai to gakkō – Iwate, Miyagi, Fukushima no kyōiku gyōsei to kyōshokuin gumiai no kiryoku* (Document collection: the Great East Japan Earthquake/nuclear disaster and schools – records of educational administration and teaching unions in Iwate, Miyagi and Fukushima), Akashi Shoten.

Komatsu Kimio, 2012, *Genpatsu ni shigami tsuku hitobito no mure: genpatsu rieki kyōdōtai no himitsu ni semaru* (The nuclear gang: inside the secret world of the nuclear profiteers), Shin Nihon Shuppansha.

Komori Atsushi, 2013, 'Genpatsu iji seyo' (Maintain nuclear power!), in Asahi Shimbun Tokubetsu Hōdōbu, *Purometeusu no wana 5: Fukushima genpatsu jiko, konshin no chōsa hōdō*, Gakken Paburisshingu: 135–74.

Komori Atsushi, 2016, *Nihon wa naze datsu genpatsu dekinai no ka: 'genshiryoku mura' to iu riken* (Why can't Japan abolish nuclear power? The nuclear village as a network of interests), Heibonsha.

Kondō Michiyo, 2001, 'Shihō seido kaikaku shingikaigi no hōkōsei to genkai: "Minji shihō no arikata", hōsō jinkō mondai o chūshin ni' (The possibilities and limitations of the Justice System Reform Council: considering the administration of the civil law in terms of the size of the legal profession), in Minshushugi Kagaku Kyōkai Hōritsu Bukai (ed.), *Hō no kagaku* (Legal science), Nippon Hyōronsha, pp. 136–43.

Kōno Kei, Naka'aki Hiroshi and Hara Miwako, 2016, 'Shinsai 5 nen: Kokumin to hisaichi no ishiski: "Bōsai to enerugī ni kansuru yoron chōsa 2015" kara' (The 2015 public opinion survey on 'Disaster Prevention and Energy'), Tokyo, Hōsō Kenkyū to Chōsa (Broadcasting Research and Surveys), pp. 28–69.

Koopmans, Ruud and Paul Statham, 1999, 'Ethic and civic conceptions of nationhood and the differential success of the extreme right in

Germany and Italy', in Marco Giugni, Doug McAdam and Charles Tilly (eds), *How Social Movements Matter*, Minneapolis, MN: University of Minnesota Press, pp. 225–51.

Kriesi, Hanspeter, Ruud Koopmans, Jan Willem Duyvendak and Marco G. Giugni, 1995, *New Social Movements in Western Europe: A Comparative Analysis*, Minneapolis, MN: University of Minnesota Press.

Kubo Yoshiaki, 2015, 'Saigai to seitō seiji no sōgo kankei: 2012 nen sōsenkyo e muketa seikyoku to seisaku ronsō no kentō wo tsūjite' (The interrelationship between disasters and party politics: through examination of the political conditions and policy debates leading to the 2012 general election), in Tsujinaka Yutaka (ed.), *Higashi nihon daishinsai eno seijigakuteki apurōchi* (Political approaches to the Great East Japan Earthquake), University of Tsukuba, pp. 131–65.

Kubo Yoshiaki, 2016, '2012 nen sōsenkyo e muketa seikyoku to seisaku ronsō: seitō seijika wa shinsai to genpatsu jiko wo dō toraetanoka' (The political conditions and policy debates leading to the 2012 general election: how parties and politicians grasped the triple disaster and nuclear accident, in Tsujinaka Yutaka (ed.), *Daishinsai ni manabu shakai kagaku, dai 1 kan, seiji katei to seisaku* (Lessons for the Social Sciences from the 2011 Great East Japan Earthquake, Volume 1 Political process and policy), Tōyō Keizai, pp. 271–93.

Kushida Kenji E. and Phillip Lipscy (eds), 2013, *Japan under the DPJ: The Politics of Transition and Governance*, Stanford, CA: Walter H. Shorenstein Asia-Pacific Research Center.

Kyodo News, 2013a, 'Kankai e eikyōryoku kōshi' (Influence in the bureaucracy), *Nihon o tsukuru: genpatsu to kokka*, viewed October 21, 2014, <www.47news.jp/47topics/tsukuru/article/post_35.html>.

Kyodo News, 2013b, 'Tsubusareta shimin shizen enerugī hōan' (Breaking the renewable energy bill), *Nihon o tsukuru: genpatsu to kokka*, viewed October 14, 2014, <www.47news.jp/47topics/tsukuru/article/post_32.html>.

Kyodo News, 2013c, 'Han-genpatsu gakusha o kanshi' (Surveillance of antinuclear scholars), *Nihon o tsukuru: genpatsu to kokka*, viewed October 21, 2014, <www.47news.jp/47topics/tsukuru/article/post_38.html>.

Kyodo News, 2013d, 'Seijika o ayatsuru' (Manipulating politicians), *Nihon o tsukuru: genpatsu to kokka*, viewed October 21, 2014, <www.47news.jp/47topics/tsukuru/article/post_39.html>.

Kyodo News, 2013e, 'Shimbun, terebi ni shintō' (Permeating the newspaper and television industries), *Nihon o tsukuru: genpatsu to kokka*, viewed October 27, 2014, <www.47news.jp/47topics/tsukuru/article/post_36.html>.

Lee Chaek, 2012, 'Tebanashi de wa yorokobenai "saisei ene hō" no seiritsu: Denjiren & Nagatachō "shizen enerugī tsubushi" no teguchi' (We cannot uncritically praise passing the renewable energy law: how the FEPC and politicians crush renewable energy), in Y. Ichinomiya, H. Koide, T. Suzuki, T. Hirose et al., *Genpatsu saikadō no fukai yami*, Takarajimasha, pp. 179–97.

Leiserson, Michael Avery, 1966, Coalitions in Politics: A Theoretical and Empirical Study, PhD dissertation, Yale University.

Lewis, David E., 2003, *Presidents and the Politics of Agency Design: Political Institutions in the United States Government Bureaucracy*, 1946–1997, Stanford University Press, CA.

Lijphart, Arend, 2012, *Patterns of Democracy: Government Forms and Performance in Thirty-six Countries*, 2nd edn, New Haven, CT: Yale University Press.

Lipsky, Michael, 1980, *Street-level Bureaucracy: Dilemmas of the Individual in Public Service*, New York: Russell Sage Foundation.

McAdam, Doug, 1996, 'Conceptual origins, current problems, future directions', in Doug McAdam, John D. McCarthy and Meyer N. Zald (eds), *Comparative Perspective on Social Movements: Political Opportunities, Mobilizing Structures and Cultural Framings*, Cambridge, MA: Cambridge University Press, pp. 23–40.

McAdam, Doug, 1999, *Political Process and the Development of Black Insurgency, 1930–1970*, 2nd edn, Chicago, IL: University of Chicago Press.

McCarthy, John D., Clark McPhail and Jackie Smith, 1996, 'Images of protest: dimensions of selection bias in media coverage of Washington demonstrations, 1982 to 1991', *American Sociological Review*, 61 (3): 478–99.

McCommon, Holly J., Courtney Sanders Muse, Harmony D. Newman and Teresa M. Terrell, 2007, 'Movement framing and discursive opportunity structures: the political successes of the U.S. women's jury movements', *American Sociological Review*, 72 (5): 725–49.

McElwain, Kenneth Mori and Umeda Michio, 2011, 'Party democratization and the salience of party leaders', *Journal of Social Science*, 62 (1):173–93.

McElwain, Kenneth Mori, 2012, 'The nationalization of Japanese elections', *Journal of East Asian Studies*, 12 (3): 323–50.

Maeda Ko, 2008, 'Re-examining the contamination effect of Japan's mixed electoral system using the treatment-effects model', *Electoral Studies*, 27 (4): 723–31.

Maeda Ko, 2009, 'Has the electoral system reform made Japanese elections party-centered?', in Steven R. Reed, Kay Shimizu and Kenneth Mori McElwain (eds), *Political Change in Japan: Electoral Behavior, Party Realignment, and the Koizumi Reforms*, Stanford, CA: Walter H. Shorenstein Asia-Pacific Research Center.

Maeda Yukio, 2015, 'Minshutō seiken ni taisuru yūkensha no hyōka: getsuji yoron chōsa dēta no bunseki' (Voters' opinions of the DPJ administration: an analysis of the monthly data), in Maeda Yukio and Tsutsumi Hidenori (eds), *Tōchi no jōken: minshutō ni miru seiken unei to tōnai tōchi* (The preconditions for government: political administration and party politics in the DPJ), Chikura Shobō, pp. 291–328.

Makihara Izuru, 2014, 'Futatsu no "saigo" o tsuranuku "tochi"' in Mikuriya Takashi and Iio Jun (eds), *Bessatsu Asuteion: 'saigo' no bunmei* (AΣTEION Special edition: Civilization 'after disaster'), Hankyū Communications, pp. 46–63.

Makino Junichiro, 2015, 'Science literacy after 3.11, No. 36', *Science Journal Kagaku*, 85 (10): 937–41.

Martin, Lanny W. and Randolph T. Stevenson, 2001, 'Government formation in parliamentary democracies', *American Journal of Political Science*, 45: 33–50.

Masuyama Mikitaka, 2003, *Gikaiseido to Nihon seiji: giji unei no keiryō seijigaku* (The parliamentary system and Japanese politics: a political arithmetic approach to Diet business), Bokutakusha.

Masuyama Mikitaka, 2013, 'Shōsenkyoku hireidaihyō heiritsu sei to nidai seitōsei: chōfuku rikkōho to genshoku yūi' (Parallel voting and the two-party system: dual candidacy and the incumbency advantage), *Leviathan*, 52: 8–42.

Matsui Nozomi, 2012, 'Tōkei seido – "Shireitō" no sekkei to "Shōchō kyōdōtai" no jizoku' (Statistics systems – the maintenance of 'departmental community' and the design of 'central control'), in Akira Morita and Toshiyuki Kanai (eds), *Seisaku henyō to seido sekkei – Seikai, shōchō saihenn zengo no gyōsei* (Policy change and institutional design: governance before and after the reorganization of the political and administrative systems), Minerva Shobō.

Matsuo Makiko, 2015, 'Shokuhin chū no hōshasei busshitsu o meguru mondai no kei'i to sono gabanansu' (Radioactive food problem and its governance), in H. Shiroyama, *Fukushima genpatsu jiko to fukugō gabanansu* (The Fukushima nuclear accident and complex governance), Tōyō Keizai Shinpōsha, pp. 249–75.

Meguid, Bonnie M., 2005, 'Competition between unequals: the role of mainstream party strategy in niche party success', *American Political Science Review*, 99 (3): 347–59.

METI (Ministry of Economy, Trade and Industry), 2014, 'Hinan shiji kuiki no gainenzu to kaku kuiki no jinkō oyobi setai sū' (Map outlining the evacuation instruction zones with population and household numbers for each area), viewed August 10, 2016, <www.meti.go.jp/earthquake/nuclear/pdf/141001.pdf>.

MEXT (Ministry of Education, Culture, Sports, Science and Technology), 2008, *JCO rinkai jiko ji no genshiryoku songai baishō taiō in tsuite 'hoshō taisaku nado ichiren no ugoki'* (On nuclear damage compensation after the Tōkaimura nuclear accident), viewed May 14, 2015, <www.mext.go.jp/b_menu/shingi/chousa/kaihatu/007/shiryo/08061105/004/001.htm>.

MEXT, Genshiryoku Hisaisha Seikatsu Shien Chiimu, 2015, *Hinan shiji chi'iki no jōkyō ni tsuite* (The situation in the evacuation instruction zones), January, viewed August 10, 2016, <www.mext.go.jp/b_menu/shingi/chousa/kaihatu/016/shiryo/__icsFiles/afieldfile/2015/01/28/1354739_3_1.pdf>.

MHLW (Ministry of Health, Labour and Welfare), 2011, *The 2011 Ministry of Health, Labour and Welfare White Paper*, Nikkei Insatsu.

MHLW, n.d., 'Shokuhin Eisei Ho ni motozuku shokuhinchū no hōshasen busshitsu ni kansuru zantei kiseichi' (Provisional regulatory standards concerning radioactive materials in foods stipulated in the Food Safety Act), viewed February 24, 2017, <www.mhlw.go.jp/shinsai_jouhou/dl/shokuhin.pdf>.

Mikuriya Takashi, 2014, *Chi no kakutō: Okiteyaburi no seijigaku kōgi* (The combat of knowledge: a lesson in the political science of rule-breaking), Chikuma shobō.

Minister for Internal Affairs and Communication, 2011 (August 19), 'Communique: the adoption of legal measures regarding the special circumstances and response to those required to re-locate in connection to the evacuation of residents necessary as a result of the nuclear power plant incident that followed the Great East Japan Earthquake'.

Ministry of Agriculture, Forestry and Fisheries, 2012–16, *Shokuryō, nōgyō, nōson hakusho Heisei 23nenban–27nenban* (White Papers on food, agriculture and agricultural regions 2011–2015).

Ministry of Agriculture, Forestry and Fisheries, 2017, 'Import regulations imposed by foreign countries', viewed April 1, 2017, <www.maff.go.jp/export/e.info/pdf/kisei_all_170317.pdf>.

Ministry of Foreign Affairs, 2011, "G8 declaration renewed commitment for freedom and democracy', viewed December 13, 2013, <www.mofa.go.jp/policy/economy/summit/2011/declaration.html>.

Mirai Support Ishinomaki, 2013, 'Ishinomaki saigai fukkō shien kyōgikai katsudō hōkokusho' (Report of Ishinomaki Disaster Reconstruction Support Committee activity).

Miyagi Prefecture, 2012, 'Higashi Nihon daishinsai – Miyagi ken no 6 ka getsu kan no saigai taiō to sono kenshō' (The Great East Japan Earthquake – review of six months of disaster response in Miyagi Prefecture).

Miyake Katsuhisa, 2011, *Nihon o horobosu denryoku fuhai* (The electric utility corruption that will sink Japan), Shinjinbutsu Ōraisha.

Miyazaki Tomomi, Kimura Hideaki and Kobayashi Gō, 2013, *Fukushima genpatsu jiko taimurain 2011–2012* (Timeline of the Fukushima nuclear accident 2011–2012), Iwanami Shoten.

MOE (Ministry of the Environment), 2011, 'Josen no hanashi' (Decontamination), viewed September 10, 2015, <http://josen.env.go.jp/material/download/pdf/josen.pdf>.

MOE, 2012a, 'Josen tokubetsu chi'iki ni okeru josen no hōshin (josen rōdo mappu) ni tsuite' (Outline of decontamination policy in the special contamination areas (the decontamination roadmap)), January 26, viewed September 10, 2015, <www.env.go.jp/press/files/jp/19093.pdf>.

MOE (Environment Restoration Review Committee), 2012b, 'Kongo no shinrin josen no arikata ni kansuru tōmen no seiri ni tsuite' (Urgent measures for the decontamination of forested areas), viewed September 10, 2015, <http://josen.env.go.jp/material/session/007.html>.

MOE (Decontamination Team), 2013, 'Kuni oyobi chihō jichitai ga jisshi shita josen jigyō ni okeru josen no kōka (kūkan senryō ritsu) ni tsuite' (The results of decontamination works carried out by the national and municipal governments), December, viewed September 10, 2015, <www.env.go.jp/jishin/rmp/conf/10/ref05.pdf>.

MOE (Environmental Management Bureau), 2014, 'Josen no shinpo jōkyō ni tsuite' (Progress on decontamination), August, viewed September 10, 2015, <www.env.go.jp/Jishin/jishin/rmp/conf/12/mat06.pdf>.

MOE, 2015a, *FY2014 Decontamination Report,* viewed February 6, 2017, <http://josen.env.go.jp/en/cooperation/pdf/decontamination_report1503_full.pdf>.

MOE (Decontamination Team), 2015b, 'Josen no shinpo jōkyō ni tsuite' (Progress on decontamination), March, viewed August 10, 2016, <http://josen.env.go.jp/material/session/pdf/014/mat02_1.pdf>.

MOE, 2017, 'Progress on off-site clean-up and interim storage facility in Japan', viewed April 29, 2017, <http://josen.env.go.jp/en/pdf/progressseet_progress_on_cleanup_efforts.pdf>.

MOJ (Ministry of Justice), 2012, *Hōsō yōsei seido kentōkai, Hōsō jinkō ni kansuru kisoteki shiryō* (Basic documents on the legal workforce) (Appendix 2), September 20.

Mori Hiroki, 1997, 'Senkyokatei ni okeru gōrisei no shōtotsu: Jimintō seiken no keizoku to Shakaitō' (Crash of reasonability in the election process: continuation of the LDP administration and the JSP), *Tsukuba hōsei* (Tsukuba Journal of Law and Politics), 23: 215–32.

Mori Hiroki, 2001, *Nihon Shakaitō no kenkyū: rosen tenkan no seiji katei* (Studies of the JSP: the political process of policy change), Bokutakusha.

Mori Hiroki, 2016, '2012 nen sōsenkyo no tokuhyō bunseki: shinsaigo no kokuseisenkyo ni arawareta min'i' (Voting analysis of the 2012 general election: popular will which appeared in the national election after the great earthquake), in Tsujinaka Yutaka (ed.), *Daishinsai ni manabu shakai kagaku, dai 1 kan, seiji katei to seisaku* (Lessons for the Social Sciences from the 2011 Great East Japan Earthquake: Volume 1 political process and policy), Tōyō Keizai, pp. 295–317.

Mōri Yoshitaka, 2009, *Sutorīto no shisō: tenkanki to shite no 1990 nendai* (Philosophy in the streets: the 1990s as a turning point), NHK Publishing INC.

Morishima Akio, 2011a, 'Genshiryoku jiko no higaisha kyūsai: songai baishō to hoshō 1' (Aid for disaster victims after Fukushima: compensation and reparations, part 1), *Toki no Hōrei* (Contemporary Law), 1882: 39–47.

Morishima Akio, 2011b, 'Genshiryoku jiko no higaisha kyūsai: songai baishō to hoshō 2' (Aid for disaster victims after Fukushima: compensation and reparations, part 2), *Toki no Hōrei* (Contemporary Law), 1884: 35–42.

Morishima Akio, 2011c, 'Genshiryoku jiko no higaisha kyūsai: songai baishō to hoshō 3' (Aid for disaster victims after Fukushima: compensation and reparations, part 3), *Toki no Hōrei* (Contemporary Law), 1888: 49–57.

Moury, Catherine and Elisabetta De Giorgi, 2015, 'Introduction: conflict and consensus in parliament during the economic crisis', *Journal of Legislative Studies*, 21 (1): 1–13.

Muramatsu Michio, 1994, *Nihon no gyōsei: katsudōkei kanryōsei no henbō* (Governance in Japan: transformation of activity-style bureaucracy), Chūkō shinsho.

Muramatsu Michio, 1997, *Local Power in the Japanese State*, Betsey Scheiner and James White (trans.), Berkeley, CA: University of California Press.

Muramatsu Michio, 2006, 'Kihan, seido, insentibu kōzō no hen'yō' (Transformation of norms, systems and incentive structures), in Muramatsu Michio and Kume Ikuo (eds), *Nihon seiji hendō no 30 nen:*

seijika, kanryō, dan'tai chōsa ni miru kōzō hen'yō (Japanese politics, 30 years of transformation: structural transformation as seen in surveys of politicians, bureaucrats and organizations), Tōyō Keizai Inc.

Muramatsu Michio, Farrukh Iqbal and Ikuo Kume (eds), 2001, *Local Government Development in Post-war Japan*, New York: Oxford University Press.

Nagai Kōju, 2012, 'Higashi Nihon daishinsai de no bengoshi-kai no hisaisha shien katsudō' (Great East Japan Earthquake disaster victim support activity by Bar Associations), *NBL*, 974: 12–20.

NAIIC (National Diet of Japan Fukushima Nuclear Accident Independent Investigation Commission), 2012, Official Report of the National Diet of Japan Fukushima Nuclear Accident Independent Investigation Commission, viewed February 6, 2017, <http://warp.da.ndl.go.jp/info:ndljp/pid/3856371/naiic.go.jp/en/report/>.

Nakajima Hajime, 2013, *Genpatsu baishō chūkan shishin no kangaekata* (The philosophy behind the interim guidelines on the determination of the scope of nuclear damage), Shōjihōmu.

Nakanishi Junko, 2014, *Genpatsu jiko to hōshasen to risuku gaku* (Nuclear accidents and the study of radiation risk), Nippon Hyōronsha.

Nakazawa Hideo, 2005, *Jūmin Tōhyō Undō to Rōkaru Rejiimu: Niigataken Maki machi to kongenteki minshushugi no hosomichi* (Local referendum movement and local regime: Maki, Niigata and the narrow road to true democracy), Harvest-sha.

Namie Town, 2011, 'Fukkō ni kansuru chōmin ankēto shūkei kekka' (Results of a survey of municipal residents concerning reconstruction), viewed August 10, 2016, <www.town.namie.fukushima.jp/uploaded/attachment/71.pdf>.

Natori Ryota, 2014, '2012 nen shuinsen ni okeru seitō tōhyō to kōhosha tōhyō' (Party vote and personal vote in the 2012 House of Representatives election), *Jōhō kenyū* (Information Studies, Kansai University, Faculty of Informatics Bulletin), 41: 71–84.

Network for Radiation-Exposed Work, 2012, *Nuclear Power Station Accident and Radiation-exposed Work*, San-ichi Shobo.

NGA (National Governors Association), 2013, *Higashi Nihon daishinsai ni okeru Zenkoku Chijikai no torikumi* (The National Governors Association's response to the Great East Japan Earthquake).

NHK, 2012, 'NHK Supesharu 3.11 ano hi kara 1 nen "chōsa hōkoku genpatsu manē": 3 chō en wa chiiki o dō kaeta no ka' (NHK special: one year since 3.11, research report on nuclear power stations and money: how did 3 trillion yen change the host communities?), broadcast March 8.

NHK ETV Tokushū Shuzaihan, 2013, *Genpatsu merutodaun e no michi: genshiryoku seisaku kenkyūkai 100 jikan no shōgen* (The road to a nuclear meltdown: 100 hours of testimony from the council on nuclear policy), Shinchōsha.

Nihei Norihiro, 2012, '3.11 borantia no "teitai" mondai o saikō suru – 1995 nen no paradaimu wo koete' (Reconsidering the issue of the 3.11 volunteer 'backlog' – beyond the 1995 paradigm), in Hasebe Toshiharu and Funabashi Harutoshi (eds), *Jizoku kanōsei no kiki – jishin, tsunami, genpatsu jiko higai ni mukiatte* (Sustainability crisis in the face of an earthquake, tsunami and nuclear disaster), Ochanomizu shobō, pp. 159–88.

Nihon Genshiryoku Bunka Shinkō Zaidan (Japan Atomic Energy Relations Organization), 2014, *Heisei 25 Nendo Genshiryoku Riyō no Chishiki Fukyū Keihatsu ni Kansuru Yoron Chōsa Hōkokusho* (Report on the public opinion survey on public awareness of the use of nuclear power).

Nihon kango kyōkai shuppankai henshūbu hen (Editorial Committee of the Japanese Nursing Association Publishers (ed.), 2011, *Rupo: sono toki kango wa naasu hatsu Higashi Nihon Daishinsai repooto* (A record of nurses at that time: a report from the Great East Japan Earthquake and tsunami), Nihon kango kyōkai shuppankai (Japanese Nursing Association Publishers).

Nishikido Makoto and Harada Shun, 2013, 'Higashi Nihon daishinsai ni yoru kengai hinansha ni tai suru jichitai taiō to shien: Saitama Ken no jichitai o jirei toshite' (The support and response from local government to evacuees from outside the prefecture following the Great East Japan Earthquake: the case of Saitama Prefecture municipalities), *Ningen Kankyo Ronsyu* (Human Environment Studies), 14 (1): 1–26.

Nishio Masaru, 2013, *Jichi, bunken saikō* (Rethinking municipal division of powers), Gyōsei.

NPA (National Police Agency), 2012, *The White Paper on Police 2011*, digest edn.

Nuclear Damage Compensation Facilitation Corporation, 'About the amount of general and special contributions', viewed March 7, 2017, <www.ndf.go.jp/press/at2016/20160331_02.html>.

Nuclear Emergency Response Headquarters, 2011, 'Josen ni kansuru kinkyu jisshi hōshin' (A guideline on urgent decontamination procedures), August.

Nuclear Emergency Response Headquarters, 2013, 'Genshiryoku saigai kara no Fukushima no fukkō kasoku ni mukete' (For accelerating the reconstruction of Fukushima from the nuclear disaster), December 20.

OECD (Organisation for Economic Co-operation and Development), 2015, *Government at a Glance 2015*, Paris: OECD Publishing, viewed February 23, 2017, <www.oecd-ilibrary.org/docserver/download/4215081e.pdf?expires=1487830525&id=id&accname=guest&checksum=7575D00FF47339A8178A76B713ED458C>.

Office of the Cabinet, 2012, *Heisei 24 nendo bōsai hakusho* (Heisei 24 [2012] Disaster prevention white paper).

Ogawa Akihiro, 2014, 'Civil society: past, present, and future', in Jeff Kingston (ed.), *Critical Issues in Contemporary Japan*, New York: Routledge, pp. 52–63.

Oguma Eiji, 2013, 'Mōten wo saguriateta shikō' (The thinking process which found blind spots after groping), in Oguma Eiji (ed.), *Genpatsu wo tomeru hitobito: 3.11 kara kanteimae made* (People who stop nuclear power plants: from 3.11 to the prime minister's residence), Bungei Shunjū, pp. 193–304.

Okamoto Masakatsu, 2013, 'Higashi Nihon daishinsai kara no fukkō – tamesareru seifu no nōryoku' (Reconstruction after the Great East Japan Earthquake: testing the capabilities of government), *Nenpō gyōsei kenkyū* (Journal of the Japanese Society for Public Administration),

48, Higashi Nihon daishisai ni okeru gyōsei no yakuwari (The role of government in the Great East Japan Earthquake), Gyōsei.

Okamoto Masahiro, 2013a, '"Higashi Nihon daishinsai de wa, nannin ga borantia ni itta no ka" to iu toi kara' (From the question of 'How many people came to Iwate Prefecture to volunteer at the time of the Great East Japan Earthquake?'), *Borantarizumu kenkyū* (Volunteerism Studies), 2: 3–14.

Okamoto Masahiro, 2013b, 'Higashi Nihon daishinsai ni okeru 18 shūkyō kyōdan no shien katsudō – chōsa gaiyō no hōkoku' (Relief activities of 18 religious groups during the Great East Japan Earthquake – survey overview report), *Nihon NPO gakkai nyuuzuretaa* (Japan NPO Research Association Newsletter), 15 (1).

Ōkuma Town, 'Kuni nado e no yōbō' (Our demands), viewed June 30, 2015, <http://ohkuma.maildepot.jp/fukkou/%E5%9B%BD%E7%AD%89%E3%81%B8%E3%81%AE%E8%A6%81%E6%9C%9B%E7%AD%89>.

Ōkura Sae, 2015, 'Higashi Nihon dai shinsaigo no genpatsu baishō ni kansuru chōsa: kekka hōkoku' (Report of a survey on nuclear damage compensation after the Great East Japan Earthquake), in Tsujinaka Yutaka (ed.), *Higashi Nihon daishinsai e no seijigakuteki apurōchi* (A political science approach to the Great East Japan Earthquake), ICR Research Report 2015, University of Tsukuba, pp. 245–55.

Onda Katsunobu, 2011, *Tokyo Denryoku: teikoku no ankoku* (TEPCO: the dark empire), Nanatsumori Shokan.

Ōnishi Yutaka, 1993, 'Kokka kensetsu to jūmin ha'aku: Nihon to Kankoku ni okeru jūmin ha'aku seido keisei katei no kenkyū' (Nation building and mapping resident presence: a study of system structure processes for mapping resident presence in Japan and Korea), Part 1, *Ōsaka Shiritsu Daigaku Hōgaku Zasshi* (Osaka City University Law Faculty Journal), 40 (1): 64–105.

Ōnishi Yutaka, 1994, 'Kokka kensetsu to jūmin ha'aku: Nihon to Kankoku ni okeru jūmin ha'aku seido keisei katei no kenkyū' (Nation building and mapping resident presence: a study of system structure processes for mapping resident presence in Japan and Korea) (Part 2), *Ōsaka Shiritsu Daigaku Hōgaku Zasshi* (Osaka City University Law Faculty Journal), 40 (2): 137–69.

Ōshika Yasuaki, 2013, *Merutodaun: dokyumento Fukushima Daiichi genpatsu jiko* (Meltdown: documenting the Fukushima nuclear accident), pocket edn, Kōdansha.

Ōshima Ken'ichi and Yokemoto Masafumi, 2012, *Genpatsu jiko no higai to hoshō: Fukushima 'ningen no fukkō'* (Damage and compensation after a nuclear accident: Fukushima and 'human reconstruction'), Ōtsuki Shoten.

Ōshima Ken'ichi and Yokemoto Masafumi, 2014, 'Fukushima genpatsu jiko no kosuto to kokumin, denryoku shōhisha e no futan tenka no kakudai' (The cost of the Fukushima nuclear disaster and the increase in the shift of the burden to citizens and electricity consumers), Ōsaka City University The Business Review, 65 (2): 1–24.

Ōta Masakatsu, 2015, *Nihon wa naze kaku o tebanasenai no ka: 'hikaku' no shikaku* (Why can't Japan let go of nukes? Antinuclear's blind spot), Iwanami Shoten.

Ōtake Hideo, 1996, *Gendai Nihon no seiji kenryoku keizai kenryoku* (Political and economic power in contemporary Japan), revised edn, Tokyo, Sanichi Shobō.

Ōtsuka Tadashi, 2011, 'Fukushima Daiichi genpatsu jiko ni yoru songai baishō to baishō shien kikōhō: fuhō kōi hōgaku no kanten kara' (Nuclear damage compensation after the Fukushima accident and the Nuclear Damage Compensation Facilitation Corporation Act: the tort law perspective), *Jurisuto* (Monthly Jurist), 1433: 39–44.

Peace-Boat Disaster Volunteer Center, 2014, '2013 nendo katsudō hōkokusho' (2013 activity report).

Pekkanen, Robert J., 2006, *Japan's Dual Civil Society: Members without Advocates*, Stanford, CA: Stanford University Press.

Pekkanen, Robert, Steven Reed and Ethan Scheiner (eds), 2013, *Japan Decides 2012: The Japanese General Election*, Basingstoke, Hampshire, and New York: Palgrave Macmillan.

Pekkanen, Robert, Steven Reed and Ethan Scheiner (eds), 2016, *Japan Decides 2014: The Japanese General Election*, Basingstoke, Hampshire, and New York: Palgrave Macmillan.

Peters, Guy B. and Brian W. Hogwood, 1985, 'In search of the issue-attention cycle', *Journal of Politics*, 47: 238–53.

Pierson, Paul, 1994, *Dismantling the Welfare State?: Reagan, Thatcher, and the Politics of Retrenchment*, Cambridge: Cambridge University Press.

Prefectural Oversight Committee for the Fukushima Health Management Survey, 2013, 'Shiryo 1: Kenmin kenkō chōsa "kihon chōsa" no jisshi jokyō ni tsuite' (Material 1: about the implementation of the 'Basic Survey', Fukushima Health Management Survey) (Materials for the 12th Oversight Committee Meeting), August 20.

Prefectural Oversight Committee for the Fukushima Health Management Survey, 2015a, 'Shiryo 1: Kenmin kenkō chōsa "kihon chōsa" no jisshi jokyō ni tsuite' (Materials 1: about the implementation of the 'Basic Survey', Fukushima Health Management Survey) (Materials for the 19th Oversight Committee Meeting), May 18.

Prefectural Oversight Committee for the Fukushima Health Management Survey, 2015b, 'Shiryo 3-1: Kenmin kenkō chōsa "kōjōsen kensa (senkō kensa)" kekka gaiyō (zantei ban)' (Materials 3-1: the summary results of the 'Thyroid Ultrasound Test (the Primary Test)', Fukushima Health Management Survey (provisional version)) (Materials for the 19th Oversight Committee meeting), May 18.

Prefectural Oversight Committee for the Fukushima Health Management Survey, 2016a, 'Kenmin kenkō chōsa ni okeru chūkan torimatome' (The interim report related to the Fukushima Health Management Survey), March.

Prefectural Oversight Committee for the Fukushima Health Management Survey, 2016b, 'Shiryo 2-1: Kenmin kenkō chōsa "kōjōsen kensa (honkaku kensa)" jisshi jokyō' (Materials 2-1: about the implementation of 'the Thyroid Ultrasound Test (the Full-Scale Test)', Fukushima Health Management Survey) (Materials for the 23rd Oversight Committee Meeting), June 6.

Prefectural Oversight Committee for the Fukushima Health Management Survey, 2016c, 'Shiryo 2-2: Kenmin kenkō chōsa "kōjōsen kensa (senkō kensa)" kekka gaiyō (Heisei 27 nendo tsuiho ban)' (Materials 2-2: the summary results of 'the Thyroid Ultrasound Test (the Primary Test)', Fukushima Health Management Survey) (additional version in 2015) (Materials for the 23rd Oversight Committee Meeting), June 6.

Prefectural Oversight Committee for the Fukushima Health Management Survey, Thyroid Ultrasound Examination Assessment Section, 2015, 'Kōjōsen kensa ni kansuru chūkan torimatome' (The interim report related to the Thyroid Ultrasound Test), May 18.

Public Relations Office, Government of Japan, 2014, 'Government public relations online, September 1', viewed February 24, 2017, <www.gov-online.go.jp/useful/article/201204/3.html>.

Ramseyer, J. Marc and Eric Rasmusen, 2003, *Measuring Judicial Independence: The Political Economy of Judging in Japan*, University of Chicago Press.

Reconstruction Agency, 2012, 'Total number of evacuees throughout Japan', viewed September 1, 2015, <www.reconstruction.go.jp/topics/main-cat2/sub-cat2-1/hinanshasuu.html>.

Reconstruction Agency, 2014a, '1: Residents' intention survey in the nuclear disaster stricken local government in Heisei', June 25.

Reconstruction Agency, 2014b, 'Heisei 25 nendo genshiryoku hisai jichitai ni okeru jūmin ikō chōsa' (Opinion survey of residents in municipalities affected by the nuclear disaster conducted in 2013).

Reconstruction Design Council in Response to the Great East Japan Earthquake, 2011, *Toward Reconstruction: Hope beyond the Disaster*, June 25.

Reed, Steven, 2013, 'Challenging the two-party system: third force parties in the 2012 election', in Robert Pekkanen, Steven Reed and Ethan Scheiner (eds), *Japan Decides 2012: The Japanese General Election*, Basingstoke, Hampshire, and New York: Palgrave Macmillan.

Research Center for Crisis and Contingency Management, 2012, 'Bōsai, kiki kanri shisaku ni kansuru ankēto chōsa tantōka ankēto' (Survey on disaster and crisis management), viewed February 23, 2017, <www.kisc.meiji.ac.jp/~crisishp/ja/pdf/2012/questionnaire-in_charge.pdf>.

Rhodes, R.A.W., 1997, *Understanding Governance: Policy Networks, Governance, Reflexibility and Accountability,* London, Open University Press.

Rhodes, R. A. W., 1999, *Control and Power in Central–Local Government Relations*, 2nd edn, Brookfield, VT: Ashgate.

Riker, William, 1962, *The Theory of Political Coalitions*, New Haven, CT: Yale University Press.

Riker, William, 1996, *The Strategy of Rhetoric: Campaigning for the American Constitution*, New Haven and London: Yale University Press.

Rosenbluth, Frances McCall and Michael F. Thies, 2010, *Japan Transformed: Political Change and Economic Restructuring*, Princeton, NJ: Princeton University Press.

Saitō Makoto, 2015, *Shinsai fukkō no seiji keizaigaku* (The political economy of disaster reconstruction), Tokyo, Nippon Hyōronsha.

Saitō Shinichi, Toshio Takeshita and Tetsurō Inaba, 2014, 'Shimbun no chōsa wa dokusha no taido ni eikyō suru ka: genpatsu mondai wo jirei to shite' (Does the tone of newspapers affect their subscribers' attitudes?:

A case of nuclear energy issues), *Shakai to Chōsa* (Advances in Social Research), 13: 58–69.

Saitō Takao, 2011, *Min'i no tsukurarekata* (Making the popular will), Iwanami Shoten.

Saitō Takao, 2012, *'Tokyo denryoku' kenkyū, haijo no keifu* (Research on TEPCO, a genealogy of exclusion), Kōdansha.

Sakurai Kiyoshi, 2011, *Genpatsu saiban* (Judgments in nuclear power lawsuits), Ushio Shuppansha.

Sakurai Toshio, Masaki Hiroyuki and Yanase Sho, 2012, 'Fukkō suishin taisei no seibi – fukkō tokku hō, Fukkōchō secchi hō, Fukushima tokuso hō' (Establishing reconstruction support structures: the Special Reconstruction Zone Law, the Establishment of the Reconstruction Agency Law and the Fukushima Special Measures Law), Rippō to Chōsa (Legislation and Inquiry), 329: 14–25.

Samuels, Richard J., 2013, *3.11: Disaster and Change in Japan*, Ithaca, NY: Cornell University Press.

Sasaki Kei'ichi, 2011a, '*Yomiuri Shimbun*, *Shūkan Shinchō*, *Sotokoto*, *Gekkan WiLL*, *Ushio*...shūkanshi, shimbun no "Tōden kōkoku" shukkō hindo wāsuto rankingu!' (*Yomiuri Shimbun*, *Shūkan Shinchō*, *Sotokoto*, *Gekkan WiLL*, *Ushio*...TEPCO advertising in the weekly magazines and newspapers, the worst of the worst), *Bessatsu Takarajima* 1796: 64–9.

Sasaki Kei'ichi, 2011b, 'Tōdai, Kyōdai, Handai e no jōhō kōkai seikyū de hakkaku, goyō gakusha ga uketotta genshiryoku sangyō no kyogaku kifukin!' (The enormous donations received by pronuclear academics from the nuclear energy industry as revealed in a freedom of information request to University of Tokyo, Kyoto University, Osaka University), *Bessatsu Takarajima* 1796: 102–4.

Sasaki Kei'ichi, 2011c, 'Kokumin ga kizukanai yoron kōsaku 1: "Yomiuri Shimbun" ni genpatsu suishin kōkoku o dasu "chikyū o kangaeru kai" no shikingen' (The crafting of public opinion 1: who paid for the Chikyū o Kangaeru Kai's pronuclear advertisements in the Yomirui Shimbun?), *Bessatsu Takarajima* 1821: 69–71.

Satō Eisaku, 2011, *Fukushima genpatsu no shinjitsu* (The truth about the Fukushima Nuclear Power Station), Heibonsha.

Satō Katsuharu and Abe Arata, 2013, 'Fukushima dai 1 genpatsu jiko ni yoru dojō osen no josen no genjō: Minamisōma shi, Kawauchi mura ni okeru osen jōkyō jūten chōsa kuiki no josen jirei kara' (Decontamination of radioactive soil after the Fukushima Daiichi nuclear accident: the example of the intensive contamination survey areas in Minamisōma and Kawauchi), *Kankyō keizai, seisaku kenkyū* (Research on Environmental Economics and Policy), 6 (2): 54–9.

Satō Tetsuo, Yoshida Fumiharu and Hashimoto Satoshi, 2003, 'Shihō seido kaikaku to ADR' (The Justice System Reform and ADR), in Kojima Takeshi (ed.), *ADR no jissai to riron* (ADR in theory and practice), Chūō Daigaku Shuppanbu, pp. 26–63.

Satō Toshihiro, 2012, 'Minutes of the Fukushima Kaigi Josen Bunkakai', viewed June 30, 2015, <www.fullchin.jp/f/2012bukai/2012111zyosenbu/02/bu02.html>.

Sawafuji Tōichirō, 2001, 'Minji saiban wa dō naru' (What will happen to civil trials?), in Minshushugi Kagaku Kyōkai Hōritsu Bukai (ed.), *Dare no tame no 'shihō kaikaku' ka: 'shihō seido kaikaku shingikai chūkan hōkoku' no hihanteki kentō* (Justice System Reform for whom? A critique of the interim report of the Justice System Reform Council), Nippon Hyōronsha, pp. 64–71.

Scheiner, Ethan, 2012, 'The electoral system and Japan's partial transformation: party system consolidation without policy realignment', *Journal of East Asian Studies*, 12 (3, special issue): 351–79.

Schwartz, Frank J. and Susan J. Pharr, 2003, *The State of Civil Society in Japan*, Cambridge University Press.

Sengoku Yoshito, 2013, *Enerugī, genshiryoku dai tenkan: denryokugaisha, kanryō, han-genpatsu ha to no kōshō hiroku* (The great transformation in energy and nuclear policy: a confidential record of negotiations with the electric utilities, the bureaucracy and opponents of nuclear power), Kōdansha.

Shibata Tetsuji and Tomokiyo Hiroaki, 1999, *Genpatsu kokumin yoron: yoron chōsa ni miru genshiryoku ishiki no hensen* (Public opinion on nuclear power: people's changing consciousness about nuclear power as seen through opinion polling), ERC Shuppan.

Shikata Noriyuki, 2014, *Higashi Nihon dai shinsai no kantei kara no kokusai kōhō katsudō to paburikku dipuromashii* (International public relations and public diplomacy after the Great East Japan Earthquake), in M. Kaneko and M. Kitano (eds), 2014, *Paburikku dipuromashii senryaku: imēji o kisou kokka kan gēmu ni ika ni shōri suru ka* (Public diplomacy strategy: how to win the image war between nations), PHP Kenkyūjo, pp. 227–46.

Shimoyama Kenji, 2014, 'Genshiryoku songai to kisei kengen fukōshi no kokka baishō sekinin' (Nuclear damage and state liability), *Hōgaku Jihō* (Legal News) 86 (2): 62–7.

Shimura Kaichirō, 2011, *Tōden teikoku, sono shippai no honshitsu* (Imperial TEPCO, the truth about its fall), Bungei Shunjū.

Shinada Yutaka, 1998, 'Senkyo kōyaku seisaku deita ni tsuite' (Making the data of Japanese politicians' policy preferences at the 1990's general elections), *Kōbe hōgaku zasshi* (Kobe Law Journal), 48 (2): 541–72.

Shinada Yutaka, 2006, 'Senkyo kōyaku seisaku deita ni tsuite' (On campaign promise policy data), *Nihon seiji kenkyū* (Japanese Political Studies), 3 (2): 63–91.

Shindō Muneyuki, 2012, *Shihō yo! omae ni mo tsumi ga aru: genpatsu soshō to kanryō saibankan* (Hey judges! Your hands are dirty too: antinuclear lawsuits and the bureaucratic judiciary), Kōdansha.

Shindō Muneyuki and Tamura Yasuhiro, 2014, 'Intabyū: Miyamoto Kōichi, zenkoku barabara ni sundeitemo jūminhyō wa Tomioka ni' (Even if we are scattered all over the country, we are residents of Tomioka: interview with Miyamoto Kōichi), *Toshi Mondai* (Urban Issues), March.

Shiozaki Yasuhisa, 2012, *Gabanansu o seiji no te ni: 'Genshiryoku kisei iinkai' sōsetsu e no tatakai* (Bringing governance under political control: the fight to establish the Nuclear Regulation Authority), Tokyo Puresu Kurabu.

Shushō Kantei Kokusai Kōhō Shitsu, 2012, 'Higashi Nihon dai shinsai to kantei kokusai kōhō katsudō' (The Great East Japan Earthquake and international public relations by the Prime Minister's Office), PowerPoint paper provided by Noriyuki Shikata of the Ministry of Foreign Affairs.

Soeda Takashi, 2014, *Genpatsu to ōtsunami, keikoku o hōmutta hitobito* (Nuclear power and the great tsunami, the people who buried the warning signs), Iwanami Shoten.

Soga Kengo, 2016, 'Gyōsei – Higashi Nihon daishisai ni taisuru chūō fushō no taiō' (Governance: central agency response to the Great East Japan Earthquake), in Yutaka Tsujinaka (ed.), *Daishinsai ni manabu shakaikagaku* (Lessons for the Social Sciences from the 2011 Great East Japan Earthquake)*, Volume 1, Political Process and Policy*, Tōyō Keizai.

Strom, Kaare, 1990, *Minority Government and Majority Rule*, Cambridge, New York and Melbourne: Cambridge University Press.

Supreme Court of Japan (ed.), 2002, *Saibansho dētabukku 2002* (Court Data 2002), Hanrei Chōsa.

Suzuki Hiroshi, 2013, 'Fukushima Daiichi Genpatsu jiko no fukkō no kadai' (The Fukushima nuclear accident and the open questions about reconstruction), *Keikaku Gyōsei* (Planning and Public Management), 36 (3).

Suzuki Manami, 2014, *Nihon wa naze genpatsu o yushutsu suru no ka* (Why does Japan export nuclear energy?), Heibonsha.

Suzuki Tatsuru, 1983, *Denryoku gaisha no atarashii chōsen: gekidō no 10 nen o norikoete* (A new challenge for the utilities: overcoming a turbulent decade), Nihon Kōgyō Shimbunsha.

Suzuki Yoshinori, 2013, 'Fukushima ken ni okeru josen no genjō to kadai' (Current issues in the decontamination of radioactive substances in Fukushima Prefecture), *Refarensu* (Reference), 63 (3): 97–108.

Tahara Sōichirō, 2011, *Dokyumento Tokyo Denryoku: Fukushima genpatsu tanjō no uchimaku* (Documenting TEPCO: the inside scoop about the construction of the Fukushima nuclear power station), Bungei Shunjū.

Takahashi Atsushi, 2011, 'Rōdō undō no moto tōshi kara bōryokudan ni tsunagaru kai'insei jōhōshi shusaisha made: Tōden no "ura masukomi taisaku" ni an'yaku shita gyōkaijin tachi' (From former union fighters to the president of a subscriber-only publication with links to the mafia: the business people who secretly engaged in TEPCO's 'hidden media strategy'), *Bessatsu Takarajima* 1796: 70–6.

Takahashi Atsushi, 2012, 'Genshiryoku bunka shinkō zaidan, Denryoku chūō kenkyūjo hoka, genshiryoku mura no kōeki hōjin ni "amakudari" shita shimbun sha kanbu tachi no jitsumei' (Japan Atomic Energy Relations Organization, Central Research Institute of the Electric Power Industry (CRIEPI) etc., the names of the newspaper managers who 'descended' to the nuclear village's public-interest corporations), in Y. Ichinomiya, H. Koide, T. Suzuki, T. Hirose et al., *Genpatsu saikadō no fukai yami*, Takarajimasha, pp. 95–112.

Takahashi Shigeru, 2014, 'Genshiryoku songai baishō hō no hōteki shomondai' (Legal issues of the Act on Compensation for Nuclear Damage), *Kōkyō Seisaku Kenkyū* (Journal of Public Policy Studies) 14: 86–98.

Takahashi Yasufumi, 2012, *Kaisetsu: Genshiryoku songai baishō shien kikō hō* (Understanding the Nuclear Damage Compensation Facilitation Corporation Act), Shōjihōmu.

Takayasu Kensuke, 2009, *Shushō no kenryoku: nichi-ei hikaku kara miru seikentō tono dainamizumu* (The power of prime ministers: dynamism of their relationships with the governing party comparing Japan and Great Britain), Sobunsha.

Takenaka Harukata 2004, '"Nihon gata bunkatsu seifu" to sangi'in no yakuwari' ('Separation of powers, Japanese style' and the role of the House of Councillors), *Nenpō seijigaku* (The Annals of Japanese Political Science Association), 55: 99–125.

Takenaka Yoshihiko, 2010, 'Renritsu seiken to seisaku kūkan' (The space of policy under the coalition government), *Seikatsu keizai seisaku* (Economic Policy Institute for Quality Life), May: 14–18.

Takeuchi Keiji, 2013, *Denryoku no shakaishi: nani ga Tokyo Denryoku o unda no ka* (A social history of electricity: how did TEPCO come to be?), Asahi Shimbun Shuppan.

Takeuchi Naoto, 2015, 'Shinsai fukkō ni okeru hisaisha jūtaku saiken shien seido no tenkai' (The development of a support system for the reconstruction of homes of disaster victims during the Great East Japan Earthquake restructure), in Kohara Takaharu and Inatsugu Hiroaki (eds), *Daishinsai ni manabu shakaikagaku dai 2 kan: shinsaigo no jichitai gabanansu* (Lessons for the Social Sciences from the 2011 Great East Japan Earthquake, vol. 2: Municipal governance after the disaster), Tōyō keizai, pp. 287–308.

Tanami Hisae, 2010, 'Hanshin Awaji daishinsai no kengai hisasha no ima – shinsai kara 15 nen' (The present circumstances of Hanshin–Awaji Great Earthquake victims outside the prefecture – 15 years after the disaster), *Saigai fukkō kenkyū* (Disaster Recovery Studies), 2: 143–59.

Tanami Hisae, 2011, 'Ikigai hinansha ni taisuru jōhō teikyō: Miyakejima funka saigai no hinansha chōsa o chūshin ni' (Providing information for evacuees outside the area: focusing on a survey of evacuees from the Miyake Island volcanic explosion), *Saigai fukkō kenkyū* (Disaster Recovery Studies), 3: 167–75.

Tanami Hisae, 2012, 'Higashi Nihon daishinsai ni okeru kengai hinansha eno shien' (The support to evacuees from outside the prefecture following the Great East Japan Earthquake), *Saigai fukkō kenkyū* (Disaster Recovery Studies), 4: 15–24.

Tanigaki Sadakazu, 2013, 'Yatō Jimintō o dō tatenaoshita ka' (Reforming the LDP in opposition), in Mikuriya Takashi, Makihara Izuru and Satō Shin, *Seiken kōtai o koete: seiji kaikaku no 20 nen* (Beyond regime change: 20 years of political reform), Iwanami Shoten, pp. 77–87.

Tarrow, Sidney, 1998, *Power in Movements, Collective Action and Politics,* 2nd edn, New York: Cambridge University Press.

TEPCO (Tokyo Electric Power Company), 2015, 'Fukushima fukkō honsha ni okeru baishō, josen, fukkō suishin ni kansuru torikumi jōkyō: Fukushima fukkō e no sekinin o hatasu tame ni' (Fukushima Revitalization Headquarters' handling of compensation, decontamination, and revitalization: taking responsibility for revitalizing

Fukushima), viewed April 5, 2017, <www.tepco.co.jp/fukushima_hq/images/150128_01-j.pdf>.

TEPCO (Tokyo Electric Power Co. Inc.), 2016a (April 28), 'About the distribution of exposure doses', in 'About the assessment status of the exposure doses by the workers of the Fukushima Daiichi Nuclear Power Plant', attached materials for press release, viewed June 1, 2016, <www.tepco.co.jp/press/release/2016/1280695_8626.html>.

TEPCO (Tokyo Electric Power Co. Inc.), 2016b, 'The situations of compensation payment', viewed September 15, 2016, <www.tepco.co.jp/fukushima_hq/compensation/results/index-j.html>.

Tokyo Denryoku Kaikau, 1F Mondai Iinkai, 2016 (December 20), *Tōdēn kaikaku sengen* (A manifesto for TEPCO reform), viewed March 1, 2017, <www.meti.go.jp/committee/kenkyukai/energy_environment/touden_1f/pdf/161220_teigen.pdf>.

Tokyo Metropolitan Office, 2012, Tōkyō-to shien katsudō hōkokusho (Tokyo Municipal Office outreach support report), March.

Tokyo Metropolitan Office, 2013, Tōkyō-to shien katsudō hōkokusho (Tokyo Municipal Office outreach support report), March.

Tokyo Metropolitan Office, 2014, Tōkyō-to shien katsudō hōkokusho (Tokyo Municipal Office outreach support report), March.

Tokyo Shimbun Genpatsu Jiko Shuzaihan, 2012, *Reberu 7: Fukushima genpatsu jiko, kakusareta Shinjitsu* (Level 7: the hidden truth about the Fukushima nuclear accident), Gentōsha.

Tōno City, 2013, '3.11 Higashi Nihon daishinsai: Tōno-shi kōhō shien katsudō kenshō kiroku-shi' (3.11 Great East Japan Earthquake: records of review of Tōno City back-up aid activities).

Torfing, Jacob, 2007, 'Introduction: democratic network governance', in Martin Marcussen and Jacob Torfing (eds), *Democratic Network Governance in Europe*, Basingstoke: Palgrave Macmillan.

Tsuda Toshihide, Tokinobu Akiko, Yamamoto Eiji and Suzuki Etsuji, 2016, 'Thyroid cancer detection by ultrasound among residents ages 18 years and younger in Fukushima, Japan: 2011 to 2014', *Epidemiology*, 27 (3): 316–22.

Tsugane Shōichirō, 2014, 'Fukushima ken ni okeru kōjōsen gan yūbyō sha sū no suii' (Estimation of the prevalence [of thyroid carcinoma] in the Fukushima prefecture), Prefectural Oversight Committee for the Fukushima Health Management Survey Thyroid Ultrasound Examination Assessment Section, 2014, Materials 5 (The 4th meeting of Thyroid Assessment Section), November 11.

Tsujinaka Yutaka (ed.), 2016, *Seiji hendōki no atsuryoku dantai* (Japanese pressure groups in a time of flux), Tokyo, Yūhikaku.

Tsujinaka Yutaka, Robert J. Pekkanen and Hidehiro Yamamoto, 2009, *Gendai Nihon no jichikai, chōnaikai* (Neighborhood associations and local governance in Japan), Tokyo, Bokutakusha.

Tsunekawa Keiichi 2016a, 'Toward a balanced assessment of Japan's responses to the triple disaster', in Keiichi Tsunekawa (ed.), *Five Years After: Reassessing Japan's Responses to the Earthquake, Tsunami, and the Nuclear Disaster*, Tokyo, University of Tokyo Press, pp. 1–32.

Tsunekawa Keiichi, 2016b, 'Crisis communication and foreign response', in K. Tsunekawa (ed.), *Five Years After: Reassessing Japan's Responses to the Earthquake, Tsunami, and the Nuclear Disaster*, Tokyo: University of Tokyo Press, pp. 193–222.

Tsunekawa Keiichi (ed.), 2016, *Five Years After: Reassessing Japan's Responses to the Earthquake, Tsunami, and the Nuclear Disaster,* Tokyo, University of Tokyo Press.

Tsutsumi Hidenori, 1998, '1996 shūgiin senkyo ni okeru kōhosha no kōyaku to tōhyō kōdō' (Candidate policy commitments and voting behavior in the 1996 election for the House of Representatives), *Senkyo kenkyū* (Japanese Journal of Electoral Studies), 13: 89–99.

Tsutsumi Hidenori, 2002, 'Senkyo seido kaikaku to kōhosha no seisaku kōyaku: shōsenkyoku hireidaihyo heiritsusei dōnyu to kōhosha no senkyo senryaku' (The effect of the electoral reform on the pledge in Japan), *Kagawa hōgaku* (Kagawa Law Review), 22 (2): 90–120.

Tsutsumi Hidenori and Uekami Takayoshi, 2007, '2003 sōsenkyo ni okeru kōhosha reberu kōyaku to seitō no rieki shūyaku kinō' (Party policy coherence in Japan: evidence from 2003 candidate-level electoral platforms), *Shakai kagaku kenkyū* (Journal of Social Science), 58 (5 and 6, combined issue): 33–48.

Tsutsumi Hidenori and Uekami Takayoshi, 2011, 'Minshutō no seisaku: keizokusei to henka' (The DPJ policy: continuity and change), in Uekami Takayoshi and Tsutsumi Hidenori (eds), *Minshutō no soshiki to seisaku* (The organization and the policy of the DPJ), Tōyō Keizai Inc., pp. 225–53.

Uekami Takayoshi, 2008, 'Tōshu senshutsu katei no minshuka: Jimintō to Minshutō no hikaku kentō' (The democratization of party leadership selection in Japan), *Nenpō seiji gaku* (The Annuals of the Japanese Political Science Association), 2008 (1): 220–40.

Uekami Takayoshi, 2010, 'Senkyo seido kaikaku to Jimintō sosai senshutsu katei no henyō: riidaa shippu wo umidasu kōzō to kosei no sōkoku' (Electoral reform and the transformation of LDP presidential elections: leadership, structure and personality), *Senkyo kenkyū* (Japanese Journal of Electoral Studies), 26 (1): 26–37.

Uekami Takayoshi, 2011, 'Senkyo seido kaikaku to Minshutō daihyō senshutsu katei ni okeru hunsō kanri: shintō ga chokumen suru kōzō mondai' (Election system reform and conflict management in the DPJ leadership election process: structural problems confronted by new parties), in Uekami Takayoshi and Tsutsumi Hidenori (eds), *Minshutō no soshiki to seisaku* (The organization and the policy of the DPJ), Tōyō Keizai Inc., pp. 71–98.

Uekami Takayoshi, 2015, 'Minshutō seikenki ni okeru daihyō senkyo' (Party leadership elections during the DPJ administration period), in Maeda Yukio and Tsutsumi Hidenori (eds), *Tōchi no jōken: Minshutō seiken ni miru seiken un'ei to tōnai tōchi* (The preconditions for government: political administration and party politics in the DPJ), Chikura Shobō, pp. 79–111.

Uekami Takayoshi and Tsutsumi Hidenori, 2015, 'Minshutō seiken ni okeru seisaku keisei to manifesuto' (Policy formation and manifestos in the DPJ administration), in Maeda Yukio and Tsutsumi Hidenori (eds), *Tōchi no jōken: Minshutō seiken ni miru seiken un'ei to tōnai tōchi*

(The preconditions for government: political administration and party politics in the DPJ), Chikura Shobō, pp. 113–46.

Ueta Kazuhiro, 2016, *Hiyō no hōkatsu teki ha'aku* (A comprehensive overview of the damages and costs of the Great East Japan Earthquake), Tōyō Keizai Shinpōsha.

United Nations Scientific Committee on the Effects of Atomic Radiation, 2008, 'Sources and effects of radiation', report (Japanese version), vol. 2, Independent Administrative Institution, National Institute of Radiological Sciences.

Wada Akiko, 2015, 'Kengai hinansha ukeire jichitai no taiō' (Response of local authorities who accepted disaster victim evacuees from outside the prefecture), in Kohara Takaharu and Inatsugu Hiroaki (eds), *Daishinsai ni manabu shakaikagaku dai 2 kan: shinsaigo no jichitai gabanansu* (Lessons for the Social Sciences from the 2011 Great East Japan Earthquake, vol. 2: Municipal governance after the disaster), Tōyō Keizai, pp. 191–212.

White Paper on Local Public Finance 2016, visual edn, <www.soumu.go.jp/iken/zaisei/28data/index.html>.

WHO (World Health Organization), 2011, 'Impact on seafood safety of the nuclear accident in Japan – 9 May 2011', viewed December 21, 2013, <www.who.int/foodsafety/impact_seafood_safety_nuclear_accident_japan_090511.pdf>.

Wolferen, Karel van, 1989, *The Enigma of Japanese Power*, New York: Alfred A. Knopf.

Yaginuma Mitsuhiko, 2009, 'Genshiryoku songai baishō hō nado no ichibu kaiseian: genshiryoku songai ni okeru higai hoshō no jūjitsu' (Partial amendments to the Act on Compensation for Nuclear Damage: adequate compensation for victims of nuclear damage), *Rippō to Chōsa* (House of Councillors Research Reports), 291, pp. 16–23.

Yamada Kenta, 2013, *3.11 to media* (3.11 and the media), transview.

Yamada Masahiro, 2016, *Seiji sanka to minshu seiji* (Political participation and democratic politics), University of Tokyo Press.

Yamagata City, 2014, 'Population and household statistics', viewed August 1, 2016, <www.city.yamagata-yamagata.lg.jp/kakuka/kikaku/kikaku/sogo/gazoufile/toukei/suikei/H28suikei/p5mken.pdf >.

Yamagata Prefecture, 2011a, 'Number of evacuees in evacuation centers in Yamagata Prefecture, 2011', published 5.00 pm, December 1.

Yamagata Prefecture, 2011b, 'Overview of implementing rental accommodation for evacuees'.

Yamagata Prefecture, 2011c, 'Rental administration strategies regarding accommodation for evacuees'.

Yamagata Prefecture, 2012, 'Number of evacuees in evacuation centers in Yamagata Prefecture, 2012', published 5.00 pm, January 26.

Yamagata Prefecture, 2014, 'Yamagata Prefecture population and household statistics (past data)', viewed August 1, 2016, <www.pref.yamagata.jp/ou/kikakushinko/020052/tokei/jinkel.html>.

Yamagata Prefecture, 2015 (June 18), 'Concerning the extension of the accommodation period for evacuees from Fukushima Prefecture', viewed August 1, 2016, <www.pref.yamagata.jp/kurashi/bosai/kikikanri/7020072kariage-jyutaku.html>.

Yamagata Prefecture-wide Support Measures Headquarters, 2011, 'On accepting voluntary evacuees from Fukushima Prefecture', June 16.

Yamakawa Yukio, 2013, 'Genpatsu jiko higaisha kyūsaisaku no mondaiten: kodomo hisaisha shienhō o megutte' (Issues regarding support policies for nuclear accident victims: concerning the Support for Children and Victims Law), *Chihō jichi shokuin kenshū* (Local Government Staff Training), December: 18–20.

Yamamoto Hidehiro, 2014, 'International comparison of protest norms between Japan, South Korea and Germany', paper presented at XVIII International Sociological Association World Congress of Sociology, Yokohama, July.

Yamamoto Hidehiro, 2016, 'Datsu genpatsu to mini no yukue: genshiryoku hatsuden wo meguru sōten kanshin no purosesu' (The will of the people toward nuclear-power-free society: issue-attention process of nuclear power policy on reconstruction in response to the Great East Japan Earthquake), inTsujinaka Yutaka (ed.), *Daishinsai ni manabu shakai kagaku 1: seiji katei to seisaku* (Lessons for the Social Sciences from the 2011 Great East Japan Earthquake, vol 1, political process and policy), Tōyō Keizai Inc., pp. 245–68.

Yamaoka Jun'ichiro, 2015, *Nihon denryoku sensō: shigen to ken'eki, genshiryoku o meguru tōsō no keifu* (Japan's electricity wars: resources and profits, a genealogy of the struggle over nuclear power), Sōshisha.

Yamashita Yūsuke, Ichimura Takashi and Satō Akihiko, 2013, *Ningen naki fukō: genpatsu hinan to kokumin no furikai o megutte* (Reconstruction without people: nuclear evacuees and the lack of public understanding), Akashi Shoten.

Yanase Sho, 2013, 'Fukushima no fukkō, saisei ni muketa taisei no kyōka – Tokyo, Fukishima 2 honsha taisei no kōchiku to Fukushima fukkōsaisei tokubetsu sochi hō no kaisei' (Strengthening structures of Fukushima reconstruction and recovery: the structure of the Tokyo/Fukushima 2 headquarters system and amendments to the Reconstruction and Recovery Special Measures Law), *Rippō to chōsa* (Legislation and Inquiry), 341: 1–13.

Yasutaka Tetsuo and Naitō Wataru, 2013, 'Analysis of the cost and effectiveness of decontamination', July, viewed June 30, 2015, <www.aist-riss.jp/wp-content/uploads/2014/12/5d22c054334dc5ff4f6d1e14e6d636c1.pdf>.

Yoda Hiroshi, Kume Ikuo, Sakano Tomokazu, Kariya Hisao and Ōtsuki Emi, 2000, *Hanshin Awaji daishinsai: gyōsei no shōgen, soshite shimin* (The Hanshin–Awaji earthquake: testimony by government, and the people), Kunpuru.

Yokemoto Masafumi, 2013a, *Genpatsu baishō o tou: aimai na sekinin, honrō sareru hinansha* (Rethinking nuclear liability: shirking responsibility, trifling with evacuees), Iwanami Shoten.

Yokemoto Masafumi, 2013b, 'Genpatsu jiko higai no kaifuku to baishō, hoshō wa dō aru beki ka: 'furusato' no sōshitsu o chūshin ni' (What form should reconstruction and compensation for nuclear damage take? On the loss of the hometown), *Kankyō to Kōgai* (Research on Environmental Disruption), 43 (2).

Yokemoto Masafumi and Watanabe Toshihiko (eds), 2015, *Genpatsu saigai wa naze fukintō na fukkō o motarasu no ka* (Why is the nuclear disaster producing unequal reconstruction outcomes?), Mineruva Shobō.

Yomiuri Shimbun Seijibu, 2011, *Bōkoku no saishō: kantei kinō teishi no 180 nichi* (Prime minister of a ruined country: 180 days in a dysfunctional prime minister's office), Shinchōsha.

Yonemori Yasumasa (ed.) 2015, *Fukkōchō meikan – 2015 nenpan* (Reconstruction Agency directory: 2015 edition), Jihyōsha.

Yonezawa City, 2014, 'Yonezawa city population (estimated population and household numbers)', viewed September 1, 2015, <www.city.yonezawa.yamagata.jp/3548.htm>.

Yoshida Kōhei and Harada Shun, 2012, 'Gaisetsu: genpatsu shūhen jichitai no hinan no keii' (Overview: an account of evacuation in municipalities in the nuclear power plant vicinity), in Yamashita Yūsuke and Kainuma Hiroshi (eds), 'Genpatsu hihan' ron: hinan no jitsuzō kara sekandotaun, kokyō saisei made' (A theory of 'nuclear accident evacuees': from the real image of the evacuee to a second-town, hometown revival), Akashi Shoten, pp. 365–89.

Yoshioka Hitoshi, 2011, *Genshiryoku no shakaishi: sono Nihon teki tenkai* (A social history of nuclear energy: its development in Japan), new edn, Asahi Shimbun Shuppan.

Websites

Iwate Prefecture Social Welfare Council Volunteer and Citizen Activity Center, viewed December 28, 2014, <www.iwate-shakyo.or.jp/soshiki/volun/>.

Miyagi Prefecture Disaster Volunteer Center, viewed December 28, 2014, <http://svc.miyagi.jp>.

Fukushima Prefecture Everyday Life Reconstruction Volunteer Center (August 2014 archives), viewed December 28, 2014, <www.pref-f-svc.org/wp-content/uploads/2014/08/32595c8d27290e8cec7a764eaaf04320.pdf>.

ADR Kentōkai, <www.kantei.go.jp/jp/singi/sihou/kentoukai/03adr.html>.

Agency for Natural Resources and Energy, a, 'Enerugī kihon keikaku' (Strategic energy plan), viewed February 28, 2016, <www.enecho.meti.go.jp/category/others/basic_plan/pdf/140411.pdf>.

Agency for Natural Resources and Energy, b, 'Sōgō shigen enerugī chōsakai kihon seisaku bunkakai chōki enerugī jukyū mitōshi shō iinkai' (Advisory Committee for Natural Resources and Energy, Strategic Policy Committee, Long-term Energy Supply and Demand Outlook Subcommittee), viewed March 26, 2016, <www.enecho.meti.go.jp/committee/council/basic_policy_subcommittee/>.

Cabinet Secretariat, a, 'Enerugī, kankyō no sentakushi ni kansuru iken chōshukai' (Meeting to hear opinions on energy and environmental policy options), viewed February 24, 2016, <www.cas.go.jp/jp/seisaku/npu/kokumingiron/>.

Cabinet Secretariat, b, 'Enerugī, kankyō no sentakushi ni kansuru tōrongata yoron chōsa' (Deliberative Poll on Energy and Environmental Policy Options), viewed February 24, 2016, <www.cas.go.jp/jp/seisaku/npu/kokumingiron/dp/index.html>.

Japan National Press Club, 'The 46th general election of the members of the House of Representatives debate and press conference with 11 leaders of political parties', viewed April 7, 2017, <s3-ap-northeast-1.amazonaws.com/jnpc-prd-public/files/2012/11/7eba6ba77c605322983a133a5b95bf46.pdf>.

Japanese Trade Union Confederation, 'Rengō kōsei soshiki ichiran' (Rengō organizational chart), viewed October 5, 2014, <www.jtuc-rengo.or.jp/rengo/shiryou/kousei/>.

Justice System Reform Council, <www.kantei.go.jp/jp/sihouseido/index.html>.

Liberal Democratic Party, 'Dai 46 kai Shūgiin Senkyo, Jimintō Senkyo Kōyaku' (The 46th House of Representatives election, Liberal Democratic Party's election promise), viewed February 24, 2016, <https://jimin.ncss.nifty.com/pdf/seisaku_ichiban24.pdf>.

National Antinuclear Litigation Directory, <www.datsugenpatsu.org/bengodan/list>.

National Policy Unit, a, 'Kakushinteki enerugī, kankyō senryaku enerugī, kankyō kaigi' (Innovative energy and environmental strategy, Energy and Environment Council), viewed February 24, 2016, <www.cas.go.jp/jp/seisaku/npu/policy09/archive01.html>.

National Policy Unit, b, 'Kakushinteki enerugī, kankyō senryaku, enerugī, kankyō kaigi' 'Dai 13 kai enerugī kankyō kaigi giji shidai haifu shiryō' (Innovative energy and environmental strategy, Energy and Environment Council, the 13th Energy and Environment Council, Agenda, distribute materials), viewed February 24, 2016, <www.cas.go.jp/jp/seisaku/npu/policy09/archive01_13.html#haifu>.

Nuclear Damage Compensation Dispute Resolution Center, <www.mext.go.jp/a_menu/genshi_baisho/jiko_baisho/detail/adr-center.htm>.

Nuclear Regulation Authority, 'Genshiryoku kisei iinkai iin no rinri nado ni kakawaru kōdō kihan ni motozuku kōhyō suru jōhō nado' (Published information based on a code of conduct on ethics for NRA Commissioners)', viewed September 1, 2016, <www.nsr.go.jp/data/000068989.pdf>.

Reconstruction Agency, 'Hisaisha shien' (Disaster victim support), <www.reconstruction.go.jp/topics/main-cat2/>.

Yamagata City, 'Saigai taiō hinan' (Disaster response and evacuation), <www.city.yamagata-yamagata.lg.jp/shimin/sub9/saigaitaiou/>.

Yamagata Prefecture, 'Yamagata Ken nai ni hinan sarete iru mina sama e' (To those evacuees who are in Yamagata Prefecture), <www.pref.yamagata.jp/ou/kankyoenergy/020072/fukkou/hinansha.html>.

Yonezawa City, 'Higashi Nihon Daishinsai kankei' (Related to the Great East Japan Earthquake), </www.city.yonezawa.yamagata.jp/4351.htm>.

Yonezawa City Disaster Evacuee Support Center (Yonezawa shi hinansha shien sentaa), 'Oide' (Welcome), <http://yonezawanet.jp/oide/>.

Name Index

Subject Index